Springer Tracts in Nature-Inspired Computing

Series Editors

Xin-She Yang, School of Science and Technology, Middlesex University, London, UK

Nilanjan Dey, Department of Information Technology, Techno India College of Technology, Kolkata, India

Simon Fong, Faculty of Science and Technology, University of Macau, Macau, Macao

D1826907

The book series is aimed at providing an exchange platform for researchers to summarize the latest research and developments related to nature-inspired computing in the most general sense. It includes analysis of nature-inspired algorithms and techniques, inspiration from natural and biological systems, computational mechanisms and models that imitate them in various fields, and the applications to solve real-world problems in different disciplines. The book series addresses the most recent innovations and developments in nature-inspired computation, algorithms, models and methods, implementation, tools, architectures, frameworks, structures, applications associated with bio-inspired methodologies and other relevant areas.

The book series covers the topics and fields of Nature-Inspired Computing, Bio-inspired Methods, Swarm Intelligence, Computational Intelligence, Evolutionary Computation, Nature-Inspired Algorithms, Neural Computing, Data Mining, Artificial Intelligence, Machine Learning, Theoretical Foundations and Analysis, and Multi-Agent Systems. In addition, case studies, implementation of methods and algorithms as well as applications in a diverse range of areas such as Bioinformatics, Big Data, Computer Science, Signal and Image Processing, Computer Vision, Biomedical and Health Science, Business Planning, Vehicle Routing and others are also an important part of this book series.

The series publishes monographs, edited volumes and selected proceedings.

More information about this series at http://www.springer.com/series/16134

Nilanjan Dey
Editor

Applications of Firefly Algorithm and its Variants

Case Studies and New Developments

 Springer

Editor
Nilanjan Dey
Department of Information Technology
Techno India College of Technology
Kolkata, West Bengal, India

ISSN 2524-552X   ISSN 2524-5538   (electronic)
Springer Tracts in Nature-Inspired Computing
ISBN 978-981-15-0308-5   ISBN 978-981-15-0306-1   (eBook)
https://doi.org/10.1007/978-981-15-0306-1

This Springer imprint is published by the registered company Springer Nature Singapore Pte Ltd.
The registered company address is: 152 Beach Road, #21-01/04 Gateway East, Singapore 189721, Singapore

Preface

Swarm intelligence optimization algorithm has been a hotspot study in the field of computational intelligence and artificial intelligence in latest years, and its application has reached many areas already. The firefly algorithm (FA) is a comparatively new algorithm in many types of smart algorithms and demonstrates outstanding efficiency. Firefly algorithm is among the well-known swarm-based algorithms that have gained popularity in a brief period of time and have various applications. Understanding and implementing the algorithm are simple. Current studies indicate that it is susceptible to premature convergence and suggest the relaxation of having constant parameters. This book introduces FA and its applications. This book also introduces the fundamental ideas of the FA and enhanced versions of the FA by combining the features of particular issues to solve multiple issues such as numerical optimization, clustering assessment and protein complexes that are discovered on the network of protein–protein interaction. The detailed steps of implementation and the comparison of results to show the feasibility of FA and extensive use of its applications are also included in this book.

The purpose of this book is not only to help beginners with a holistic approach towards understanding firefly algorithm but also to present to the researchers new technological trends and design challenges they have to cope with, while designing systems using this algorithm. This book is concerned with supporting and enhancing the utilization of firefly algorithm in several systems and real-world activities. It provides a well-standing forum to discuss the characteristics of the firefly algorithm in different domains. This book is proposed for professionals, scientists, and engineers who are involved in the techniques using the firefly algorithm. It provides an outstanding foundation for undergraduate and postgraduate students as well. It has several features, including an outstanding basis of the firefly algorithm analysis, and it includes different applications and challenges with extensive studies for systems that have used firefly algorithms.

This book is organized as follows:

Chapter 1 gives the overview of firefly algorithm and its variants in digital image processing. The significance and requirements of digital image processing arise from two main areas of applications: the improvement of visual information for human interpretation and the encoding of scene data for the independent perception of machines. However, human is often involved in such processing for manually tuning up the parameters, which takes a long time, and it remains as an unresolved issue. Due to the efficacy and success in solving various digital image analysis problems, the firefly algorithm, inspired by fireflies' flashing behaviour in nature, is used in various image analysis optimization studies. This chapter is dedicated to a comprehensive review of the firefly algorithm to solve optimization problems in various steps of digital image analysis, like image preprocessing, segmentation, compression, feature selection, and classification. Various applications of the firefly algorithm in image analysis are also discussed in this chapter. Key issues and future research directions are also highlighted in this chapter.

Chapter 2 deals with the development of firefly algorithm interface for parameter optimization of electrochemical-based machining processes. Manufacturing of micro-complex profiles is a challenging task, which is time-consuming and expensive. The optimum parameter setting of machining process is essential to obtain a desired profile on the workpiece with higher material removal rate. To achieve this purpose, a technique to create a graphical user interface (GUI) that associates metaheuristic technique, the firefly algorithm (FA), is highlighted in this chapter. The developed GUI can be used effectively by the user for obtaining the optimal solution of non-traditional machining (NTM) processes. The GUI is tested on three NTM processes, namely electrochemical machining (ECM), electro-chemical micromachining (EMM) and electrochemical turning (ECT), and the results show the effectiveness of interface accompanying the considered FA.

Chapter 3 is devoted to a firefly algorithm-based approach for identifying coalitions of energy providers that best satisfy the energy demand. Smart grids aim to ensure the network reliability by providing electricity when it is demanded, while lowering the impact they have on the surrounding environment. Due to the great advances of renewable technologies, the current trend is to encourage the integration of renewable energy providers that could participate in satisfying the requested demand. The grid reliability can be achieved by identifying in advance the coalitions of heterogeneous electricity providers that could satisfy the forecast demand. This problem can be seen as an optimization problem aiming to identify the coalition that best satisfies the electricity demand curve over a particular period of time, by taking into account the energy supply forecast for each provider. In this context, this chapter proposes an optimization methodology for identifying the coalition of providers that satisfies the electricity demand curve over a time interval using an adapted version of the firefly algorithm. Additionally in this chapter, the fitness function has introduced two penalty components that penalize a solution's adherence to the requested provider heterogeneity and maximum desired price. To apply the firefly algorithm in this context, the firefly movement strategy has using the genetic crossover and mutation operators. The proposed optimization methodology has been evaluated in this chapter on an in-house developed data set consisting of forecast energy values for energy supply and demand.

Chapter 4 discusses the structural damage identification using adaptive hybrid evolutionary firefly algorithm. A recently developed adaptive hybrid evolutionary firefly algorithm (AHEFA) as a cross-breed of differential evolution (DE) approach and firefly algorithm (FA) is utilized to address inverse optimization problems in two-stage damage detection of truss structures. In the first step, the most potentially damaged elements are recognized utilizing a modal strain energy-based index (MSEBI). In the remaining one, AHEFA is employed as an optimizer to estimate the real damage severity relied on the afore-collected information in the first stage. The effectiveness and correctness of the present algorithm are demonstrated via three numerical examples. The results given by the current paradigm are validated with those solved by DE and FA. The outcomes indicate that AHEFA can perform well in precisely recognizing the locations and extents of multi-damage trusses with a lower computational attempt.

Chapter 5 focuses on an automated approach for developing a convolutional neural network using a modified firefly algorithm for image classification. The design phase of CNNs requires potential from non-experts of machine learning. In this chapter, a fully automated algorithm to develop CNN was proposed based on firefly optimization. The proposed method can design a CNN structure with any number of layer depth without any limitation on the depth value. The proposed method employed the skip connection as a fundamental building block of CNN. A modified firefly algorithm was presented based on the k-nearest neighbour attraction model to reduce the computational complexity of the firefly. The CIFAR-10 and CIFAR-100 were used for the training and validation of the proposed method to perform image classification. The proposed method provided high accuracy when compared to the cutting-edge approaches.

Chapter 6 provides an overview of the enhanced firefly algorithm for optimum steel construction design. The philosophy of structural engineering is the consideration of safety, economy and aesthetics in fundamentally meeting the requirement for sheltering. So, the design of the structures is to be safe and at the same time economical is the main aim of the designers. Thus, not only the construction is to be designed as safely carrying the calculated design loads limited to the structural provisions, but also the minimum level of material is to be used. So then, the steel structural engineers perform minimum weighted designs in order to select the optimum one among the feasible designs within the provision limits. Hence, a variety of structural optimization methods have been developed to solve these complex engineering problems. The traditional gradient-based mathematical methods are not sufficient to solve those tedious design problems. To this end, metaheuristics have been effectively utilized as an instrument of achieving the optimal structural designs. In this chapter, an enhanced version of firefly algorithm, which is based on the social behaviours of fireflies while communicating with each other, to prevent it from confinement in a local optima is presented in order to obtain the optimum design of various types of steel constructions under code provisions that such problem is categorized as discrete nonlinear programming problem.

Chapter 7 deals with the application of firefly algorithm for face recognition. Face recognition is steadily making its way into commercial products. As such, the accuracy of face recognition systems is becoming extremely crucial. In the firefly algorithm, the brightness of fireflies is used to measure attraction between a pair of unisex fireflies. The firefly with higher brightness attracts the less bright firefly. The objective function is defined in proportion to the brightness, to define a maximization problem. This chapter aims to present the promising application of the firefly algorithm for face recognition. The firefly algorithm is used in a hyper-dimensional feature space to select features that maximize the recognition accuracy. This chapter delineates how the firefly algorithm is a suitable algorithm for selection of the features in a face recognition model. The firefly algorithm is then applied to this feature space to identify and select the best features. Fireflies are arbitrarily placed on various focal points of the image under consideration. The advantage of this approach is its fast convergence in selecting the best features. The gamma parameter controls the movement of fireflies in this feature space and can be tuned for gaining an improvement in the performance of the face recognition model. This chapter aims to evaluate the performance and viability of using the firefly algorithm for face recognition.

Chapter 8 provides the application of chaos-based firefly algorithm optimized controller for automatic generation control of two-area interconnected power system with energy storage unit and UPFC. Firefly algorithm (FFA) optimization technique is a popular continuous optimization technique and also widely applied to tune controller gain values of proportional–integral–derivative (PID) controller. In chaos-based firefly algorithm, the chaotic maps are included to improve the randomness when generating new solutions and thereby to increase the diversity of the population. It increases global search mobility for robust global optimization. The investigated power system incorporates two thermal power systems, and two areas are interconnected via tie line. Also, this work system is equipped with a secondary (PID) controller. The gain value of the controller is optimized by implementing chaos-based firefly algorithm (CFA) by applying step load in area 1. The effectiveness and supremacy of proposed optimization technique-tuned controller performance are verified by comparing genetic algorithm (GA), particle swarm optimization (PSO) technique and firefly algorithm-tuned PID controller performance in the same interconnected power system. The supremacy of the proposed technique is evaluated by considering time domain specification parameters of the proposed optimization technique-based controller response. Further, the effectiveness of recommended optimization technique is verified by adding nonlinearities (generation rate constraint (GRC) and governor dead band (GDB)) to the investigated power system with connecting Unified Power Flow Controller (UPFC) is applied, and it is a flexible alternating current transmission system (FACTS) controller family device connected parallel to the tie line. Further, hydrogen aqua electrolyser (HAE) energy storage unit with a fuel cell is connected in area 2 for improving the performance of the system during sudden load demand. Finally, simulation result proved that proposed CFA technique gives better controller performance compared with GA, PSO and FFA technique-tuned controller performance in terms of fast settled response with a minimum peak overshoot and undershoot, damping oscillations during unexpected load demand conditions.

Chapter 9 focuses on plant biology-inspired genetic algorithm: superior efficiency to firefly optimizer. This chapter analytically compares the efficiency of the recent plant biology-inspired genetic algorithm (PBGA) and the firefly algorithm (FA) optimizer. The comparison is over a range of well-known critical benchmark test functions. Through statistical comparisons over the benchmark functions, the efficiency of PBGA has been evaluated versus FA as a well-known accurate metaheuristic optimizer. Through a considerable number of Mont Carlo runs of searching for a solution by both optimizers, their performance has been statistically measured by several valid indices. In addition, the convergence curves give a visual comparison of both techniques where the stability, speed and accuracy dominance of PBGA are clearly observable. However, in the case of benchmark function with smooth nature like Rosenbrock, Sphere and Dixon–Price, FA has better performance in average, while PBGA performance is still comparable to FA.

Chapter 10 deals with the firefly algorithm-based Kapoor's thresholding and Hough transform to extract leukocyte section from hematological images. Computerized disease examination techniques are widely adopted in the literature to evaluate a considerable number of medical images ranging from the RGB scale to greyscale. This chapter proposes a novel image extraction method by combining Kapur's thresholding and Hough transform (HT) to extract the leukocyte segment from the RGB-scaled blood smear image (BSI). Automated mining of the leukocyte region is always preferred in medical clinics for fast disease examination and treatment planning process. This study aims to implement a hybrid procedure to extort the leukocyte segment. Kapur's thresholding is considered to enhance the RGB-scaled test image, and HT is used to detect and extract the circle section from the image. In this work, the hematological images of leukocyte images for segmentation and classification (LISC) database are adopted for the examination. The extracted leukocyte picture is then evaluated with ground truth, and the essential image performance parameters (IPPs) are then computed. This work is then validated against the semi-automated approaches, such as Shannon's entropy-based Chan–Vese and level set segmentation techniques existing in the literature. The outcome of the proposed techniques confirms that proposed procedure gives better IPP values compared to the existing semi-automated techniques.

Chapter 11 provides an overview of the effect of population size over parameterless firefly algorithm. Nature-inspired optimization algorithms have proved their efficacy for solving highly nonlinear problems in science and engineering. Firefly algorithm (FA), which is one of them, has been taken to be broadly discussed in this chapter. However, practical studies demonstrate that the proper setting of FA's parameters is the key difficulty. FA associates with some sensitive parameters, and proper setting of their values is extremely time-consuming. Therefore, parameterless variants of FA have been proposed in this study by incorporating the adaptive formulation of the control parameters which are free from the tuning procedure. Population size greatly influences the convergence of any nature-inspired optimization algorithm. Generally, lower population size enhances convergence speed but may trap into local optima, while larger population size maintains the diversity but slow down the convergence rate. Therefore, the crucial effect of different population sizes over FA's efficiency has also been investigated here. The population size (of considered) FA is varied within the size [10, 1280] where it gets doubled in each run starting with $n = 10$. Therefore, eight instances of FA have been executed and the best one is considered by the user over different classes of functions. Experimental results also show that the proposed parameterless FA with a population size between 40 and 80 provides the best results by considering optimization ability and consistency over any classes of functions.

West Bengal, India   Nilanjan Dey

Contents

1   **Firefly Algorithm and Its Variants in Digital Image Processing: A Comprehensive Review**   1
Nilanjan Dey, Jyotismita Chaki, Luminița Moraru, Simon Fong
and Xin-She Yang

2   **Development of Firefly Algorithm Interface for Parameter Optimization of Electrochemical-Based Machining Processes**   29
D. Singh and R. S. Shukla

3   **A Firefly Algorithm-Based Approach for Identifying Coalitions of Energy Providers that Best Satisfy the Energy Demand**   53
Cristina Bianca Pop, Viorica Rozina Chifu, Eric Dumea
and Ioan Salomie

4   **Structural Damage Identification Using Adaptive Hybrid Evolutionary Firefly Algorithm**   75
Qui X. Lieu, Van Hai Luong and Jaehong Lee

5   **An Automated Approach for Developing a Convolutional Neural Network Using a Modified Firefly Algorithm for Image Classification**   99
Ahmed I. Sharaf and El-Sayed F. Radwan

6   **Enhanced Firefly Algorithm for Optimum Steel Construction Design**   119
S. Carbas

7   **Application of Firefly Algorithm for Face Recognition**   147
Jai Kotia, Rishika Bharti, Adit Kotwal and Ramchandra Mangrulkar

8   **Application of Chaos-Based Firefly Algorithm Optimized**
     Controller for Automatic Generation Control of Two Area
     Interconnected Power System with Energy Storage Unit
     and UPFC   173
     K. Jagatheesan, B. Anand, Soumadip Sen and Sourav Samanta

9   **Plant Biology-Inspired Genetic Algorithm: Superior Efficiency**
     to Firefly Optimizer   193
     Neeraj Gupta, Mahdi Khosravy, Om Prakash Mahela and Nilesh Patel

10   **Firefly Algorithm-Based Kapur's Thresholding and Hough**
     Transform to Extract Leukocyte Section from Hematological
     Images   221
     Venkatesan Rajinikanth, Nilanjan Dey, Ergina Kavallieratou
     and Hong Lin

11   **Effect of Population Size Over Parameter-less Firefly**
     Algorithm   237
     Krishna Gopal Dhal, Samarendu Sahoo, Arunita Das and Sanjoy Das

About the Editor

Nilanjan Dey is an Assistant Professor in Department of Information Technology at Techno India College of Technology, Kolkata, India. He is a visiting fellow of the University of Reading, UK. He was an honorary Visiting Scientist at Global Biomedical Technologies Inc., CA, USA (2012-2015). He was awarded his PhD. from Jadavpur University in 2015. He has authored/edited more than 50 books with Elsevier, Wiley, CRC Press and Springer, and published more than 300 papers. He is the Editor-in-Chief of International Journal of Ambient Computing and Intelligence, IGI Global, Associated Editor of IEEE Access and International Journal of Information Technology, Springer. He is the Series Co-Editor of Springer Tracts in Nature-Inspired Computing, Springer, Series Co-Editor of Advances in Ubiquitous Sensing Applications for Healthcare, Elsevier, Series Editor of Computational Intelligence in Engineering Problem Solving and Intelligent Signal processing and data analysis, CRC. His main research interests include Medical Imaging, Machine learning, Computer Aided Diagnosis, Data Mining etc. He is the Indian Ambassador of International Federation for Information Processing – Young ICT Group. Recently, he has been awarded as one among the top 10 most published academics in the field of Computer Science in India (2015-17).

Chapter 1
Firefly Algorithm and Its Variants in Digital Image Processing: A Comprehensive Review

Nilanjan Dey, Jyotismita Chaki, Luminiţa Moraru, Simon Fong and Xin-She Yang

1 Introduction

The image processing consists of many tasks and steps such as visualization, image sharpening, and restoration, and one of the main tasks in image processing is to extract feature and recognize patterns and ultimately to make sense of the actual image content using a vast array of image processing techniques [1]. For many tasks such as image enhancement, clustering, and classification, feature extraction, or more advanced image interpretation, either supervised or unsupervised learning or their combinations may be used [2, 3]. Image processing nowadays sometimes has to deal with time-dependent, complex, dynamic, noisy images and to make sense of image data with good accuracy at the minimum computational cost, often with the minimum probable energy consumption and in a practically acceptable time scale.

N. Dey
Department of Information Technology, Techno India College of Technology, Kolkata, West Bengal 700156, India
e-mail: neelanjandey@gmail.com

J. Chaki (✉)
School of Information Technology and Engineering, Vellore Institute of Technology, Vellore, India
e-mail: jyotismita.c@gmail.com

L. Moraru
Department of Chemistry, Physics and Environment, Faculty of Sciences and Environment, Dunarea de Jos University of Galati, 47 Domneasca Str., 800008 Galati, Romania
e-mail: Luminita.Moraru@ugal.ro

S. Fong
Department of Computer and Information Science, University of Macau, Taipa, Macau
e-mail: ccfong@umac.mo

X.-S. Yang
School of Science and Technology, Middlesex University, London, UK
e-mail: xy227@cam.ac.uk

© Springer Nature Singapore Pte Ltd. 2020
N. Dey (ed.), *Applications of Firefly Algorithm and its Variants*,
Springer Tracts in Nature-Inspired Computing,
https://doi.org/10.1007/978-981-15-0306-1_1

These are all a series of very challenging and interesting issues concerning computer vision and optimization [4].

Optimization is almost everywhere in every application, particularly in engineering designs, whether to reduce energy consumption and cost, or to boost profits, output, sustainability, overall performance, and effectiveness [5, 6]. An optimization issue pertains to maximizing or minimizing an objective function by setting the appropriate values from many to almost infinite choices of viable values for the design variables. Not only in complex scientific studies, but also in our daily activities, these problems appear [7, 8]. One of the main optimization issues in image processing is an algorithm's reliability that can be influenced by several factors, such as the algorithm's intrinsic structure, how new solutions are generated, and setting of its algorithm-dependent parameters [9, 10]. The essence of an optimizer is a properly designed search or optimization procedure to perform the required search capabilities so as to find the better, ideally, the optimal solutions to an optimization problem under consideration. Usually, starting with some initial solution vector, often as a random solution or an educated guess, an algorithm starts to explore the search space so as to generate better solution vectors [11, 12]. The ultimate goal is to try to find the best solution as fast as probable, either in terms of the minimum computing costs or the minimum number of steps [13, 14]. Usually, this is not attainable in practice. In many situations, it may not be easy to find even a single feasible solution.

Several tasks of image processing can be redesigned as problems of optimization. For example, image thresholding [15–18] and segmentation [19–23] in terms of the threshold design parameter can clearly be constructed as optimization problems, while the objective functions can be properly addressed. In this study, some metaheuristic algorithms are presented, which are needed to optimize the enhancement [24–26] and denoise [27–30] function to preprocess images. Several segmentation, compression [15, 16, 31, 32], and optimization techniques are also discussed.

By imitating some successful features scenario from nature, various optimization algorithms [33–35] have been introduced. For example, Darwin's theory of the survival of the fittest motivates the genetic algorithm [36–38], while particle swarm optimization [10, 39–43] is also another metaheuristic technique that imitates how a swarm moves by following one another. In addition, the shuffled frog-leaping algorithm [44–46] is influenced by immersive actions and exchange of leaping frogs, searching for food in a pond on discrete stones, while the cuckoo algorithm [47–51] is motivated by a bird family's lifestyle known as cuckoos. Furthermore, the bat algorithm [52–58] is inspired by the echolocation behaviour of microbats, with varying pulse rates of emission and loudness, while the flower pollination algorithm is inspired by the pollination characteristics of flowering plants [59, 60].

The ant colony optimization algorithm [61–64] is inspired by the behaviour of real ants, and the bacterial foraging optimization algorithm [65–68] is inspired by bacteria's chemotaxis behaviour that will portray chemical gradients (such as nutrients) in the climate and move towards or away from precise signals. In the similar fashion, the artificial fish swarm algorithm [69–71] is motivated by how a fish can always find food in a location where there is plenty of food; hence the more food, the more fish, while artificial bee colony algorithm [72–74] is an optimization algorithm based on honey bee intelligent scavenging behaviour, and simulated annealing [21, 23, 75–77] is inspired by the process of annealing in metallurgy; i.e., a material is heated and then slowly cooled under controlled conditions to increase the size of the crystals in the material for reducing their defects.

All these optimization techniques utilize various degrees of exploration and exploitation based on their various methods of search. For a more detailed review of these techniques, readers can refer to Kennedy et al. [43] and Yang [78].

Even with several advantages, these methods can still have some limitations. It is not straightforward to select parameters such as a number of generations, population size, and others [79–82]. Though, in most cases, optimal solutions can be found quite effectively, sometimes, it might not be difficult to find the true optimal solution to the defined problem. For example, when using particle swarm optimization [83–85], it needs to define initial design parameters, which requires parametric studies and it can converge prematurely. Another example is the limitations of shuffled frog-leaping algorithm [86–88] in terms of its initial non-uniform population, long and slow and search speed in late iterations, which can lead to being easily trapped in local extremities. In addition, bacterial foraging algorithm [89–91] for the optimization process can have the weak ability to perceive the environment and vulnerable to the perception of local extremes, while artificial fish swarm algorithm [85, 92, 93] can have higher time complexity, lower convergence speed, and lack of balance amongst global search and local search. The limitations of artificial bee colony optimization algorithm [94–96] are lack of use of secondary information, require new fitness tests on new algorithm parameters, a higher number of objective function evaluation, and slow when in sequential processing, while simulated annealing [97–99] can be a very slow process. There are other studies and discussions about various other algorithms [39–41, 100–107]. Furthermore, it still lacks the theoretical analyses for most of these algorithms, though it is important to gain some insight into the working mechanisms of all the major algorithms. However, this is not our main focus here in this paper.

Our main focus of this paper is to review the firefly algorithm (FA) in a comprehensive way [81, 108]. FA has various applications in digital image analysis [27, 28, 109, 110]. Its efficiency and simple steps allow researchers from various disciplines to use this algorithm in image preprocessing, segmentation, compression, feature selection, and classification [111–113]. FA can deal with highly nonlinear, multi-modal optimization problems naturally and effectively. It does not utilize velocities, and there is no problem as it associated with velocity as those in particle swarm optimization. The speed of convergence of this method is very high, in the probabilitic searching for the global optimality. In addition, FA has the flexibility of integration with other optimization methods to form hybrid tools, which does not need a good initial solution to start its iteration process. Different studies have been conducted to refine the basic firefly algorithm in order to increase its overall performance and make it more useful to a challenging problem at hand. Thus, a detailed survey on the firefly algorithm and its variants as well as different applications will be carried out here in this survey. A detailed discussion with several other relevant studies on extended firefly algorithms is also be given.

Therefore, the rest of this survey is organized as follows. In Sect. 2, metaheuristic algorithms in digital processing is briefly introduced, while in Sects. 3–9, the necessity of firefly algorithm in following aspects, image preprocessing, image segmentation, image compression, feature selection, classification, and different applications, respectively, is reviewed, respectively. Finally, i n Sct. 9 conclusions are briefly drawn and future directions of firefly algorithms in digital image analysis are discussed.

2   Metaheuristic Algorithms in Digital Image Processing

2.1   *Firefly Algorithm (FA)*

Firefly algorithm is a metaheuristic algorithm originally introduced by Yang [55, 56, 57, 58, 108], inspired by the patterns of flashing illumination produced by tropical fireflies in nature. From their lower stomach, they produce chemically generated light. Such bioluminescence with flashing patterns is utilized to allow communication amongst two adjacent insects searching for prey and also finding mates. The classical FA is formulated by using the following idealizations:

- All fireflies are unisex, and regardless of their gender, one firefly will be attracted to the closest firefly.
- The attraction is proportional to the luminance intensity amongst two fireflies. The firefly with the brighter luminance will attract the firefly with less luminance for any pair of flashing fireflies. The attraction amongst a pair of fireflies depends mainly on the Cartesian distance and is proportional to the brightness that reduces as the distance amongst fireflies increases. In a region, if the entire fireflies have less luminance, they will simply move in the dimensional search space in a random manner till they discover a brighter luminance firefly.
- A firefly's brightness can indeed be associated with the objective function's values so as to guide the search procedure.

A firefly's luminance can be regarded as proportional to the cost function value (i.e., luminance = objective function) for a maximization problem. The FA's performance (time of exploration, convergence speed, and precision of optimization) can depend on the parameter settings and type of problems, which will be discussed in detail later.

2.1.1   Firefly Algorithm (FA)

The main parameters that may influence the FA's efficiency are the intensity of light and attractiveness variations amongst adjacent fireflies. The scaling parameter in the distance amongst fireflies will also have some effect. The following Gaussian form can demonstrate variation in luminance analytically.

$$L(d) = L_0 e^{-\gamma r^2} \tag{1}$$

where L and L_0 are the intensity at distance r and original light intensity at $r = 0$ and γ is the coefficient of the light absorption.

The attractiveness (C) can be a direct link to the luminosity and can be abstractly defined as:

$$C = C_0 e^{-\gamma r^2} \tag{2}$$

where C_0 can be considered as the attractiveness at $r = 0$. In essence, the above equation has a critical distance characteristic $D = 1/\sqrt{\gamma}$ that can substantially alter the attractiveness from C_0 to $C_0 e^{-1}$. Any monotonically reducing functions like the following form can be the attractiveness function $C(r)$:

$$C(r) = C_0 e^{-\gamma r^n}, \ (n \geq 1) \tag{3}$$

For a static γ, the characteristic length turns out to be

$$D = \gamma^{-1/n} \to 1, \ n \to \infty \tag{4}$$

On the other hand, for a specified length scale D, the parameter γ can be utilized as initial atypical value ($\gamma = 1/D^n$).

The Cartesian distance of a pair of fireflies p and q at s_p and s_q, in the m-dimensional space, can be expressed mathematically as:

$$r_{pq}^t = \left\| S_q^t - S_p^t \right\|_2 = \sqrt{\sum_{a=1}^{m}(S_{q,a} - S_{p,a})^2} \tag{5}$$

In the FA, the light intensity obeys the inverse square law at a given distance r from the light source S_p^t. A firefly's light intensity L decreases as the distance r increases in terms of $L \propto 1/r^2$. The movement of the attracted firefly p to a brighter firefly q can be defined by the equation of the update position demarcated as follows:

$$S_p^{t+1} = S_p^t + C_0 e^{-\gamma r_{pq}^2}\left(S_q^t - S_p^t\right) + \varphi \tag{6}$$

where S_p^{t+1} and S_p^t are the firefly's updated position and initial position, respectively, $C_0 e^{-\gamma r_{pq}^2}\left(S_q^t - S_p^t\right)$ is the attraction term amongst fireflies, and φ is a random vector with a parameter controlling the step sizes.

2.1.2   The Working Principle of FA

In this section, the traditional FA's working concept is illustrated utilizing a two-dimensional problem of optimization. For simplicity and for the purpose of illustration here, it is assumed that there are six fireflies in the population. When the algorithm is initialized, the entire population of fireflies in the two-dimensional search space is randomly distributed. In this case, the search space is assumed to have two best global values and one best local value (so three best values in total).

Throughout the preliminary search, some fireflies move in the direction of the global best (G) values and some to the local best (L) value as demonstrated in Fig. 1a. From Fig. 1a, it is detected that Firefly-1 (F1) is at L1, Firefly-4 (F4) is at G, and Firefly-5 (F5) is at L2. F2 lies amongst L1 and G, Firefly-3 (F3) lies amongst G and L2, and Firefly-6 (F6) is amongst G and L2. The light intensity emitted by F4 is brighter than F1 and F5. At this situation, F2 travels towards L1 or G based on the Cartesian distance "r". In this case, the distance amongst F1 and F2 (D1) is less than (D2); thus, F2 moves towards L1. Likewise, the Cartesian distance amongst F4 and F3 (D3) is less than (D4), and F3 is more probably attracted to G than L2. The Cartesian distance amongst F6 and F5 (D5) is less than (D6), and F6 is probably attracted to L2.

3   Image Preprocessing

3.1   Image Enhancement

The process of transformation/mapping of digital images for better image analysis is known as image enhancement [114]. Contrast enhancement techniques are required for visual quality improvement. In spatial domain enhancement techniques, image pixels are controlled to attain the required enhancement using a transform function. Transform functions help to stretch the occupied greyscale range of the provided dark/blur images (e.g., logarithmic transforms, histogram equalization, and power law transform). Transformation function optimization is a nonlinear optimization problem with several constraints.

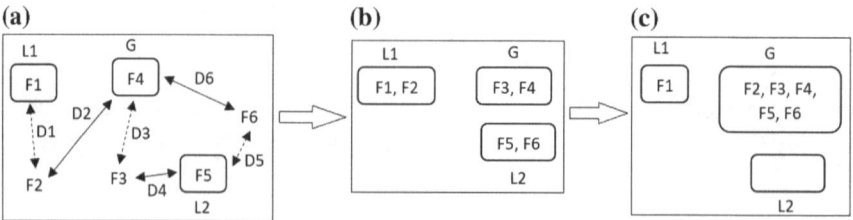

Fig. 1 Various stages of FA: **a** initial stage of the search, **b** intermediate stage, **c** final stage

3.1.1 Enhancement Criterion

Selecting criteria is essential for obtaining the objective function for the optimization process of image enhancement. Ideally, image features are determined by the fitness function which is associated with the selection criteria.

Three performance measurement parameters are reported in Ashour et al. [115] namely: (i) entropy, (ii) sum of the intensity of edge, and (iii) the number of edgels/edge pixels. Higher edge intensity and more edges help improve the visual quality of an image [116]. Homogeneous intensity allotment refers to the equalized histogram and entropy enhancement. Dhal et al. [117] proposed this following objective function (Obj (t)):

$$\text{Obj}(t) = \log_e^{E_n(I_{eh})} x \left(n_{\text{edgels}(I_{eh})} \middle/ (R \times C) \right) \times H_e(t) \tag{7}$$

where the fitness value of the improved image is represented by Obj(t). The pixel intensities' sum of Sobel edge image I_{eh} is denoted by $E_n(I_{eh})$. The number of edge pixels in the Sobel edge image whose intensity value exceeds a threshold is represented by $n_{\text{edgels}(I_{eh})}$. Entropy value $H_e(t)$ is computed from the enhanced image t by studying the histogram. C and R are the numbers of column and row of the image, respectively.

In the literature, the above-mentioned parametric optimization techniques are widely used as an enhancement criterion in the various application domains. In Samanta et al. [118], a log transform-based optimal image enhancement using the firefly algorithm was deployed in aerial photography. In Dhal et al. [117], a chaotic Lévy flight strategy for grey-level image improvement was reported in the bat and firefly algorithm. In Draa et al. [119], an opposition-based firefly algorithm for medical image contrast enhancement was carried out. In Hassanzadeh et al. [120], a nonlinear greyscale image enhancement technique using firefly algorithm was conducted.

3.2 Image Denoising

Random signals (noise and artefact) can reduce the image quality at the time of image acquisition. In recent years, wavelet thresholding-based efficient approaches for denoising have been extensively explored in the literature [121–124]. In Nasri and Nezamabadi-pour [125], the wavelet-based neural thresholding (WT-TNN) methodology for adaptive thresholding was reported. The proposed function of efficiency was further used in a new sub-band-adaptive neural thresholding network to improve denoising. The proposed scheme reduced noise irrespective of its distribution and modelling of image wavelet coefficients distribution. In that study, simultaneous learning of threshold function parameters and threshold value was confirmed in each sub-band of the wavelet transformation. The main limitation of this proposed technique is its high computational time because of the usage of the steepest descent gradient method. Achieving quick convergence of the learning procedure is difficult because of proper initialization of (i) threshold and (ii) thresholding parameters. In Bhandari et al. [19, 20], the firefly algorithm was used in wavelet transform-based thresholding neural network-based denoising methodology to overcome the above-mentioned limitations for satellite image denoising.

4 Image Segmentation

Image segmentation is a process of subdividing an image (I) into a non-overlapping finite number(s) of constituent regions (($I_1, I_2, \ldots, I_s$) with consistent characteristics based on texture, colour, position, homogeneity, etc., of image regions and can be represented as:

$$I = \bigcup_{i=1}^{s} I_i \quad \text{and} \quad I_i \bigcap I_J = \varnothing \quad \text{where } i \neq j \tag{8}$$

Image segmentation techniques can be loosely classified into three traditional categories, namely (i) threshold-based, (ii) region-based, and (iii) edge-based.

4.1 Threshold-Based Techniques

Threshold-based image segmentation techniques are simplest, robust, accurate, competent, and relatively easy to implement [126]. In a bilevel thresholding technique, the input image is divided into background and foreground (two classes), which can be further extended to multilevel thresholding by an increasing number of classes. In local thresholding, various threshold values are allocated for individual portion of the input image techniques, whereas in global thresholding, a global threshold value [127] is obtained using various parametric and nonparametric approaches.

In parametric approaches [128–130], each class's grey-level distribution has a function of probability density (acquired utilizing Gaussian distribution). These techniques attempt to search an approximation of distribution parameters that best suits the defined histogram data and generally lead to computationally expensive and initial condition-dependent nonlinear optimization problems. In nonparametric approaches, thresholds are obtained by maximizing class variance, entropy measures, etc. For example, Otsu's between-class variance method [131] and Kapur's entropy criterion [132] method are widely used multilevel image segmentation techniques. As the number of segmentation level increases, the complexity of the thresholding problem increases accordingly (a large amount of calculation). Traditional segmentation methods tend to use more computational time [133] for exhaustive search of optimal thresholds. To resolve this issue, various population-based intelligent optimization algorithms have been already explored, including genetic algorithm (GA)-based study in Yin [134], Lai and Tseng [135], particle swarm optimization in [136–138] ant colony optimization in Ye et al. [139], cuckoo search (CS) in Samantaa et al. [140], Rajinikanth et al. [141], bat algorithm (BA) in Alihodzic and Tuba [52], Satapathy et al. [53], bacterial foraging algorithm (BF) in Sathya and Kayalvizhi [67, 68], honey bee optimization (HBO) in Horng et al. [142], krill herd optimization (KHO) in Resma and Nair [143], and grey wolf optimizer (GWO) in Khairuzzaman and Chaudhury [144].

Performances of these algorithms are compared based on the incurred computational cost to obtain satisfactory results and achieve the best values of the objective function.

In earlier studies [108, 145], it is clearly reported that the firefly algorithm works well as compared to most of the existing metaheuristic algorithms. In order to find the optimal values for the thresholds for multilevel segmentation, the firefly algorithm has also been used to maximize Otsu's, Kapur's, and Tsallis's objective function.

4.1.1 Otsu's Between-Class Variance Method

In Otsu's bilevel thresholding technique ([146], October), the threshold that reduces the variance within the class is searched and can be represented as a weighted sum of the two-class variances:

$$C_P^2(T) = P_1(T)C_1^2(T) + P_2(T)C_2^2(T) \tag{9}$$

where weights P_1 and P_2 are the probabilities of the two classes segmented by threshold T, and $C_1^2(T)$ and $C_2^2(T)$ are the two-class variances. The probabilities of the classes are calculated from the H histogram bins:

$$P_1(T) = \sum_{j=0}^{T-1} p(j), \ P_2(T) = \sum_{j=T}^{H-1} p(j) \tag{10}$$

Otsu demonstrated that reducing the variance within the class was the same as increasing the variance between classes. The maximum inter-class variance can be represented by:

$$C_M^2 = C^2 - C_P^2(T) = P_1(\overline{c_1} - \overline{c_t})^2 + P_2(\overline{c_2} - \overline{c_t})^2$$
$$= P_1(T)P_2(T)[\overline{c_1}(T) - \overline{c_2}(T)]^2 \tag{11}$$

which can be expressed in terms of probabilities P and means $\overline{c}$ of the classes. The class mean $\overline{c_{1,2,t}}(T)$ can be represented by:

$$\overline{c_1}(T) = \frac{\sum_{j=0}^{T-1} jp(j)}{P_1(T)} \tag{12}$$

$$\overline{c_2}(T) = \frac{\sum_{j=T}^{H-1} jp(j)}{P_2(T)} \tag{13}$$

$$\overline{c_t}(T) = \sum_{j=0}^{H-1} jp(j) \tag{14}$$

Thus, the optimal threshold is the maximum of $C_M^2(T)$.

In the Otsu multilevel threshold problem, an image is classified into i classes $(C_1, C_2, \ldots, C_i)$ with threshold set $(T_1, T_2, \ldots, T_{i-1})$. The optimal threshold is the maximum of C_{Mul}^2

$$C_{\text{Mul}}^2 = \sum_{k=1}^{i} P_k (\overline{c_k} - \overline{c_t})^2 \tag{15}$$

Multilevel thresholds searching for Otsu's between-class variance method is a time-consuming task and involves extensive computation. However, metaheuristic algorithms can be effective alternatives. In fact [147, 148], reported efficient results for solving the above-mentioned problem using the firefly algorithm.

4.1.2 Kapur's Entropy Criterion Method

To use Kapur's technique [149] to obtain the threshold with $T = \begin{bmatrix} t_1, t_2, \ldots, t_{l-1} \end{bmatrix}$ as an image threshold vector, Kapur's entropy is estimated by

$$H_{\max} = \sum_{i=1}^{l} P_i^C \tag{16}$$

In general, each entropy is independently calculated based on the specific t value. For a multilevel thresholding problem, they can be represented as:

$$P_1^C = \sum_{i=1}^{t_1} \frac{p_i^C}{o_0^C} \ln\left(\frac{p_i^C}{o_0^C}\right) \tag{17}$$

$$P_2^C = \sum_{i=t_1+1}^{t_2} \frac{p_i^C}{o_1^C} \ln\left(\frac{p_i^C}{o_1^C}\right) \tag{18}$$

$$\vdots$$

$$P_3^C = \sum_{i=t_l+1}^{G} \frac{p_i^C}{o_{l-1}^C} \ln\left(\frac{p_i^C}{o_{l-1}^C}\right) \tag{19}$$

where P_i^C is the probability distribution of G intensity levels, C is unity for grey-level images, and $O_0^C, O_1^C, \ldots, O_{l-1}^C$ is the probability occurrence for l levels. The search randomly fine-tunes the threshold values, until $H_{\max}$ is reached.

Kapur's entropy criterion-based multilevel thresholding was reported in [150, 151] where the firefly algorithm was used to obtain $H_{\max}$.

In 1968, Kullback [152] proposed the concept of cross entropy. Cross entropy is defined as:

$$CE = \sum_{j=1}^{M} r_j \log r_j / s_j \qquad (20)$$

where $R = \{r_1, r_2, \ldots, r_j\}$ and $S = \{s_1, s_2, \ldots, s_j\}$ be two probability distributions on the same set.

For multilevel image thresholding [153], minimizing the cross-entropy techniques can often provide accurate measurement in threshold selection and the technique is also simple, but computation time rises exponentially as the segmentation level rises [34, 35]. The reduction in the computation time by the firefly algorithm was carried out by Horng [73] to select several thresholds by reducing the cross entropy amongst the object and the background (via a minimum cross-entropy thresholding technique).

4.1.3 Tsallis Entropy Criterion Method

Tsallis entropy is a recent advancement of statistical mechanics, based on the concept of non-extensive entropy and originally derived from Shannon's theory [154, 155]. Shannon's entropy can be represented as:

$$S_r = \frac{1 - \sum_{k=1}^{\text{Thr}} (P_k)^r}{r - 1} \qquad (21)$$

The above equation satisfies Shannon's entropy, when entropic index $(\mathbf{r}) \to 1$. Considering 1 grey levels $(\{0, 1, \ldots, (l-1)\})$ with probability distributions $\mathbf{P_k} = \mathbf{P_0}, \mathbf{P_1}, \ldots, \mathbf{P_{l-1}}$ of a greyscale image where $\sum \mathbf{P_k} = 1$, Tsallis multi-thresholding can be represented as (deploying additivity rule for the entropy) and considering $f(\mathbf{Thr})$ as the optimal threshold [156]:

$$
\begin{aligned}
\mathbf{K_{max}} = \mathbf{f(Thr)} &= [\mathbf{Thr_1}, \mathbf{Thr_2}, \ldots, \mathbf{Thr_m}] \\
&= \mathbf{argmax}\big[\mathbf{S_r^A(Thr)} + \mathbf{S_r^B(Thr)} + \cdots + \mathbf{S_r^M(Thr)} \\
&\quad + (1 - \mathbf{r}).\mathbf{S_r^A(Thr)}.\mathbf{S_r^B(Thr)} \ldots \mathbf{S_r^M(Thr)} \big]
\end{aligned}
\qquad (22)
$$

where

$$S_r^A(\text{Thr}) = \frac{1 - \sum_{k=0}^{t_1-1} \left(\frac{P_k}{P^A}\right)^r}{r-1}, \quad P^A = \sum_{k=0}^{t_1-1} P_k$$

$$S_r^B(\text{Thr}) = \frac{1 - \sum_{k=t_1}^{t_2-1} \left(\frac{P_k}{P^B}\right)^r}{r-1}, \quad P^B = \sum_{k=t_1}^{t_1-1} P_k$$

$$S_r^M(\text{Thr}) = \frac{1 - \sum_{k=t_m}^{l-1} \left(\frac{P_k}{P^M}\right)^r}{r-1}, \quad P^M = \sum_{k=t_k}^{l-1} P_k$$

They are subject to these following constraints:

$$|P^A + P^B| - 1 < S < 1 - |P^A - P^B|$$

$$|P^B + P^C| - 1 < S < 1 - |P^B - P^C|$$

$$|P^M + P^{l-1}| - 1 < S < 1 - |P^M - P^{l-1}|$$

To maximize the value of Thr (optimal threshold), the search randomly alters the values of the selected threshold (having a certain probability value) until is K_{max} reached.

Tsallis entropy criterion-based multilevel thresholding was reported in Rajinikanth et al. [157] where the firefly algorithm was used to obtain K_{max}.

From the obtained simulated results in Manic et al. [158], author(s) reported that Tsallis entropy delivered better performance measure values, whereas Kapur's entropy method provided faster convergence and lower CPU time duration.

5   Image Compression

In multimedia applications, image compression with excellent reconstructed image quality is a crucial and challenging task for researchers. Vector quantization (VQ) is a powerful and efficient tool for lossy image compression because the compression technique is non-transformed. In Chiranjeevi and Jena [159], it was reported that the prime objective of VQ was to design codebook that comprised of a set of codewords assigned to the input image vector based on the Euclidean shortest distance. Linde–Buzo–Gray (LBG) [160] is the most common VQ technique that designs a local optimal codebook for image compression.

Vector quantization is a block coding technique for loss data compression [151]. Codebook generation is known as VQ's most important process. Let the original image size $I = \{i_{pq}\}$ be $N \times N$ pixels divided into multiple blocks of $m \times m$ pixels. In other words, there are $M_b = \left[\frac{M}{m}\right] \times \left[\frac{M}{m}\right]$ blocks that signified by an assortment of input vectors $Y = (y_p, p = 1, 2, \ldots, M_b)$. Let S be $m \times m$. The input vector $y_p, y_p \in \Re^S$

is S-dimensional Euclidean space. A codebook B includes M_B is S-dimensional codewords, i.e., $B = \{B_1, B_2, \ldots, B_{M_B}\}$, $B_q \in \Re^S$, $\forall q = 1, 2, \ldots, M_B$. Every input vector is characterized by a row vector $y_p = (y_{p1}, y_{p2}, \ldots, y_{pS})$, and the qth codebook codeword is signified as $B_q = (B_{q1}, B_{q2}, \ldots, B_{qS})$. The VQ methods essentially allocate a related codeword to each input vector, and the codeword will probably replace the input vectors associated with acquire the compression target.

By reducing the distortion function F, optimizing B in terms of mean square errors (MSE) can be formulated. Generally, the lower the F value, the better the quality of B will be.

$$F(B) = \frac{1}{M_b} \sum_{q=1}^{M_B} \sum_{p=1}^{M_b} \omega_{pq} \cdot \|y_p - B_q\|^2 \tag{23}$$

where

$$\sum_{q=1}^{M_B} \omega_{pq} = 1, \forall p \in \{1, 2, \ldots, M_b\}$$

$$\omega_{pq} = \begin{cases} 1 & \text{if } y_p \text{ is in the } q\text{th cluster,} \\ 0, & \text{otherwise} \end{cases}$$

and

$$S_a \leq C_{qa} \leq W_a, a = 1, 2, \ldots, S$$

Here, S_a and W_a are the smallest and largest of the ath components in the entire training and input vectors. The $\|y - B\|$ is the Euclidean distance amongst the vector y and codeword B. For an optimal vector quantizer, there are two necessary conditions.

- The partition T_q, $q = 1, \ldots, M_B$ must satisfy:

$$T_q \supset \{y \in Y : d(y, B_q) < d(y, B_a), \forall a \neq q\} \tag{24}$$

- The codeword B_q must be specified by the centre of gravity of T_q:

$$B_q = \frac{1}{M_q} \sum_{p=1}^{M_q} y_p, \ y_p \in T_q \tag{25}$$

where M_q is the overall number of vectors belonging to T_q.

Horng [161] proposed a firefly algorithm-based technique to construct the codebook of vector quantization. LBG method [162] used as the starting of FA to design the VQ algorithm. The following fitness function has been used to achieve the nearly global optimal vector quantization codebook:

$$\text{fitness}(B) = \frac{1}{F(B)} = \frac{M_b}{\sum_{q=1}^{M_B} \sum_{p=1}^{M_b} \omega_{pq} \cdot \left\| y_p - B_q \right\|^2} \tag{26}$$

The results obtained were compared to those by particle swarm optimization, optimization of quantum particle swarm, and optimization of honey bee mating to explain its expertise. In Severo et al. [163], a modified firefly algorithm was used for vector quantization codebook design, which attempted to assure a stronger influence of the training set in codebook design.

6 Other Imaging Techniques

In Gupta et al. [164], modified FA was reported for edge detection, whereas in Nikolic et al. [165, 166], FA was used in adjusted Canny edge detection algorithm for medical imaging. In Chakraborty et al. [167], FA was used in nonrigid demons registration to optimize the velocity-smoothing kernels of the demons registration. The correlation coefficient was considered as a fitness function in that study. In [168], a rigid image registration technique was explored in which input variables contain two translational parameters and one rotational parameter as well as normalized cross-correlation as the fitness function. In Lin [169], an adaptive two-pass median filter design based on support vector machines was proposed. A region-based image retrieval using the firefly algorithm and support vector machine was proposed in Kanimozhi [170]. Proposed effective suppression of impulse noise for image restoration was used to preserve more image details. In Darwish [171], automatic multilabel image annotation using Bayesian classifier coupled with firefly algorithm was reported. Obviously, these results are just preliminary results, and further investigation is needed to study the hybridization with various imaging techniques to prove the efficacy and robustness of the algorithm.

7 Feature Selection

Selection of features is a challenging task in pattern classification, as it involves searching through a high-dimensional space. An extensive search is computationally unfeasible, particularly when there are a huge number of features. This has directed to a broad range of techniques for selecting features (reduction of dimensionality). The most commonly used feature selection techniques incorporate (or eliminate) variables in a step-by-step fashion [172]. For example, direct sequential selection begins by choosing the finest feature amongst all the obtainable ones. It helps to keep the newest subgroup of features selected at each and every step and then adds the finest feature amongst the residual features to it. This technique halts when its performance is not improved significantly by introducing a new variable. Unlike sequential direct selection, sequential backward selection begins with choosing the full set of features and rejecting a single feature at a time. The feature that induces the smallest possible performance reduction is removed permanently at a specified step. The process ends when a high-performance reduction is caused by removing each of the remaining features.

The main limitation of these sequential techniques is that at an initial stage of the search process, they may dismiss a variable that is then inaccessible for probable future performance enhancement. This implies that during the search process, this type of feature selection techniques can lock at a local minimum. A promising alternative is the utilization of population-based optimization algorithms, like genetic algorithms [173, 174], particle swarm optimization [175, 176], ant colony optimization [177, 178], cuckoo search [179, 180], and firefly algorithm [181, 182]. These studies show that such approaches can explore, in an effective manner, high-dimensional spaces for function optimization.

In Zhang et al. [182], a modified firefly algorithm was used for feature optimization. This work proposed a FA variant for the selection of features to minimize the early convergence issue of the original FA model. It dealt with a variety of simulated annealing—improved global and local signals, chaotic attractiveness movements, dispersing weedy solutions strategies, and the worst and best recollections to lead the search and boost search diversity for global optimism. The objective function represented in the equation below is used to assess the fitness of each and every chosen subset of features.

$$\text{Fitness}(f) = \omega_1 * \text{accuracy}_f + \omega_2 * \left(\text{no.\_of\_features}_f\right)^{-1}$$

where ω_1 and ω_2 represent the classification accurateness weights and the number of chosen features, correspondingly, with $\omega_1 + \omega_2 = 1$. Because classification correctness is more vital than the number of carefully chosen features, the value of $\omega_1 = 0.9$ and $\omega_2 = 0.1$ is assigned based on the reference of connected studies [181]. Proposed method is assessed with 11 regression and 29 classification problems.

In Mistry et al. [181], a new facial expression identification system was used with modified local Gabor binary patterns for extraction of feature and a feature optimization FA variant was proposed.

8   Classification

In Hongjun [183, 184], extreme learning machine (ELM) and band selection for hyperspectral image recognition using FA were proposed. For the reduction of complexity of the ELM network, a subset of original bands was selected using FA. ELM parameters, namely (i) regularization coefficient (c), (ii) Gaussian kernel (σ), and

(iii) hidden number of neurons (L), are optimized using FA where the classification accuracy was considered as an objective function. In Aadit et al. [185, 186], best firefly algorithm joined with ISO-FLANN recognition was used to recognize spontaneous micro-expression for possible philanthropical applications (lie detection, national security, improving relationships, psychiatry, and clinical therapy, etc.). In Shamshirband et al. [187], support vector machine was optimized using the firefly algorithm for lens system which was an important factor in image quality.

9 Application Areas

9.1 Medical Imaging

Firefly algorithm has been widely used in medical imaging for segmentation of various anatomic structures, registration, edge detection, filtering, classification, etc. In Rahebi and Hardalaç [188], an automated system for optic disc detection from retinal images (on DRIVE, STARE, and DiaRetDB1 dataset) was proposed using the firefly algorithm. The reported accuracy and the reported time for detection of all these datasets are significantly high. In Rajinikanth et al. [157], tumour extraction from a two-dimensional (2D) magnetic resonance image (MRI) was reported. In their work, multilevel thresholding using the firefly algorithm was used for preprocessing. In Alomoush et al. [189], firefly-based fuzzy C-means (FCM) were used to find optimal initial cluster centres, thus improving all applications related to fuzzy clustering for image segmentation. In Alsmadi [190], a similar type of study was reported where a hybrid firefly algorithm with the fuzzy C-mean algorithm used for MRI brain segmentation. In Goethe [191], a hybrid supervised feature selection algorithm, namely tolerance rough set firefly-based quick reduct, was MRI brain imaging. In Roopini et al. [192], a segmentation technique for brain tumour detection from axial, sagittal, and coronal views' brain MRI images was reported using fuzzy entropy and distance regularized level set. In Senapati and Dash [193], optimized FA-based local linear wavelet neural network was used to classify breast cancer tumour detection. In Filipczuk et al. [194], a computer-aided system was designed to perform automatic marking of nuclei for breast cancer diagnosis by analysing microscopic images of fine needle biopsy material. FA was used for this extraction of nuclei from cytological image processing. In Sohani [195], support vector machines (SVMs) coupled with firefly algorithm were used to forecast the malaria incidences. Appropriate choice of SVM parameters was reported using FA.

In Nikolic et al. [165, 166], the Canny edge detection algorithm was adjusted using FA, and in Kaur and Singh [196], FA-based sharpening technique was proposed for ultrasound imaging. In Boscolo et al. [197], a knowledge-based segmentation system with sophisticated active contour model-based medical image segmentation was proposed. In Xiaogang et al. [198], a multi-resolution medical image registration algorithm (mutual information registration) based on firefly algorithm and Powell approach was deployed.

9.2 Object Tracking

In Gao et al. [199], arbitrary target tracking in various challenging conditions using a FA-based tracking was reported. In this work, the parameters' sensitivity and adjustment of the FA were also discussed. In Gao et al. [103], particles in the particle filter were optimized for enhanced visual tracking utilizing the firefly algorithm prior to resampling. The number of significant particles was increasable, and the particles can more accurately approximate the true state of the target.

9.3 Satellite Imaging

In Bhandari et al. [19, 20], an edge preserved satellite image denoising using adaptive differential evolution algorithm is discussed (using optimal sub-band adaptive thresholding). Learning the adaptive thresholding function parameters necessary for optimal performance using various optimization algorithms including FA-based approach was reported. In Suresh and Lal [200], multilevel thresholding for segmentation of satellite images [113, 201] was reported. In that study, it was shown that an improved version of the firefly algorithm [202] outperformed DPSO, differential evolution (DE), and FA algorithm for benchmark images, but obtained results were poor in higher thresholding levels [203]. An hourly global solar radiation prediction using FA was reported in Ibrahim and Khatib [204].

9.4 Watermarking and Steganography

In Agarwal [205], an optimized greyscale image watermarking using DWT–SVD and firefly algorithm was reported. Authors successfully demonstrated the use of multiple scaling factors to achieve the best results and the robustness of the proposed technique. However, in Ali [206], a fundamental flaw of Mishra's work was reported. Authors tried to address the false positive detection for SVD-based image watermarking. In Dey et al. [38], scaling factors during embedding of manifold medical information were optimized and applied on ophthalmic imaging. In Dong et al. [207], gray-scale watermarking using DWT-DCT-SVD and Chaotic Firefly algorithm and in Guo et al. [208], DWT-QR transform domain and FA was reported. In Kazemivash and Moghaddam [209], an LWT-FA-based watermark technique was discussed.

In Chhikara [210], a discrete firefly optimization technique with dynamic alpha and gamma parameters and t-test filter technique for hybrid feature selection algorithm was reported to improve detectability of hidden message for blind image steganalysis, whereas an improved dynamic discrete firefly algorithm was reported by the same author in Chhikara [211]. Optimal parameter selection using variable step size firefly algorithm and lifting wavelet transform was used for quick response code-based image steganography in Raja [212].

Apart from the above-mentioned area, the firefly algorithm has also used in many other areas (imaging applications) like key-points' search [213], biometric authentication [214] using fingerprint (Al-Ta'l 201), and hand vein [215].

10   Conclusions

This work presents a comprehensive survey on various imaging techniques in combination with the FA, including enhancement, denoising, segmentation, registration, and others. Application of FA-based imaging techniques has been also discussed. The efficiency of FA is already well established in the literature. Most of the cases FA outperforms other metaheuristic algorithms and significantly improves the obtained results of image processing-based applications. An insignificant number of colour image processing has been reported in the literature.

There are some research topics that demand for further research. First, a more detailed parametric study may be needed to identify the parameter ranges and more fine-tuning of parameters of FA so as to maximize its performance. Second, a more diverse range of applications can be explored so as to solve many real-world image processing problems. Third, some theoretical analysis of the firefly algorithm and other algorithms will be useful to gain greater insights into these algorithms. Fourth, hybridization with both conventional image processing techniques and other metaheuristics can be explored to see if further improvements can be achieved. Finally, these methods can be extended to study more complex image processing tasks in applications related to machine learning and artificial intelligence.

References

1. Pitas I (2000) Digital image processing algorithms and applications. Wiley, New York
2. Vandenbroucke N, Macaire L, Postaire JG (2000) Color image segmentation by supervised pixel classification in a color texture feature space. Application to soccer image segmentation. In: Proceedings 15th IEEE international conference on pattern recognition. ICPR-2000, vol 3, pp 621–624 (September)
3. Vandenbroucke N, Macaire L, Postaire JG (2000) Color image segmentation by supervised pixel classification in a color texture feature space. Application to soccer image segmentation.

In: Proceedings IEEE 15th international conference on pattern recognition. ICPR-2000, vol 3, pp 621–624 (September)
4. Daly S (1994) A visual model for optimizing the design of image processing algorithms. In Proceedings of 1st IEEE international conference on image processing, vol 2, pp 16–20 (November)
5. Ruiz JE, Paciornik S, Pinto LD, Ptak F, Pires MP, Souza PL (2018) Optimization of digital image processing to determine quantum dots' height and density from atomic force microscopy. Ultramicroscopy 184:234–241
6. Grangetto M, Magli E, Martina M, Olmo G (2002) Optimization and implementation of the integer wavelet transform for image coding. IEEE Trans Image Process 11:596–604
7. Dalrymple B, Smith J (2018) Forensic digital image processing: optimization of impression evidence. CRC Press
8. Diamond S, Sitzmann V, Boyd S, Wetzstein G, Heide F (2017) Dirty pixels: optimizing image classification architectures for raw sensor data. arXiv preprint arXiv:1701.06487
9. Wang D, Li G, Jia W, Luo X (2011) Saliency-driven scaling optimization for image retargeting. Vis Comput 27:853–860
10. Shao P, Wu Z, Zhou X, Tran DC (2017) FIR digital filter design using improved particle swarm optimization based on refraction principle. Soft Comput 21:2631–2642
11. George EB, Karnan M (2012) MR brain image segmentation using bacteria foraging optimization algorithm. Int J Eng Technol (IJET) 4:295–301
12. Precht H, Gerke O, Rosendahl K, Tingberg A, Waaler D (2012) Digital radiography: optimization of image quality and dose using multi-frequency software. Pediatr Radiol 42:1112–1118
13. Vahedi E, Zoroofi RA, Shiva M (2012) Toward a new wavelet-based watermarking approach for color images using bio-inspired optimization principles. Digit Signal Proc 22:153–162
14. Loukhaoukha K, Chouinard JY, Taieb MH (2011) Optimal image watermarking algorithm based on LWT-SVD via multi-objective ant colony optimization. J Inf Hiding Multimedia Sig Process 2:303–319
15. Tuba E, Alihodzic A, Tuba M (2017) Multilevel image thresholding using elephant herding optimization algorithm. In: 2017 IEEE 14th international conference on engineering of modern electric systems (EMES), pp 240–243 (June)
16. Tuba E, Tuba M, Simian D, Jovanovic R (2017) JPEG quantization table optimization by guided fireworks algorithm. In: International workshop on combinatorial image analysis. Springer, Cham, pp 294–307 (June)
17. Moallem P, Razmjooy N (2012) Optimal threshold computing in automatic image thresholding using adaptive particle swarm optimization. J Appl Res Technol 10:703–712
18. Ye Z, Wang M, Jin H, Liu W, Lai X (2015) An image thresholding approach based on ant colony optimization algorithm combined with genetic algorithm. Dimensions 15:16
19. Bhandari AK, Kumar D, Kumar A, Singh GK (2016) Optimal sub-band adaptive thresholding based edge preserved satellite image denoising using adaptive differential evolution algorithm. Neurocomputing 174:698–721
20. Bhandari AK, Kumar A, Chaudhary S, Singh GK (2016) A novel color image multilevel thresholding based segmentation using nature inspired optimization algorithms. Expert Syst Appl 63:112–133
21. Li H, He H, Wen Y (2015) Dynamic particle swarm optimization and K-means clustering algorithm for image segmentation. Optik 126:4817–4822
22. Mahalakshmi S, Velmurugan T (2015) Detection of brain tumor by particle swarm optimization using image segmentation. Indian J Sci Technol 8:1
23. Li Z, Cao J, Zhao X, Liu W (2015) Atmospheric compensation in free space optical communication with simulated annealing algorithm. Opt Commun 338:11–21
24. Dhal KG, Ray S, Das A, Das S (2018) A survey on nature-inspired optimization algorithms and their application in image enhancement domain. Arch Comput Methods Eng 1–32
25. Song KS, Kim MS, Kang MG (2016) Image enhancement algorithm using dynamic range optimization. J Inst Electron Inf Eng 53:101–109

26. Mahapatra PK, Ganguli S, Kumar A (2015) A hybrid particle swarm optimization and artificial immune system algorithm for image enhancement. Soft Comput 19:2101–2109
27. Zhang C, Qin Q, Zhang T, Sun Y, Chen C (2017) Endmember extraction from hyperspectral image based on discrete firefly algorithm (EE-DFA). ISPRS J Photogram Remote Sens 126:108–119
28. Zhang L, Zhou X, Wang Z, Tan C, Liu X (2017) A nonmodel dual-tree wavelet thresholding for image denoising through noise variance optimization based on improved chaotic drosophila algorithm. Int J Pattern Recogn Artif Intell 31:1754015
29. Krishnaveni M, Subashini P, Dhivyaprabha TT (2016) A new optimization approach-SFO for denoising digital images. In: 2016 IEEE international conference on computation system and information technology for sustainable solutions (CSITSS), pp 34–39 (October)
30. Kockanat S, Karaboga N (2017) Medical image denoising using metaheuristics. Metaheuristics for medicine and biology. Springer, Berlin, pp 155–169
31. Ahmadi K, Javaid AY, Salari E (2015) An efficient compression scheme based on adaptive thresholding in wavelet domain using particle swarm optimization. Sig Process Image Commun 32:33–39
32. Emara ME, Abdel-Kader RF, Yasein MS (2017) Image compression using advanced optimization algorithms. J Commun 12
33. Shrivastava P, Shukla A, Vepakomma P, Bhansali N, Verma K (2017) A survey of nature-inspired algorithms for feature selection to identify Parkinson's disease. Comput Methods Programs Biomed 139:171–179
34. Pare S, Kumar A, Bajaj V, Singh GK (2017) An efficient method for multilevel color image thresholding using cuckoo search algorithm based on minimum cross entropy. Appl Soft Comput 61:570–592
35. Pare S, Bhandari AK, Kumar A, Singh GK (2017) An optimal color image multilevel thresholding technique using grey-level co-occurrence matrix. Expert Syst Appl 87:335–362
36. Gholami A, Bonakdari H, Ebtehaj I, Mohammadian M, Gharabaghi B, Khodashenas SR (2018) Uncertainty analysis of intelligent model of hybrid genetic algorithm and particle swarm optimization with ANFIS to predict threshold bank profile shape based on digital laser approach sensing. Measurement 121:294–303
37. Hamid MS, Harvey NR, Marshall S (2003) Genetic algorithm optimization of multidimensional grayscale soft morphological filters with applications in film archive restoration. IEEE Trans Circuits Syst Video Technol 13:406–416
38. Dey N, Ashour A, Beagum S, Pistola D, Gospodinov M, Gospodinova E, Tavares J (2015) Parameter optimization for local polynomial approximation based intersection confidence interval filter using genetic algorithm: an application for brain MRI image de-noising. J Imaging 1:60–84
39. Wang GG, Gandomi AH, Yang XS, Alavi AH (2016) A new hybrid method based on krill herd and cuckoo search for global optimisation tasks. Int J Bio-Inspired Comput 8:286–299
40. Wang Q, Zhou D, Nie R, Jin X, He K, Dou L (2016). Medical image fusion using pulse coupled neural network and multi-objective particle swarm optimization. In: Eighth international conference on digital image processing (ICDIP 2016), International society for optics and photonics, vol 10033, p 100334K (August)
41. Wang Q, Zhou D, Nie R, Jin X, He K, Dou L (2016, August) Medical image fusion using pulse coupled neural network and multi-objective particle swarm optimization. In Eighth international conference on digital image processing (ICDIP 2016), International society for optics and photonics, vol 10033, p 100334K
42. Zheng Z, Saxena N, Mishra KK, Sangaiah AK (2018) Guided dynamic particle swarm optimization for optimizing digital image watermarking in industry applications. Future Gener Comput Syst 88:92–106
43. Kennedy J, Eberhart R, Shi Y (2001) Swarm intelligence. Academic Press
44. Ladgham A, Hamdaoui F, Sakly A, Mtibaa A (2015) Fast MR brain image segmentation based on modified Shuffled Frog Leaping Algorithm. SIViP 9:1113–1120

45. Amiri B, Fathian M, Maroosi A (2009) Application of shuffled frog-leaping algorithm on clustering. Int J Adv Manuf Technol 45:199–209
46. Wang N, Li X, Chen XH (2010) Fast three-dimensional Otsu thresholding with shuffled frog-leaping algorithm. Pattern Recogn Lett 31:1809–1815
47. Brajevic I, Tuba M (2014) Cuckoo search and firefly algorithm applied to multilevel image thresholding. Cuckoo search and firefly algorithm. Springer, Cham, pp 115–139
48. Senthilnath J, Das V, Omkar SN, Mani V (2013) Clustering using levy flight cuckoo search. In: Proceedings of seventh international conference on bio-inspired computing: theories and applications (BIC-TA 2012). Springer, India, pp 65–75
49. Tiwari V (2012) Face recognition based on cuckoo search algorithm. Image 7:9
50. Yang XS, Deb S (2009) Cuckoo search via Lévy flights. In: 2009 IEEE world congress on nature & biologically inspired computing (NaBIC), pp 210–214 (December)
51. Yang XS, Deb S (2010) Engineering optimisation by cuckoo search. arXiv preprint arXiv: 1005.2908
52. Alihodzic A, Tuba M (2014) Improved bat algorithm applied to multilevel image thresholding. Sci World J
53. Satapathy SC, Raja NSM, Rajinikanth V, Ashour AS, Dey N (2018) Multi-level image thresholding using Otsu and chaotic bat algorithm. Neural Comput Appl 29:1285–1307
54. Cai X, Wang H, Cui Z, Cai J, Xue Y, Wang L (2018) Bat algorithm with triangle-flipping strategy for numerical optimization. Int J Mach Learn Cybern 9:199–215
55. Yang XS (2010) Nature-inspired metaheuristic algorithms. Luniver Press
56. Yang XS (2010) Firefly algorithm, Levy flights and global optimization. Research and development in intelligent systems XXVI. Springer, London, pp 209–218
57. Yang XS (2010) Firefly algorithm, stochastic test functions and design optimisation. arXiv preprint arXiv:1003.1409
58. Yang XS (2010) A new metaheuristic bat-inspired algorithm. Nature inspired cooperative strategies for optimization (NICSO 2010). Springer, Berlin, pp 65–74
59. Yang XS (2012) Flower pollination algorithm for global optimization. International conference on unconventional computing and natural computation. Springer, Berlin, pp 240–249 (September)
60. Yang XS, Karamanoglu M, He X (2013) Multi-objective flower algorithm for optimization. Procedia Comput Sci 18:861–868
61. Tian J, Yu W, Xie S (2008) An ant colony optimization algorithm for image edge detection. In: 2008 IEEE congress on evolutionary computation (IEEE world congress on computational intelligence), pp 751–756 (June)
62. Dorigo M (1992) Optimization, learning and natural algorithms. Ph.D. thesis, Politecnico di Milano
63. Cinsdikici MG, Aydın D (2009) Detection of blood vessels in ophthalmoscope images using MF/ant (matched filter/ant colony) algorithm. Comput Methods Programs Biomed 96:85–95
64. Tao W, Jin H, Liu L (2007) Object segmentation using ant colony optimization algorithm and fuzzy entropy. Pattern Recogn Lett 28:788–796
65. Hanmandlu M, Verma OP, Kumar NK, Kulkarni M (2009) A novel optimal fuzzy system for color image enhancement using bacterial foraging. IEEE Trans Instrum Meas 58:2867–2879
66. Sanyal N, Chatterjee A, Munshi S (2011) An adaptive bacterial foraging algorithm for fuzzy entropy based image segmentation. Expert Syst Appl 38:15489–15498
67. Sathya PD, Kayalvizhi R (2011) Optimal multilevel thresholding using bacterial foraging algorithm. Expert Syst Appl 38:15549–15564
68. Sathya PD, Kayalvizhi R (2011) Modified bacterial foraging algorithm based multilevel thresholding for image segmentation. Eng Appl Artif Intell 24:595–615
69. Neshat M, Sepidnam G, Sargolzaei M, Toosi AN (2014) Artificial fish swarm algorithm: a survey of the state-of-the-art, hybridization, combinatorial and indicative applications. Artif Intell Rev 42:965–997
70. El-Said SA (2015) Image quantization using improved artificial fish swarm algorithm. Soft Comput 19:2667–2679

71. Chu X, Zhu Y, Shi J, Song J (2010) Method of image segmentation based on fuzzy C-means clustering algorithm and artificial fish swarm algorithm. In: 2010 IEEE international conference on intelligent computing and integrated systems, pp 254–257 (October)
72. Akay B (2013) A study on particle swarm optimization and artificial bee colony algorithms for multilevel thresholding. Appl Soft Comput 13:3066–3091
73. Horng MH (2011) Multilevel thresholding selection based on the artificial bee colony algorithm for image segmentation. Expert Syst Appl 38:13785–13791
74. Zhang YD, Wu L, Wang S (2011) Magnetic resonance brain image classification by an improved artificial bee colony algorithm. Progress Electromagnet Res 116:65–79
75. Geng J, Li MW, Dong ZH, Liao YS (2015) Port throughput forecasting by MARS-RSVR with chaotic simulated annealing particle swarm optimization algorithm. Neurocomputing 147:239–250
76. Kirkpatrick S, Gelatt CD, Vecchi MP (1983) Optimization by simulated annealing. Science 220:671–680
77. Zhang Y, Yan H, Zou X, Tao F, Zhang L (2016) Image threshold processing based on simulated annealing and OTSU method. In: Proceedings of the 2015 Chinese intelligent systems conference. Springer, Berlin, pp 223–231
78. Yang XS (2014) Nature-inspired optimization algorithms. Elsevier, London
79. Bagheri M, Mirbagheri SA, Bagheri Z, Kamarkhani AM (2015) Modeling and optimization of activated sludge bulking for a real wastewater treatment plant using hybrid artificial neural networks-genetic algorithm approach. Process Saf Environ Prot 95:12–25
80. Ghosh P, Mitchell M, Tanyi JA, Hung AY (2016) Incorporating priors for medical image segmentation using a genetic algorithm. Neurocomputing 195:181–194
81. Yang XS (2008) Nature-inspired metaheuristic algorithms. Luniver Press, Bristol, UK
82. Ghamisi P, Benediktsson JA (2015) Feature selection based on hybridization of genetic algorithm and particle swarm optimization. IEEE Geosci Remote Sens Lett 12(2):309–313
83. Esmin AA, Coelho RA, Matwin S (2015) A review on particle swarm optimization algorithm and its variants to clustering high-dimensional data. Artif Intell Rev 44:23–45
84. Armano G, Farmani MR (2016) Multiobjective clustering analysis using particle swarm optimization. Expert Syst Appl 55:184–193
85. Chen Y, Zhu Q, Xu H (2015) Finding rough set reducts with fish swarm algorithm. Knowl Based Syst 81:22–29
86. Pérez-Delgado ML (2019) Color image quantization using the shuffled-frog leaping algorithm. Eng Appl Artif Intell 79:142–158
87. Ma M, Zhu Q (2017) Multilevel thresholding image segmentation based on shuffled frog leaping algorithm. J Comput Theor Nanosci 14:3794–3801
88. Sharma TK, Pant M (2017) Opposition based learning ingrained shuffled frog-leaping algorithm. J Comput Sci 21:307–315
89. Bermejo E, Cordón O, Damas S, Santamaría J (2015) A comparative study on the application of advanced bacterial foraging models to image registration. Inf Sci 295:160–181
90. Hossain MA, Ferdous I (2015) Autonomous robot path planning in dynamic environment using a new optimization technique inspired by bacterial foraging technique. Rob Auton Syst 64:137–141
91. Wan S, Chang SH, Chou TY, Shien CM (2018) A study of landslide image classification through data clustering using bacterial foraging optimization
92. Shi L, Guo R, Ma Y (2016) A novel artificial fish swarm algorithm for pattern recognition with convex optimization. In: 2016 international conference on communication and electronics systems (ICCES), pp 1–4 (October)
93. Nalluri MSR, SaiSujana T, Reddy KH, Swaminathan V (2017) An efficient feature selection using artificial fish swarm optimization and svm classifier. In 2017 international conference on networks & advances in computational technologies (NetACT), pp. 407–411
94. Bhandari AK, Kumar A, Singh GK (2015) Modified artificial bee colony based computationally efficient multilevel thresholding for satellite image segmentation using Kapur's, Otsu and Tsallis functions. Expert Syst Appl 42:1573–1601

95. Bansal JC, Gopal A, Nagar AK (2018) Stability analysis of artificial bee colony optimization algorithm. Swarm Evol Comput 41:9–19
96. Chen J, Yu W, Tian J, Chen L, Zhou Z (2018) Image contrast enhancement using an artificial bee colony algorithm. Swarm Evol Comput 38:287–294
97. Wang P, Lin JS, Wang M (2015) An image reconstruction algorithm for electrical capacitance tomography based on simulated annealing particle swarm optimization. J Appl Res Technol 13:197–204
98. Ayumi V, Rere LR, Fanany MI, Arymurthy AM (2016) Optimization of convolutional neural network using microcanonical annealing algorithm. In: 2016 IEEE international conference on advanced computer science and information systems (ICACSIS), pp 506–511 (October)
99. Dong Y, Wang J, Chen F, Hu Y, Deng Y (2017) Location of facility based on simulated annealing and "ZKW" algorithms. Math Probl Eng
100. Perez J, Melin P, Castillo O, Valdez F, Gonzalez C, Martinez G (2017) Trajectory optimization for an autonomous mobile robot using the Bat Algorithm. In: North American fuzzy information processing society annual conference, Springer, Cham, pp 232–241 (October)
101. Sameen MI, Pradhan B, Shafri HZ, Mezaal MR, bin Hamid H (2017) Integration of ant colony optimization and object-based analysis for LiDAR data classification. IEEE J Sel Top Appl Earth Obs Remote Sens 10:2055–2066
102. Gao ML, Shen J, Yin LJ, Liu W, Zou GF, Li HT, Fu GX (2016) A novel visual tracking method using bat algorithm. Neurocomputing 177:612–619
103. Gao ML, Li LL, Sun XM, Yin LJ, Li HT, Luo DS (2015) Firefly algorithm (FA) based particle filter method for visual tracking. Optik 126:1705–1711
104. Katiyar S, Patel R, Arora K (2016) Comparison and analysis of cuckoo search and firefly algorithm for image enhancement. International conference on smart trends for information technology and computer communications. Springer, Singapore, pp 62–68 (August)
105. Tabakhi S, Moradi P (2015) Relevance–redundancy feature selection based on ant colony optimization. Pattern Recogn 48:2798–2811
106. Dao TP, Huang SC, Thang PT (2017) Hybrid Taguchi-cuckoo search algorithm for optimization of a compliant focus positioning platform. Appl Soft Comput 57:526–538
107. Ye Z, Yang J, Wang M, Zong X, Yan L, Liu W (2018) 2D Tsallis entropy for image segmentation based on modified chaotic bat algorithm. Entropy 20:239
108. Yang XS (2009) Firefly algorithms for multimodal optimization. International symposium on stochastic algorithms. Springer, Berlin, pp 169–178 (October)
109. Wang H, Cui Z, Sun H, Rahnamayan S, Yang XS (2017) Randomly attracted firefly algorithm with neighborhood search and dynamic parameter adjustment mechanism. Soft Comput 21:5325–5339
110. Asl PF, Monjezi M, Hamidi JK, Armaghani DJ (2018) Optimization of flyrock and rock fragmentation in the Tajareh limestone mine using metaheuristics method of firefly algorithm. Eng Comput 34:241–251
111. Dey N (2019) Uneven illumination correction of digital images: a survey of the state-of-the-art. Optik 183:483–495
112. Dey N (ed) (2017) Advancements in applied metaheuristic computing. IGI Global, Hershey
113. Dey N, Ashour AS (2018) Meta-heuristic algorithms in medical image segmentation: a review. In: Advancements in applied metaheuristic computing. IGI Global, Hershey, pp 185–203
114. Mustafi A, Mahanti PK (2009) An optimal algorithm for contrast enhancement of dark images using genetic algorithms. In: Computer and information science 2009. Springer, Berlin, pp. 1–8
115. Ashour AS, Samanta S, Dey N, Kausar N, Abdessalemkaraa WB, Hassanien AE (2015) Computed tomography image enhancement using cuckoo search: a log transform based approach. J Signal Inf Process 6:244
116. Gorai A, Ghosh A (2009) Gray-level image enhancement by particle swarm optimization. In: 2009 IEEE world congress on nature & biologically inspired computing (NaBIC), pp 72–77 (December)

117. Dhal KG, Quraishi IM, Das S (2015) A chaotic Lévy flight approach in bat and firefly algorithm for gray level image enhancement. IJ Image Graph Sig Process 7:69–76

118. Samanta S, Mukherjee A, Ashour AS, Dey N, Tavares JMR, Abdessalem Karâa WB, Hassanien AE (2018) Log transform based optimal image enhancement using firefly algorithm for autonomous mini unmanned aerial vehicle: An application of aerial photography. Int J Image Graph 18:1850019

119. Draa A, Benayad Z, Djenna FZ (2015) An opposition-based firefly algorithm for medical image contrast enhancement. Int J Inf Commun Technol 7:385–405

120. Hassanzadeh T, Vojodi H, Mahmoudi F (2011) December) Non-linear grayscale image enhancement based on firefly algorithm. International conference on swarm, evolutionary, and memetic computing. Springer, Berlin, pp 174–181

121. Achim A, Bezerianos A, Tsakalides P (2001) Novel Bayesian multiscale method for speckle removal in medical ultrasound images. IEEE Trans Med Imaging 20:772–783

122. Argenti F, Alparone L (2002) Speckle removal from SAR images in the undecimated wavelet domain. IEEE Trans Geosci Remote Sens 40:2363–2374

123. Xie H, Pierce LE, Ulaby FT (2002) SAR speckle reduction using wavelet denoising and Markov random field modeling. IEEE Trans Geosci Remote Sens 40:2196–2212

124. Chang SG, Yu B, Vetterli M (2000) Adaptive wavelet thresholding for image denoising and compression. IEEE Trans Image Process 9:1532–1546

125. Nasri M, Nezamabadi-pour H (2009) Image denoising in the wavelet domain using a new adaptive thresholding function. Neurocomputing 72:1012–1025

126. Agrawal S, Panda R, Bhuyan S, Panigrahi BK (2013) Tsallis entropy based optimal multilevel thresholding using cuckoo search algorithm. Swarm Evol Comput 11:16–30

127. Sezgin M, Sankur B (2004) Survey over image thresholding techniques and quantitative performance evaluation. J Electron Imaging 13:146–166

128. Zahara E, Fan SKS, Tsai DM (2005) Optimal multi-thresholding using a hybrid optimization approach. Pattern Recogn Lett 26:1082–1095

129. Kittler J, Illingworth J (1986) Minimum error thresholding. Pattern Recogn 19:41–47

130. Pun T (1981) Entropy thresholding: a new approach. Comput Vision Graph Image Proc 16:210–239

131. Otsu N (1979) A threshold selection method from gray-level histograms. IEEE Trans Syst Man Cybern 9:62–66

132. Kapur JN, Sahoo PK, Wong AK (1985) A new method for gray-level picture thresholding using the entropy of the histogram. Comput Vis Graph Image Process 29:273–285

133. Song JH, Cong W, Li J (2017) A fuzzy C-means clustering algorithm for image segmentation using nonlinear weighted local information. J Inf Hiding Multimedia Sig Process 8:1–11

134. Yin PY (1999) A fast scheme for optimal thresholding using genetic algorithms. Sig Process 72:85–95

135. Lai CC, Tseng DC (2004) A hybrid approach using Gaussian smoothing and genetic algorithm for multilevel thresholding. Int J Hybrid Intell Syst 1:143–152

136. Maitra M, Chatterjee A (2008) A hybrid cooperative–comprehensive learning based PSO algorithm for image segmentation using multilevel thresholding. Expert Syst Appl 34:1341–1350

137. Shi Y, Eberhart R (1998) A modified particle swarm optimizer. In: 1998 IEEE international conference on evolutionary computation proceedings. IEEE world congress on computational intelligence (Cat. No. 98TH8360), pp 69–73 (May)

138. Gao H, Xu W, Sun J, Tang Y (2010) Multilevel thresholding for image segmentation through an improved quantum-behaved particle swarm algorithm. IEEE Trans Instrum Measur 59:934–946

139. Ye Z, Zheng Z, Yu X, Ning X (2006) Automatic threshold selection based on ant colony optimization algorithm. In: International conference on neural networks and brain, Beijing, pp 728–732

140. Samantaa S, Dey N, Das P, Acharjee S, Chaudhuri SS (2013) Multilevel threshold based gray scale image segmentation using cuckoo search. arXiv preprint arXiv:1307.0277

141. Rajinikanth V, Raja NSM, Satapathy SC (2016) Robust color image multi-thresholding using between-class variance and cuckoo search algorithm. Information systems design and intelligent applications. Springer, New Delhi, pp 379–386

142. Horng MH, Jiang TW, Chen JY (2009) Multilevel minimum cross entropy threshold selection based on honey bee mating optimization. In: Proceedings of the international multi conference of engineers and computer scientists, Hong Kong, pp 978–988

143. Resma KB, Nair MS (2018) Multilevel thresholding for image segmentation using Krill Herd Optimization algorithm. J King Saud Univ Comput Inf Sci

144. Khairuzzaman AKM, Chaudhury S (2017) Multilevel thresholding using grey wolf optimizer for image segmentation. Expert Syst Appl 86:64–76

145. Łukasik S, Żak S (2009) Firefly algorithm for continuous constrained optimization tasks. International conference on computational collective intelligence. Springer, Berlin, pp 97–106 (October)

146. Shah-Hosseini H (2011) Otsu's criterion-based multilevel thresholding by a nature-inspired metaheuristic called galaxy-based search algorithm. In 2011 IEEE third world congress on nature and biologically inspired computing, pp 383–388 (October)

147. Zhou C, Tian L, Zhao H, Zhao K (2015) A method of two-dimensional Otsu image threshold segmentation based on improved firefly algorithm. In: 2015 IEEE international conference on cyber technology in automation, control, and intelligent systems (CYBER), pp 1420–1424 (June)

148. Raja N, Rajinikanth V, Latha K (2014) Otsu based optimal multilevel image thresholding using firefly algorithm. Model Simul Eng 2014:37

149. Bhandari AK, Singh VK, Kumar A, Singh GK (2014) Cuckoo search algorithm and wind driven optimization based study of satellite image segmentation for multilevel thresholding using Kapur's entropy. Expert Syst Appl 41:3538–3560

150. He L, Huang S (2017) Modified firefly algorithm based multilevel thresholding for color image segmentation. Neurocomputing 240:152–174

151. Horng MH (2010) Multilevel minimum cross entropy threshold selection based on the honey bee mating optimization. Expert Syst Appl 37:4580–4592

152. Kullback S (1968) Information theory and statistics. Dover, New york

153. Li CH, Tam PKS (1998) An iterative algorithm for minimum cross entropy thresholding. Pattern Recognit Lett 19(8):771–776

154. Tsallis C (1988) Possible generalization of Boltzmann-Gibbs statistics. J Stat Phys 52:479–487

155. Tsallis C, Rajagopal AK, Plastino AR, Andricioaei I, Stranb JE, Abe S, Klos J (2001) Nonextensive statistical mechanics and its applications. In: Abe S, Okamoto Y (eds) Series lecture notes in physics. Springer, Berlin

156. Ramírez-Reyes A, Hernández-Montoya A, Herrera-Corral G, Domínguez-Jiménez I (2016) Determining the entropic index q of Tsallis entropy in images through redundancy. Entropy 18:299

157. Rajinikanth V, Raja NSM, Kamalanand K (2017) Firefly algorithm assisted segmentation of tumor from brain MRI using Tsallis function and Markov random field. J Control Eng Appl Inform 19:97–106

158. Manic KS, Priya RK, Rajinikanth V (2016) Image multithresholding based on Kapur/Tsallis entropy and firefly algorithm. Indian J Sci Technol 9:89949

159. Chiranjeevi K, Jena UR (2016) Image compression based on vector quantization using cuckoo search optimization technique. Ain Shams Eng J

160. Linde Y, Buzo A, Gray R (1980) An algorithm for vector quantizer design. IEEE Trans Commun 28:84–95

161. Horng MH (2012) Vector quantization using the firefly algorithm for image compression. Expert Syst Appl 39:1078–1091

162. Lloyd SP (1957) Least square quantization in PCM's. Bell Telephone Laboratories Paper, Murray Hill, NJ

163. Severo V, Leitão HAS, Lima JB, Lopes WTA, Madeiro F (2016) Modified firefly algorithm applied to image vector quantisation codebook design. Int J Innov Comput Appl 7:202–213
164. Gupta M, Tazi SN, Jain A (2014) Edge detection using Modified Firefly Algorithm. In: 2014 IEEE international conference on computational intelligence and communication networks, pp 167–173 (November)
165. Nikolic M, Tuba E, Tuba M (2016) Edge detection in medical ultrasound images using adjusted Canny edge detection algorithm. In: 2016 IEEE 24th telecommunications forum (TELFOR), pp 1–4 (November)
166. Nikolic M, Tuba E, Tuba M (2016, November) Edge detection in medical ultrasound images using adjusted Canny edge detection algorithm. In: 2016 IEEE 24th telecommunications forum (TELFOR), pp 1–4
167. Chakraborty S, Dey N, Samanta S, Ashour AS, Balas VE (2016) Firefly algorithm for optimized nonrigid demons registration. In: Bio-inspired computation and applications in image processing. Academic Press, pp 221–237
168. Zhang Y, Wu L (2012) A novel method for rigid image registration based on firefly algorithm. Int J Res Rev Soft Intell Comput (IJRRSIC) 2:141–146
169. Lin TC, Yu PT (2004) Adaptive two-pass median filter based on support vector machines for image restoration. Neural Comput 16(2):333–354
170. Kanimozhi T, Latha K (2015) An integrated approach to region based image retrieval using firefly algorithm and support vector machine. Neurocomputing 151:1099–1111
171. Darwish SM (2016) Combining firefly algorithm and Bayesian classifier: new direction for automatic multilabel image annotation. IET Image Process 10:763–772
172. Siedlecki R, Sklansky J (1988) On automatic feature selection. Int J Pattern Recog Artificial Intell 2:197–220
173. Tsai CF, Eberle W, Chu CY (2013) Genetic algorithms in feature and instance selection. Knowl Based Syst 39:240–247
174. Chtioui Y, Bertrand D, Barba D (1998) Feature selection by a genetic algorithm. Application to seed discrimination by artificial vision. J Sci Food Agric 76:77–86
175. Xue B, Zhang M, Browne WN (2014) Particle swarm optimisation for feature selection in classification: novel initialisation and updating mechanisms. Appl Soft Comput 18:261–276
176. Xue B, Zhang M, Browne WN (2013) Particle swarm optimization for feature selection in classification: a multi-objective approach. IEEE Trans Cybern 43(6):1656–1671
177. Neagoe VE, Neghina EC (2016) Feature selection with ant colony optimization and its applications for pattern recognition in space imagery. In: 2016 IEEE international conference on communications (COMM), pp 101–104 (June)
178. Kanan HR, Faez K, Taheri SM (2007) Feature selection using ant colony optimization (ACO): a new method and comparative study in the application of face recognition system. In: Industrial conference ondata mining. Springer, Berlin, Heidelberg, pp 63–76
179. Rodrigues D, Pereira LA, Almeida TNS, Papa JP, Souza AN, Ramos CC, Yang XS (2013) BCS: a binary cuckoo search algorithm for feature selection. In: 2013 IEEE international symposium on circuits and systems (ISCAS2013), pp 465–468 (May)
180. Reddi KK, Enireddy V (2016) Cuckoo search framework for feature selection and classifier optimization in compressed medical image retrieval. i-manager's J Image Process 3:1
181. Mistry K, Zhang L, Sexton G, Zeng Y, He M (2017) Facial expression recongition using firefly-based feature optimization. In: 2017 IEEE congress on evolutionary computation (CEC), pp 1652–1658 (June)
182. Zhang L, Mistry K, Lim CP, Neoh SC (2018) Feature selection using firefly optimization for classification and regression models. Decis Support Syst 106:64–85
183. Su H, Cai Y, Du Q (2016) Firefly-algorithm-inspired framework with band selection and extreme learning machine for hyperspectral image classification. IEEE J Sel Topics Appl Earth Observ Remote Sens 10(1):309–320
184. Su H, Tian S, Cai Y, Sheng Y, Chen C, Najafian M (2017) Optimized extreme learning machine for urban land cover classification using hyperspectral imagery. Front Earth Sci 11(4):765–773

185. Aadit MNA, Mahin MT, Juthi SN (2017) Spontaneous micro-expression recognition using optimal firefly algorithm coupled with ISO-FLANN classification. In: 2017 IEEE region 10 humanitarian technology conference (R10-HTC), pp 714–717 (December)
186. Aadit MNA, Mahin MT, Juthi SN (2017, December) Spontaneous micro-expression recognition using optimal firefly algorithm coupled with ISO-FLANN classification. In: 2017 IEEE region 10 humanitarian technology conference (R10-HTC), pp 714–717
187. Shamshirband S, Petković D, Pavlović NT, Ch S, Altameem TA, Gani A (2015) Support vector machine firefly algorithm based optimization of lens system. Appl Opt 54:37–45
188. Rahebi J, Hardalaç F (2016) A new approach to optic disc detection in human retinal images using the firefly algorithm. Med Biol Eng Comput 54:453–461
189. Alomoush WK, Abdullah SNHS, Sahran S, Hussain RI (2014) Segmentation of MRI brain images using FCM improved by firefly algorithms. J Appl Sci 14:66–71
190. Alsmadi MK (2014) A hybrid firefly algorithm with fuzzy-C mean algorithm for MRI brain segmentation. Am J Appl Sci 11:1676–1691
191. Jothi G (2016) Hybrid tolerance rough set–firefly based supervised feature selection for MRI brain tumor image classification. Appl Soft Comput 46:639–651
192. Roopini IT, Vasanthi M, Rajinikanth V, Rekha M, Sangeetha M (2018) Segmentation of tumor from brain MRI using fuzzy entropy and distance regularised level set. Computational signal processing and analysis. Springer, Singapore, pp 297–304
193. Senapati MR, Dash PK (2013) Local linear wavelet neural network based breast tumor classification using firefly algorithm. Neural Comput Appl 22:1591–1598
194. Filipczuk P, Wojtak W, Obuchowicz A (2012) Automatic nuclei detection on cytological images using the firefly optimization algorithm. Information technologies in biomedicine. Springer, Berlin, pp 85–92
195. Ch S, Sohani SK, Kumar D, Malik A, Chahar BR, Nema AK et al (2014) A support vector machine-firefly algorithm based forecasting model to determine malaria transmission. Neurocomputing 129:279–288
196. Kaur G, Singh R (2014) Sharpening enhancement of ultra sound images using firefly algorithm. Int J 4(8)
197. Boscolo R, Brown MS, McNitt-Gray MF (2002) Medical image segmentation with knowledge-guided robust active contours. Radiographics 22:437–448
198. Xiaogang D, Jianwu D, Yangping W, Xinguo L, Sha L (2013) An algorithm multi-resolution medical image registration based on firefly algorithm and Powell. In: 2013 IEEE third international conference on intelligent system design and engineering applications, pp 274–277 (January)
199. Gao ML, He XH, Luo DS, Jiang J, Teng QZ (2013) Object tracking using firefly algorithm. IET Comput Vis 7:227–237
200. Suresh S, Lal S (2016) An efficient cuckoo search algorithm based multilevel thresholding for segmentation of satellite images using different objective functions. Expert Syst Appl 58:184–209
201. Borra S, Thanki R, Dey N (2019) Satellite image analysis: clustering and classification. Springer, Singapore
202. Wang GG, Guo L, Duan H, Wang H (2014) A new improved firefly algorithm for global numerical optimization. J Comput Theor Nanosci 11:477–485
203. Chen K, Zhou Y, Zhang Z, Dai M, Chao Y, Shi J (2016) Multilevel image segmentation based on an improved firefly algorithm. Math Probl Eng
204. Ibrahim IA, Khatib T (2017) A novel hybrid model for hourly global solar radiation prediction using random forests technique and firefly algorithm. Energy Convers Manage 138:413–425
205. Agarwal C, Mishra A, Sharma A, Bedi P (2014) Optimized gray-scale image watermarking using DWT–SVD and firefly algorithm. Expert Syst Appl 41(17):7858–7867
206. Ali M, Ahn CW (2015) Comments on "Optimized gray-scale image watermarking using DWT-SVD and firefly algorithm". Expert Syst Appl 42(5):2392–2394
207. Dong H, He M, Qiu M (2015) Optimized gray-scale image watermarking algorithm based on DWT-DCT-SVD and chaotic firefly algorithm. In: 2015 IEEE international conference on cyber-enabled distributed computing and knowledge discovery, pp 310–313 (September)

208. Guo Y, Li BZ, Goel N (2017) Optimised blind image watermarking method based on firefly algorithm in DWT-QR transform domain. IET Image Process 11:406–415
209. Kazemivash B, Moghaddam ME (2017) A robust digital image watermarking technique using lifting wavelet transform and firefly algorithm. Multimedia Tools Appl 76:20499–20524
210. Chhikara RR, Singh L (2015) An improved discrete firefly and t-test based algorithm for blind image steganalysis. In: 2015 6th international conference on intelligent systems, modelling and simulation, pp 58–63
211. Chhikara RR, Sharma P, Singh L (2018) An improved dynamic discrete firefly algorithm for blind image steganalysis. Int J. Mach Learn Cyb 9(5):821–835
212. Raja PM, Baburaj E (2016) Optimal parameter selection for quick response code based image steganography via variable step size firefly algorithm and lifting wavelet transform. J Comput Theor Nanosci 13(11):8742–8759
213. Woźniak M, Marszałek Z (2014) An idea to apply firefly algorithm in 2d image key-points search. In: International conference on information and software technologies, pp 312–323
214. Chaki J, Dey N, Shi F, Sherratt RS (2019) Pattern mining approaches used in sensor-based biometric recognition: a review. IEEE Sens J 19:3569–3580
215. Honarpisheh Z, Faez K (2013) An efficient dorsal hand vein recognition based on firefly algorithm. Int J Electr Comput Eng 3(1):2088–8708

Chapter 2
Development of Firefly Algorithm Interface for Parameter Optimization of Electrochemical-Based Machining Processes

D. Singh and R. S. Shukla

1 Introduction

Non-traditional machining (NTM) process has great potential due to its wide variety of applications in modern industries. In literature, attempts are made to obtain the optimum parameter setting for traditional machining operation. Significant advances were seen in the last four decades in electrochemical machining (ECM) and its variants, i.e., electrochemical micro-machining (EMM) and electrochemical turning (ECT) in the manufacturing industries for producing precise components. ECM works on the principle of electrolysis [1, 2]. This process uses electrolytes which dissolve reaction products formed on the workpiece by electrochemical action. ECM process is similar to a reverse electroplating process. The workpiece hardness is not an issue; make ECM appropriate for machining hard and "difficult–to–machine" materials [1]. ECM has some benefits over other traditional machining processes in terms of performance characteristics such as "material removal rate (MRR)" and "surface roughness (Ra)." All the processes have different process parameters and performance parameters. Therefore, to select an optimal set of these parameters is essential to enhance the performance parameter of ECM and its variant. The engineering area is a wide domain of applications and researchers have attempted work for monitoring, controlling, and diagnosis of machining processes [3, 4]. In this chapter, a graphical user interface (GUI) is proposed that mimics metaheuristic algorithm,

D. Singh (✉) · R. S. Shukla
Mechanical Engineering Department, Sardar Vallabhbhai National Institute of Technology, Ichchhanath, Surat 395007, India
e-mail: dineshsinghmed@gmail.com

R. S. Shukla
e-mail: rajkamalshukla2013@gmail.com

© Springer Nature Singapore Pte Ltd. 2020
N. Dey (ed.), *Applications of Firefly Algorithm and its Variants*,
Springer Tracts in Nature-Inspired Computing,
https://doi.org/10.1007/978-981-15-0306-1_2

i.e., firefly algorithm (FA). An attempt is made to obtain the optimum parameter setting for three processes: ECM, electrochemical micro-machining (EMM), and electrochemical turning (ECT).

2 Related Works

In the last decade, the applicability and application of the NTM processes are increased considerably and several researchers have attempted to obtain the effective parameter setting for different machining processes. Jain and Jain [2] proposed a model of "MRR" in some NTM processes and developed semi-empirical models for NTM processes. Another work was reported by Jain and Jain [5] in which an attempt was made to optimize the ECM parameters (i.e., "tool feed rate," "electrolyte flow velocity," and "applied voltage") using real-coded genetic algorithm (RCGA) to enhance geometrical inaccuracy. Chakradhar and Gopal [6] attempted optimization of ECM process parameters for "EN-31 steel" using "gray relation analysis (GRA)" for the performance parameters MRR, SR, overcut (OC), and cylindricity error (CE). Das and Chakraborty [7] have developed methodology of selecting NTM processes for complex and intricate profile machining application using analytic network method.

Das et al. [8] conducted investigation on ECM to determine the significant effects of the process parameters (i.e., "electrolyte concentration," "voltage," "feed rate," and "inter-electrode gap") on MRR and Ra and used analysis of variance (ANOVA) to obtain contribution of every single process parameter on the considered responses. Bhattacharyya and Sorkhel [9] investigated the influence of ECM process parameters on machining performance using mathematical models based on response surface methodology (RSM). Schulze and Schatzing [10] proposed influences of factor contamination on selected erosion and chemical-based micro-machining to achieve high process accuracy. Senthilkumar et al. [11] attempted to obtain the effects of various ECM process parameters (i.e., "applied voltage," "electrolyte concentration," and "electrolyte flow rate") on the responses such as MRR and Ra and used RSM for optimizing the considered process parameters. Hinduja and Kunieda [12] proposed a model of chemical- and thermal-based advanced processes to simulate it in selected domain with its applications. Tang and Yang [13] conducted an experimental investigation on stainless steel 00Cr12Ni9Mo4Cu2 grade material and used GRA method to obtain the optimum process parameters (i.e., "voltage," "feed speed," and "electrolyte pressure") on MRR and Ra simultaneously. Ayyappan and Sivakumar [14] attempted micro-structure study of surfaces of the steel specimen 20MnCr5 produced during machining using ECM to determine the effects of electrolyte on the specimen.

When the ECM process used in the micrometer range, it is termed as electrochemical micro-machining (EMM). It is used for micro-range machining to produce complex profiles. Munda and Bhattacharya [15] attempted an investigation of the EMM process using RSM-based approach by developing a mathematical model to correlate the various machining parameters. Chae et al. [16] made an attempt for an investigation of micro-cutting operation and a survey was carried out to define areas that should be inspected for the quality improvement of the components during machining using micro-cutting processes. Dornfeld et al. [17] proposed wide review of recent developments in mechanical micro-machining for the future improvements and developments in the considered process. Samanta and Chakraborty [18] have attempted the problems on optimization to obtain optimized machining parameters of selected NTM processes such as ECM, ECDM, and EMM using artificial bee colony (ABC) algorithm. He et al. [19] have studied the microstructures of tungsten-machined surface using wire-EMM to enhance the machining efficiency. Goud et al. [20] have reviewed the "material removal mechanism" in electrochemical discharge machining process (ECDM) and used fuzzy logic to improve the performance of the process. Hajian et al. [21] have studied the effects of electrolyte concentration and magnetic field in ECDM process on channel depth and Ra for glass material. Kao and Hocheng [22] attempted multiple performance characteristics optimization for the performance parameter such as Ra and passivation strength using GRA to optimize several process parameters of electropolishing.

Fister et al. [23] have solved selected examples of various domains using FA. Furthermore, Gandomi et al. [24] have proposed FA with chaos that was applied to different benchmark problems and found that it outperforms the other algorithms. Zaki and El-Sawy [25] proposed hybridization of ACO with FA for unconstrained optimization problems and tested on several benchmark problems and the results have proved its dominance for obtaining the optimal solution.

Past researcher has applied several metaheuristic techniques, like "genetic algorithm (GA)" [26] "artificial bee colony (ABC) algorithm" [18, 26], "biogeography-based (BBO) algorithm" [26], Gauss–Jordan [9], and weighted- based multi-objective optimization [26] to optimize the operating parameters of the ECM process. El- Taweel and Gouda [27] evaluated the performance parameter, i.e., MRR, Ra, and roundness error (RE) of ECT processes using the RSM technique to obtain the optimum process parameter combination to maximize the MRR and to minimize Ra and RE. Similarly, Malapati and Bhattacharyya [28] investigated the influences of process parameters such as "pulse frequency," "machining voltage," "duty ratio," "electrolyte concentration" and "micro tool feed rate" on "MRR," "width overcut (WOC)," "length overcut (LOC)," and "linearity" in the EMM process using RSM approach.

It is observed in the literature that, in most of the cases, the sub-optimal solution has been obtained. Further, as far as my knowledge, no attempts have been made to develop GUI for the user in the field parameter optimization of NTM processes. This interface would be utilized by operator to have the optimum range of parameters in the considered machining processes. Ultimately, these values of process parameter will increase the cutting rate which reduces the time consumption for machining. The surface roughness of the specimen will be improved. The characteristics like width overcut, linear overcut, surface roughness and roundness error will be improved. So, precision machining will be obtained at reduce cost of machining which results in reduction of the production cost of the final product. For this purpose, a GUI is

proposed based on metaheuristic technique, i.e., FA which is applied for the improvement of performance parameters of three considered NTM processes. An attempt is
made to apply GUI having features like, flexible, and simple but yet powerful for
parameter optimization to obtain the set of solutions in the considered machining
processes.

The detail of the FA with user interface is described in the next section. In Sect. 4,
the examples considered on machining processes like, ECM, EMM, and ECT are
described and solved for parameter optimization using developed FA-GUI.

3   Problem Solution

This section reports the developed FA-GUI for the considered metaheuristic technique. The developed GUI is user-friendly and can be effectively used for continuous
domain problems.

3.1   *Firefly Algorithm-Graphical User Interface (FA-GUI)*

The researcher Yang in 2008 has developed firefly algorithm (FA)—A well-known
creatures. These lightning bugs are wonderful god creations whose lifestyle of living
is quite dissimilar from other animals. Normally, they exist in a warm environment
and work in summer nights. They are characterized by their flashes which have two
purposes, one to attract the breeding partners and second to deter the enemy [23, 29].
The flash light obeys rules of physics, i.e., the strength of light reduces with increase of
path distance. The closeness of the fireflies mating partners affects the strength of the
intensity value. The value of the intensity defines fitness value of the problems. The
fitness value is significantly affected by the intensity, absorption factor, randomness,
and attractiveness parameter. These factors govern the firefly algorithm to obtain the
solution. These factors and governing equation of the FA reported in the literature
[23, 29] are merged in single interface named as FA-GUI. The demonstration steps
of firefly algorithm to the machining problems are reported in the literature [30,
31]. The FA can deal with highly non-linear, multi-modal optimization problems
effectively. The speed of convergence of FA is very high in probability of finding the
global optimized solution. It has the flexibility of integration with other optimization
techniques. It does not require initial good solution to start its iteration process.
There is still need more research balance exploitation and exploration with fine
tuning the randomness method in order to produce a good result and performance of
the algorithm.

Every GUI is developed with a certain objective and it is an end display for
accomplishing specified task that can be easily executed by the user. The "user-
friendly" characteristic of GUI is significant as it decides its effectiveness. GUI can
be developed in several languages such Java, Python, Ruby, PHP, C++, MATLAB,

etc. A MATLAB feature "graphical user interface development environment" is used to develop the considered GUI that has all the features to generate the layout with push buttons, menus, static text, pop-up menu, etc. These are the uni-controls that are used by the end-user for the handling of GUI. An application-based two files are generated while launching a GUI that has to save as "*FIG-file*" and "*M-file*" [32, 33]. These files are used in computing of the outputs for the inputs given by the user.

The metaheuristic techniques are efficient to obtain the optimum process parameters for enhancing the responses. The researchers have proposed topological model, wireless Holon network, metaheuristic algorithms, and expert system for controlling the factors like, machining operation, job-shop scheduling, optimization of machining parameters, neural-based computing, and selection of cutting tools in manufacturing industries [34–36]. Further researcher has proposed the work in the area of automatic generation control of multi-area power thermal systems [37], multi-depot vehicle routing problem, [38] and robot arm path planning [39]. The present study attempts to connect metaheuristic technique, i.e., FA in a single interface as shown in Fig. 1. This concept of developing user interface provides flexibility to the user in obtaining the solutions of the optimization problem in the continuous domain. The user can communicate with GUI without bothering about the programming computation.

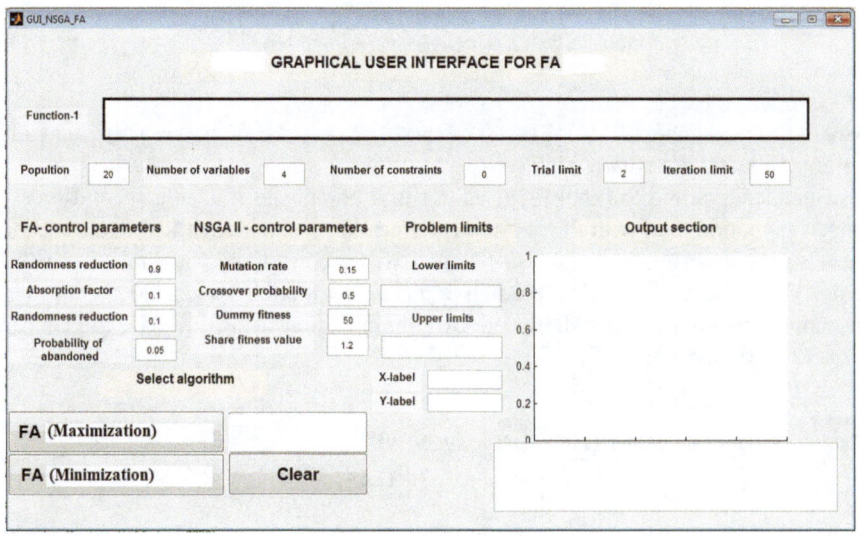

Fig. 1 FA-based graphical user interface

4 Application of FA-GUI in NTM Processes

This section demonstrates the simulation results of FA-GUI to attain the optimum values of the operating parameters for ECM, EMM, and ECT processes. For these considered processes, single objective at a time approach is considered to solve problems using FA-GUI-based optimization.

4.1 Example 1: Electrochemical Machining (ECM)

Bhattacharya and Sorkhel [9] have conducted experimentation on an ECM setup for machining cylindrical workpiece made of "EN-8 steel" using brass tool and considered four process parameters, i.e., "electrolyte concentration (g/l)," "electrolyte flow rate (l/min)," "applied voltage (V)," and "inter-electrode gap (mm)," and two performance parameter MRR (g/min) and overcut (mm). These process parameters were set at five different levels. The actual values and coded values of selected process parameters are shown in Table 1. The parametric levels for any chosen coded (X_i) need to be uncoded actual variable (X) using the Eq. (1).

$$\text{coded value } (X_i) = \frac{2X - (X_{max} + X_{min})}{\frac{X_{max} - X_{min}}{2}} \tag{1}$$

where coded value for X_i is $-2, -1, 0, 1$, and 2; X_{max} and X_{min} are maximum and minimum value of actual variable.

Bhattacharya and Sorkhel [9] used "central composite rotatable second-order" experimentation plan with thirty-one experimental runs, and then RSM was used to develop the mathematical regression equations for considered performance parameters. The same mathematical predictive regression models are considered for performance parameter (i.e., MRR and OC) optimization using FA-GUI is given in Eqs. (2)–(3), respectively.

Table 1 Process factors and their bounds for ECM [9]

Parameters	Symbols	Levels				
		−2	−1	0	1	2
Electrolyte concentration (g/l)	x_1	15	30	45	60	90
Electrolyte flow rate (l/min)	x_2	10	11	12	13	14
Applied voltage (V)	x_3	10	15	20	25	30
Inter-electrode gap (mm)	x_4	0.4	0.6	0.8	1	1.2

$$MRR = 0.6244 + 0.1523x_1 + 0.0404x_2 + 0.1519x_3$$
$$- 0.1169x_4 + 0.0016x_1^2 + 0.0227x_2^2 + 0.0176x_3^2$$
$$- 0.0041x_4^2 + 0.0077x_1x_2 + 0.0119x_1x_3 - 0.0203x_1x_4$$
$$+ 0.0103x_2x_3 - 0.0095x_2x_4 + 0.0300x_3x_4 \tag{2}$$

$$OC = 0.3228 + 0.0214x_1 - 0.0052x_2 + 0.0164x_3$$
$$+ 0.0118x_4 - 0.0041x_1^2 - 0.0122x_2^2 + 0.0027x_3^2$$
$$+ 0.0034x_4^2 - 0.0059x_1x_2 - 0.0046x_1x_3 - 0.0059x_1x_4$$
$$+ 0.0021x_2x_3 - 0.0053x_2x_4 - 0.0078x_3x_4 \tag{3}$$

The ECM problem is considered for process parameter optimization using FA in developed GUI. The results for MRR and overcut obtained are shown in Fig. 2a, b for

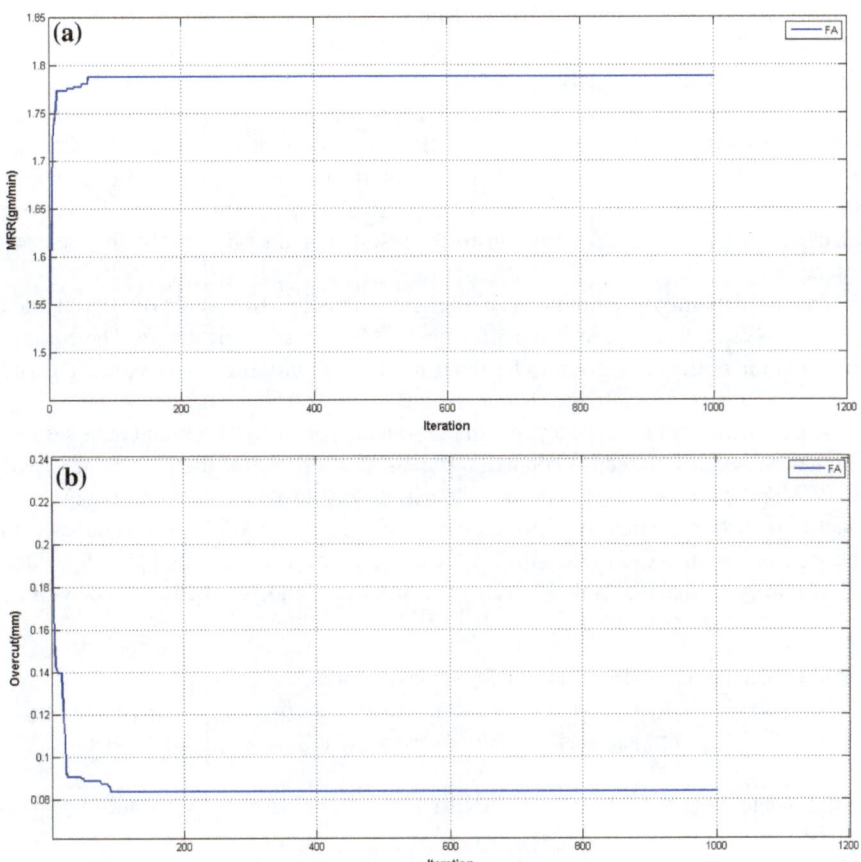

Fig. 2 **a, b** Convergence of MRR and overcut results for the ECM process

Table 2 Comparison of FA-GUI results for ECM

	MRR (g/min)	Overcut (mm)
Gauss–Jordan [9]	0.8230	0.2706
GA [26]	1.1603	0.2369
ABC [26]	1.3077	0.2067
BBO [26]	1.5069	0.1320
FA-GUI	1.7880	0.0836

the ECM process. The effectiveness of FA optimization is measured by employing Eqs. (2) and (3) in GUI. Here, the performance parameters MRR and overcut are to maximize and minimize, respectively, using FA separately. Bhattacharya and Sorkhel [9] used Gauss–Jordon algorithm and obtained performance parameters are given in Table 2. These same RSM-based equations were also attempted by Mukherjee and Chakroborty [26] using GA, ABC, and BBO and the obtained results are given in Table 2. It shows the optimal ECM performance parameters that are obtained using FA-GUI. While comparison with the results of Bhattacharya and Sorkhel [9] for the Gauss–Jordan method, it is found that the results of considered FA are better for MRR and overcut. These same results of FA-GUI compared with the results of Mukherjee and Chakroborty [26] for GA, ABC, and BBO; it is observed that the results obtained using FA-GUI are found better. The results obtained for MRR and overcut using FA are 1.7880 (g/min) and 0.0836 (mm), respectively. The results shown in Table 2 clearly indicate that the FA outperforms the results that are obtained by the previous researcher.

The corresponding optimum process parameters for MRR and overcut obtained using FA-GUI are shown in Table 3 for the ECM process, respectively. The computational time of the FA algorithm for determining the optimum solution of MRR and overcut results is 0.3738 s and 0.4038 s, respectively.

The influence of ECM process parameters on the performance parameter is studied using obtained graph trends depicted in Fig. 3a–d. As shown in the Figs. 3a–d, MRR in ECM process increases with an increase in electrolyte concentration Fig. 3a, electrolyte flow rate (Fig. 3b), applied voltage (Fig. 3c), but MRR value reduces with the increase of inter-electrode gap (Fig. 3d). Therefore, the maximum possible value of "electrolyte concentration," "electrolyte flow rate" and "applied voltage" will be

Table 3 Optimum process parameter for MRR and overcut

Parameter	Electrolyte concentration (g/l)	Electrolyte flow rate (l/min)	Applied voltage (V)	Inter-electrode gap (mm)
MRR (g/min) (1.7880)	74.5073	13.9902	29.5557	0.4029
Overcut (mm) (0.0836)	15.0000	10.0000	10.0000	0.4000

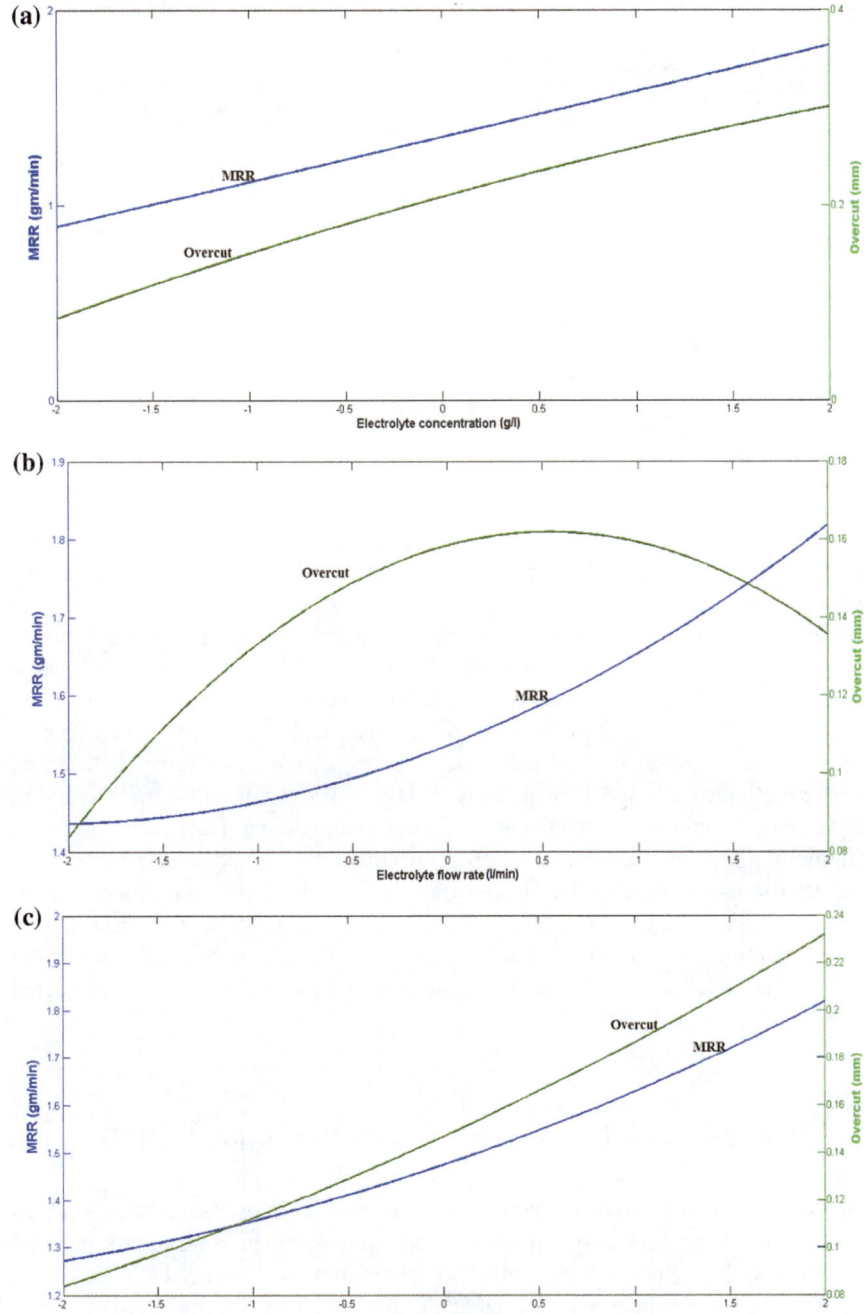

Fig. 3 a–d Variations of performance parameter w.r.t. ECM process parameter

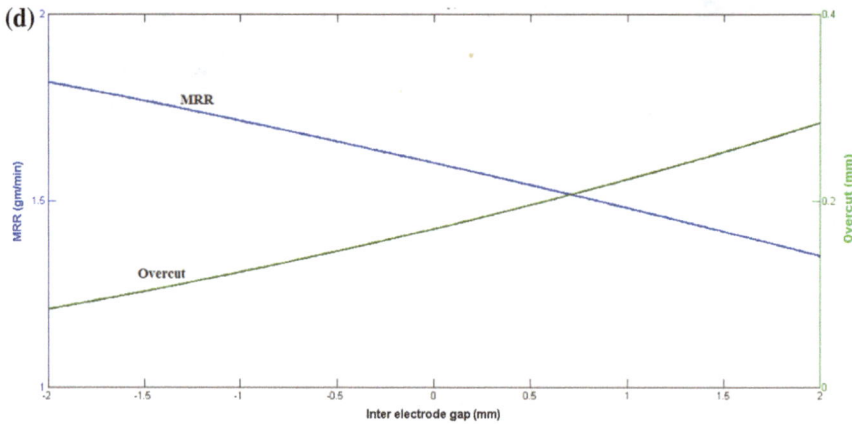

Fig. 3 (continued)

the optimum solution for the performance parameter MRR. The obtained value for "electrolyte concentration," "electrolyte flow rate" and "applied voltage" is in the agreement with this fact. Furthermore, MRR decreases with an inter-electrode gap so the minimum value is the optimal solution. Hence, the process parameter obtained for MRR (shown in Table 3) is the optimal solution and these solutions are in good accord with the trends of the variation graph for the ECM process using FA-GUI. While the performance parameter overcut increases with the increase of electrolyte concentration (Fig. 3a), overcut value rises with the increase of electrolyte flow rate up to a definite limit and then decreases (Fig. 3b), overcut value increases with the increase of applied voltage (Fig. 3c), inter-electrode gap (Fig. 3d). Therefore, minimum values of the considered process parameters should be used to get the optimal results of overcut. The values obtained for these process parameters for overcut using FA is in good agreement with the trends of the overcut with respect to the process parameters. These trends of the performance parameters MRR and overcut confirm the optimality of the solution that obtained using FA-GUI for considered ECM process.

4.2   Example 2: Electrochemical Micro-Machining (EMM)

The process EMM is used for fabrication of microstructures and in thin-film processing. The 3D process of EMM involves masking less and through-mask material removal [28]. In this process, controlled computer numerical controlled (CNC) operations can be done by linking tool to a machine to produce more complex forms structure components. This problem of EMM is taken from Malapati and Bhattacharya [28] for parameter optimization. Malapati and Bhattacharya [28] developed an EMM setup to determine the influence of the process parameters, i.e., "pulse frequency

Table 4 Process factors and their bounds for EMM [28]

Parameters	Symbol	Levels				
		−2	−1	0	1	2
Pulse frequency (kHz)	x_1	40	50	60	70	80
Machining voltage (V)	x_2	7	8	9	10	11
Duty ratio (%)	x_3	30	40	50	60	70
Electrolyte concentration (g/l)	x_4	50	60	70	80	90
Micro tool feed rate (μm/s)	x_5	150	175	200	225	250

(kHz)," "machining voltage (V)," "duty ratio (%)," "electrolyte concentration (g/l)," and "macro tool feed rate (μm/s)" on the MRR and machining accuracy factors such as WOC, LOC, and linearity. Bare copper plates and Tungsten were considered as workpieces and micro tool, respectively. In this chapter, for parameter optimization EMM, the same process parameters and performance parameters are considered. The bounds of these parameters are given in Table 4.

Malapati and Bhattacharya [28] conducted thirty-two experimental trials using a "central composite half-fraction rotatable second-order design and plan." The mathematical predictive regression models for MRR, WOC, LOC, and linearity considered are same as given in Malapati and Bhattacharya [28] which are given in Eqs. (4)–(7).

$$\begin{aligned} MRR = {}& 2.59446 + 0.237004x_2 + 0.426029x_3 + 0.340938x_4 \\ & - 0.385790x_1^2 + 0.295540x_2^2 - 0.299427x_3^2 \\ & + 0.224331x_1x_5 + 0.327156x_2x_3 - 0.243256x_2x_4 \end{aligned} \tag{4}$$

$$\begin{aligned} WOC = {}& 0.124856 + 0.0288208x_4 + 0.0270208x_5 \\ & - 0.0173807x_3^2 + 0.0140443x_5^2 + 0.0203563x_1x_4 \\ & - 0.0257063x_1x_5 - 0.0209313x_2x_3 \end{aligned} \tag{5}$$

$$\begin{aligned} LOC = {}& 0.403467 + 0.0526542x_2 + 0.0604563x_1x_2 \\ & + 0.0457563x_1x_3 + 0.0344187x_1x_4 - 0.166544x_2x_4 \\ & - 0.0554562x_2x_5 + 0.0495813x_3x_4 \\ & + 0.0413687x_3x_5 + 0.0534565x_4x_5 \end{aligned} \tag{6}$$

$$\begin{aligned} linearity = {}& 0.0721852 + 0.00868750x_1 + 0.0112125x_2 \\ & + 0.0171458x_4 - 0.0128688x_1x_3 + 0.0151188x_1x_4 \\ & + 0.00841875x_2x_4 - 0.0116062x_2x_5 - 0.00749375x_4x_5 \end{aligned} \tag{7}$$

Like the ECM process, here, the performance parameter MRR is maximized and performance parameter WOC, LOC, and linearity are minimized. Malapati and

Bhattacharya [28] obtained the optimal parameter setting of EMM process parameters as pulse frequency = 52.2818 kHz, machining voltage = 10.1033 V, duty ratio = 68.3890%, electrolyte concentration = 85.1515 g/l, micro tool feed rate = 208.5860 μm/s. The value of performance parameters MRR, WOC, LOC, and linearity is 3.1039 mg/min, 0.0003 mm, 0.1676 mm, and 0.0691 mm, respectively. The effectiveness of FA optimization is measured by using Eqs. (4)–(7) in FA-GUI for considered EMM process. The results for MRR, WOC, LOC, and linearity obtained using FA are shown in Fig. 4a–d for the EMM process at different iteration, respectively. It is found that the MRR is increased to 5.5467 (mg/min), WOC reduced to 1.58×10^{-5} (mm), LOC reduced to 1.09×10^{-5} and linearity reduced to 9.43×10^{-6}. The results of single objective optimization of the performance parameters and process parameters are reported in Tables 5 and 6, respectively. The computational time for determining the optimum solution using FA-GUI for MRR, WOC, LOC, and linearity are 0.404 s, 0.252 s, 0.257 s, and 0.258 s, respectively.

The effects of process parameters on the performance parameter can be studied using the obtained graph trends for EMM process as depicted in Fig. 5a–e. MRR in EMM process increases with an increase in pulse frequency up to a certain range, but then MRR reduced with an increase in pulse frequency (Fig. 5a), also MRR value increases with the increase of machining voltage (Fig. 5b), duty ratio (Fig. 5c), but MRR value reduces with the increase of electrolyte concentration (Fig. 5d) and micro tool feed rate (Fig. 5e). Therefore, the value obtained for the process parameter pulse frequency (−0.4492 kHz), machining voltage (1.9813 V), duty ratio (1.9234%), electrolyte concentration (−1.9162 g/l) and micro tool feed rate (−1.9102 μm/s) are in the agreement with the trends on the graph. Thus, the process parameters obtained for MRR shown in Table 6 are the optimal solution for EMM process.

While the performance parameter WOC reduces with the increase of pulse frequency. In Fig. 5a, machining voltage (Fig. 5b), also WOC value increases with the increase of the duty ratio up to a certain limit than WOC decreases with the increase of the duty ratio (Fig. 5c); WOC increases with increase of electrolyte concentration (Fig. 5d), but WOC decreases with increase of micro tool feed rate (Fig. 5e). The values obtained for these process parameters for WOC using FA-GUI is in worthy agreement with the trends. The performance parameter LOC reduces with the increase of pulse frequency (Fig. 5a), machining voltage (Fig. 5b), but LOC value increases with the increase of duty ratio (Fig. 5c), electrolyte concentration (Fig. 5d), and micro tool feed rate (Fig. 5e). Therefore, the maximum value of the process parameter ("pulse frequency" and "machining voltage") and a minimum value of the process parameter ("duty ratio," "electrolyte concentration," and "micro tool feed rate") is desirable for the optimum solution of LOC. Similarly, the performance parameter linearity increases with the increase of pulse frequency (Fig. 5a), machining voltage (Fig. 5b), duty ratio (Fig. 5c), but linearity reduces with increase of electrolyte concentration (Fig. 5d), also linearity increases very minute with micro tool feed rate (Fig. 5e). These trends of the performance parameters MRR, WOC, LOC, and linearity confirm the optimality of the solution that is obtained using FA-GUI for EMM process.

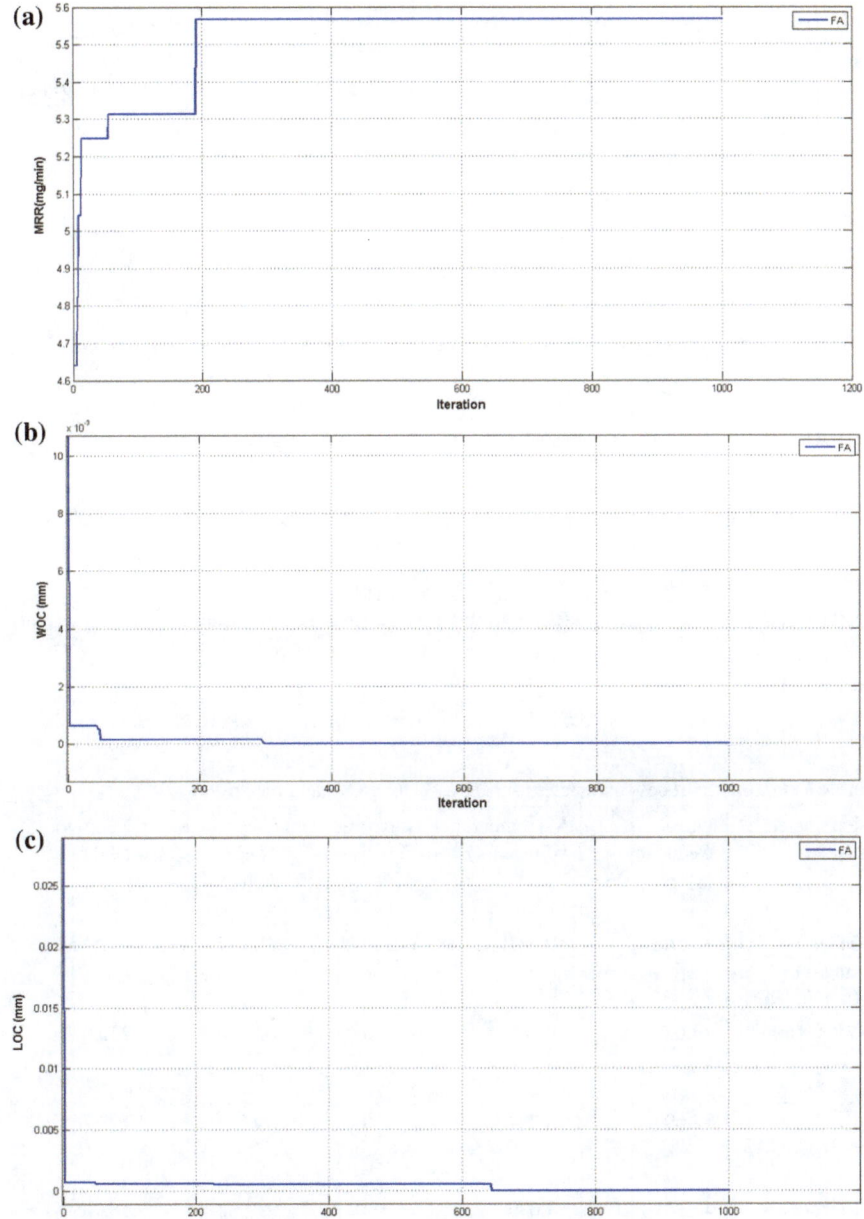

Fig. 4 a–d Convergence of MRR, WOC, LOC, and linearity results

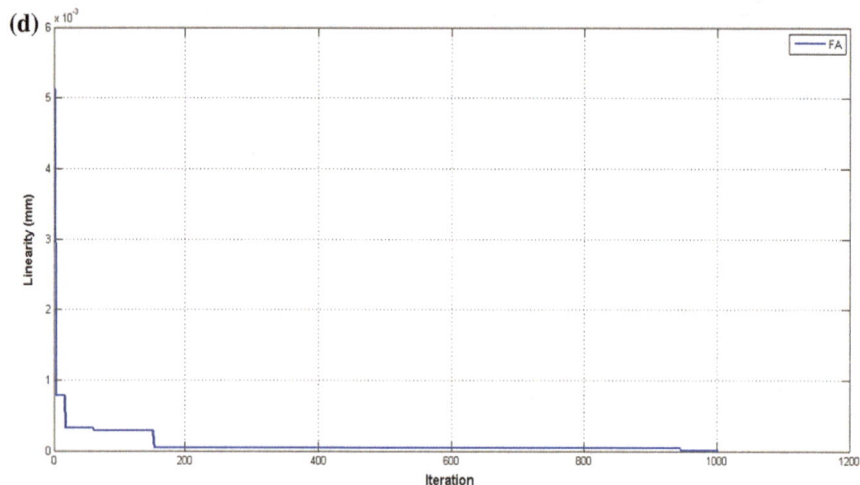

Fig. 4 (continued)

Table 5 Comparison of FA-GUI results for the EMM process

	MRR (mg/min)	WOC (mm)	LOC (mm)	Linearity (mm)
Malapati and Bhattacharyya [28]	3.1039	0.0003	0.1676	0.0691
FA-GUI	5.5467	1.58×10^{-5}	1.09×10^{-5}	9.43×10^{-6}

Table 6 Optimum process parameter for MRR, WOC, LOC, and linearity

Parameter	Pulse frequency (kHz)	Machining voltage (V)	Duty ratio (%)	Electrolyte concentration (g/l)	Micro tool feed rate (μm/s)
MRR (mg/min) (5.5467)	55.508	10.981	69.234	50.837	152.244
WOC (mm) (1.58×10^{-5})	76.627	9.669	69.353	63.758	236.009
LOC (mm) (1.09×10^{-5})	77.695	8.034	34.415	54.563	173.285
Linearity (mm) (9.43×10^{-6})	47.161	7.039	52.099	81.934	163.650

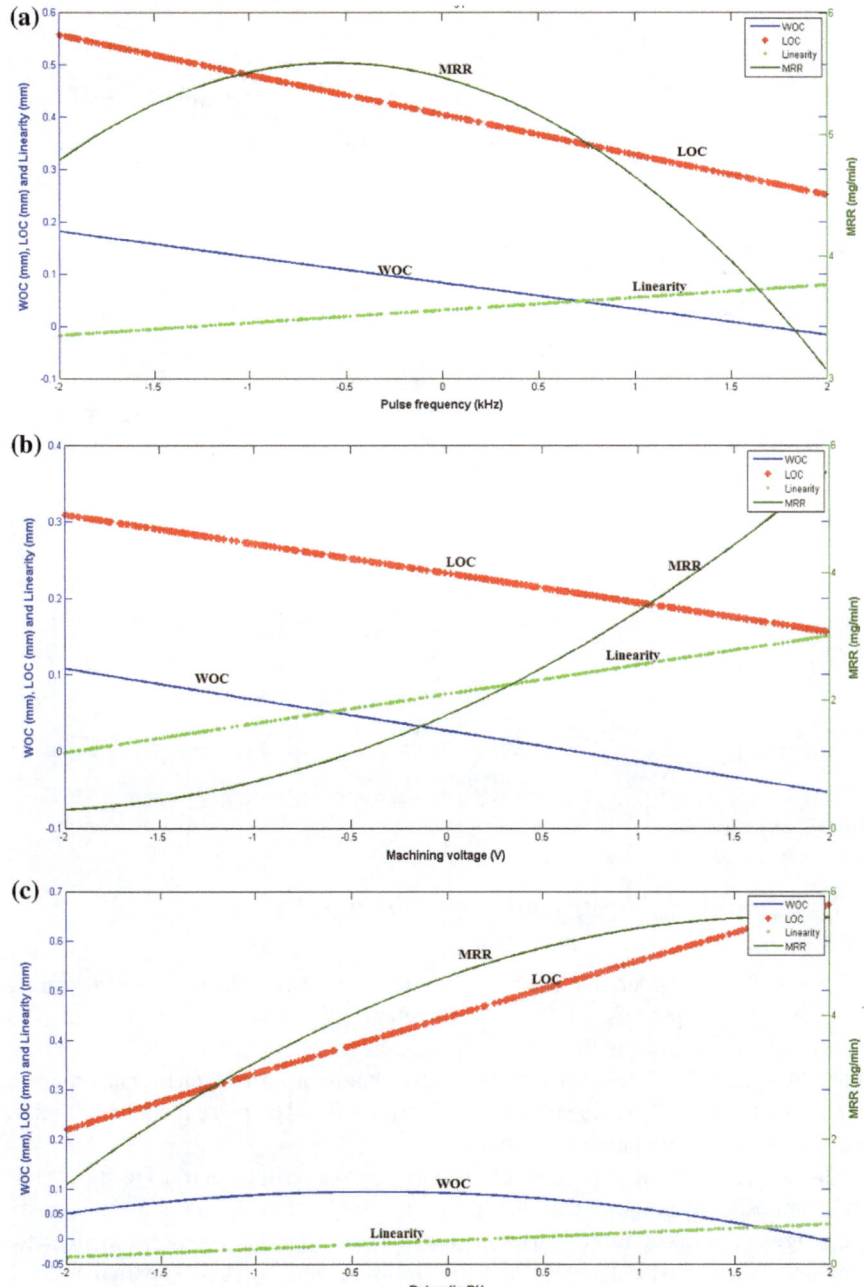

Fig. 5 a–e Variations of performance parameter w.r.t. EMM process parameter

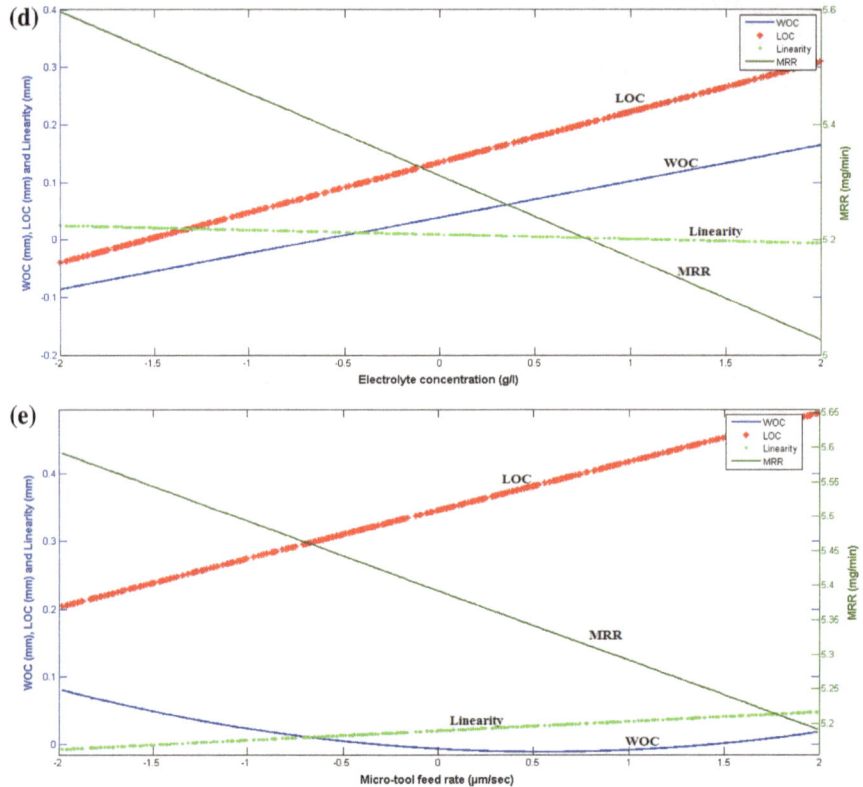

Fig. 5 (continued)

4.3 Example 3: Electrochemical Turning (ECT)

This example is considered from El-Taweel and Gouda [27]. Electrochemical turning used as finishing process. El-Taweel and Gouda [27] developed an ECT setup to determine the influence of the process parameters (i.e., "applied voltage," "wire feed rate," "wire diameter," "overlap distance," and "rotational speed") on the performance parameter such as MRR, Ra and roundness error (RE). The present study optimizes these performance parameters in FA-GUI.

El-Taweel and Gouda [27] performed thirty-two experimental trials using a "central composite second-order rotatable design" set on five levels and their working range is shown in Table 7. The mathematical predictive regression models are remodeled for the experimental results of El-Taweel and Gouda [27] with the use of "MINITAB software" for coded values of the process parameters. These RSM-based regression equations are given in Eqs. (8)–(10).

Table 7 Process factors and their bounds for ECT [27]

Parameters	Symbol	Levels				
		−2	−1	0	1	2
Applied voltage (V)	x_1	10	17.5	25	32.5	40
Wire feed rate (mm/min)	x_2	0.1	0.2	0.3	0.4	0.5
Wire diameter (mm)	x_3	0.2	0.65	1.1	1.55	2
Overlap distance (mm)	x_4	0.02	0.03	0.04	0.05	0.06
Rotational speed (rpm)	x_5	300	450	600	750	900

$$
\begin{aligned}
\text{MRR} = {} & 0.214932 + 0.050667x_1 + 0.016583x_2 \\
& + 0.006167x_3 + 0.004667x_4 + 0.010917x_5 \\
& + 0.002068x_1^2 - 0.005557x_2^2 - 0.021307x_3^2 \\
& - 0.004057x_4^2 + 0.004068x_5^2 - 0.0005x_1x_2 \\
& + 0.004375x_1x_3 - 0.010625x_1x_4 + 0.004375x_1x_5 \\
& + 0.014375x_2x_3 + 0.007625x_2x_4 + 0.011875x_2x_5 \\
& - 0.002x_3x_4 + 0.0135x_4x_5
\end{aligned} \tag{8}
$$

$$
\begin{aligned}
\text{Ra} = {} & 1.68193 + 0.16708x_1 + 0.06708x_2 - 0.20792x_3 \\
& + 0.12208x_4 - 0.13875x_5 - 0.02568x_1^2 \\
& - 0.01943x_2^2 - 0.01943x_3^2 - 0.01943x_4^2 \\
& - 0.03193x_5^2 + 0.11188x_1x_2 + 0.02438x_1x_3 \\
& + 0.27938x_1x_4 + 0.03312x_1x_5 + 0.03062x_2x_3 \\
& + 0.08563x_2x_4 - 0.01062x_2x_5 + 0.04813x_3x_4 \\
& + 0.15188x_3x_5 + 0.18188x_4x_5
\end{aligned} \tag{9}
$$

$$
\begin{aligned}
\text{RE} = {} & 7.21250 + 1.98917x_1 - 1.310x_2 + 1.71x_3 \\
& - 1.09250x_4 - 2.39250x_5 + 0.50000x_1^2 \\
& + 0.02125x_2^2 + 0.19375x_3^2 + 0.27875x_4^2 \\
& + 0.55875x_5^2 - 0.17375x_1x_2 - 0.31625x_1x_3 \\
& + 0.03250x_1x_5 - 0.41125x_2x_3 + 0.4025x_2x_4 \\
& - 0.005x_2x_5 - 0.0675x_3x_4 - 0.6225x_3x_5 + 0.91375x_4x_5
\end{aligned} \tag{10}
$$

Here, ECT problem is considered for process parameter optimization using FA-GUI. The results for MRR, Ra, and RE obtained are shown in Fig. 6a–c for the ECT process at different iterations, respectively. The effectiveness of FA optimization measured for considered ECT process by employing Eqs. (8), (9), and (10) in the developed FA-GUI. Here, the performance parameters are optimized separately. The

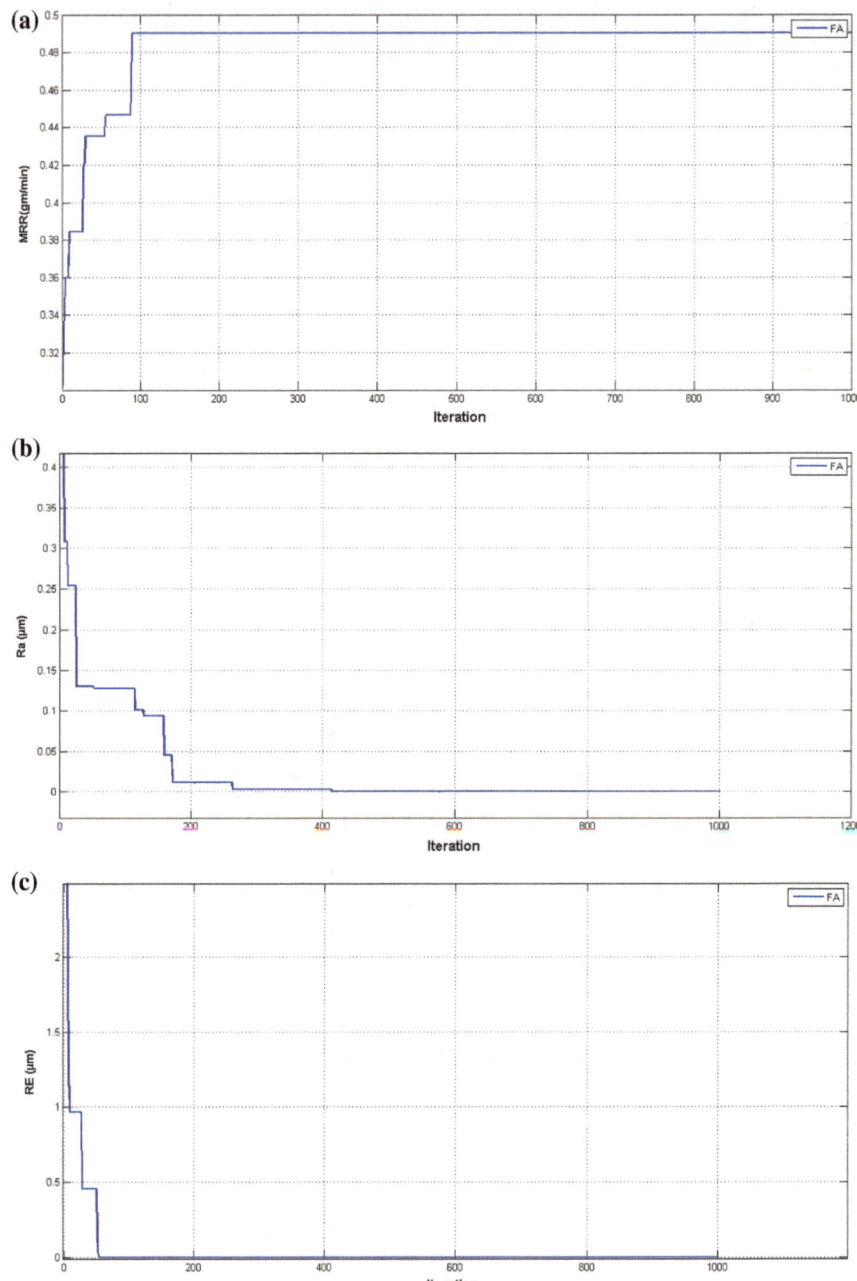

Fig. 6 **a–c** Convergence of MRR, Ra, and RE for ECT process

Table 8 Comparison of FA-GUI results for ECT

Results	MRR (g/min)	Ra (μm)	RE (μm)
El-Taweel and Gouda [27]	0.2980	1.1294	5.5432
GA [26]	0.2356	0.7656	6.7153
ABC [26]	0.3126	0.4269	4.5165
BBO [26]	0.4068	0.1081	2.7764
FA-GUI	0.4902	4.57×10^{-5}	6.42×10^{-5}

performance parameter MRR is maximized, and Ra and RE are to be minimized.
El-Taweel and Gouda [27] did not implement single objective optimization of any of
the considered performance parameters. Mukherjee and Chakroborty [26] attempted
these same problems using GA, ABC, and BBO. The obtained results of Mukherjee
and Chakroborty [26] and FA-GUI are shown in Table 8. The results obtained using
FA-GUI when compared with the results of Mukherjee and Chakroborty [26] for
GA, ABC, and BBO, it is observed that the results found using FA are significantly
better. The results obtained for MRR, Ra, and RE using FA-GUI are 0.4902 (g/min),
4.57×10^{-5} (μm), and 6.42×10^{-5} (μm), respectively. The results shown in Table 8
clearly indicate that the FA based on GUI outperforms the results that obtained by the
previous researcher. The optimum process for MRR, Ra, and RE results are shown
in Table 9. The computational time for determining the optimum solution of MRR,
Ra, and RE are 0.404 s, 0.410 s, and 0.405 s, respectively.

The effects of ECT process parameters on the performance parameter can be
studied by observing graph trends that are obtained using the optimum values of
ECT process. As shown in the (Fig. 7a–e) and using Eqs. (8)–(10), MRR in ECT
process increases with an increase in applied voltage (Fig. 7a) and wire feed rate
(Fig. 7b), MRR increases with wire diameter initially and then decreases gradually
(Fig. 7c), increase of overlap distance and workpiece rotational speed also increases
MRR as depicted in (Fig. 7d–e). The performance parameter Ra decreases with the
increase of applied voltage (Fig. 7a) but Ra surges with increase of wire feed rate
(Fig. 7b), while Ra decreases with the increase of wire diameter (Fig. 7c), also Ra
surges with increase of overlap distance (Fig. 7d), and Ra reduces with increase
of workpiece rotational speed (Fig. 7e). Similarly, the performance parameter RE
increases with the increase of applied voltage (Fig. 7a), wire feed rate (Fig. 7b), and

Table 9 Optimum process parameter for MRR, Ra, and RE

Parameter	Applied voltage (V)	Wire feed rate (mm/min)	Wire diameter (mm)	Overlap distance (mm)	Rotational speed (rpm)
MRR (g/min)	40.000	0.5000	1.518	0.060	900
Ra (μm)	39.480	0.1422	1.856	0.023	415.850
RE (μm)	13.769	0.2184	0.357	0.054	346.113

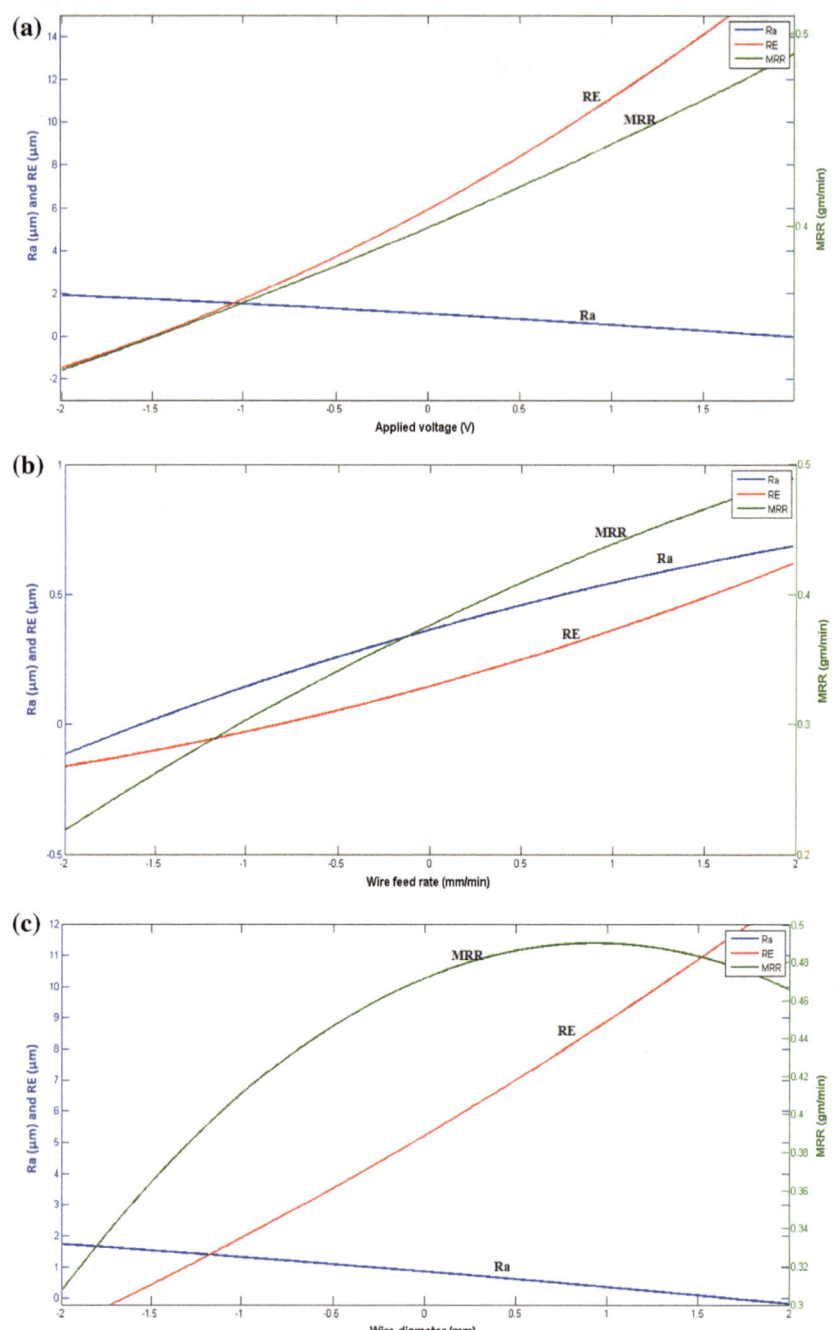

Fig. 7 **a–e** Variation of performance parameter w.r.t. ECT process parameters

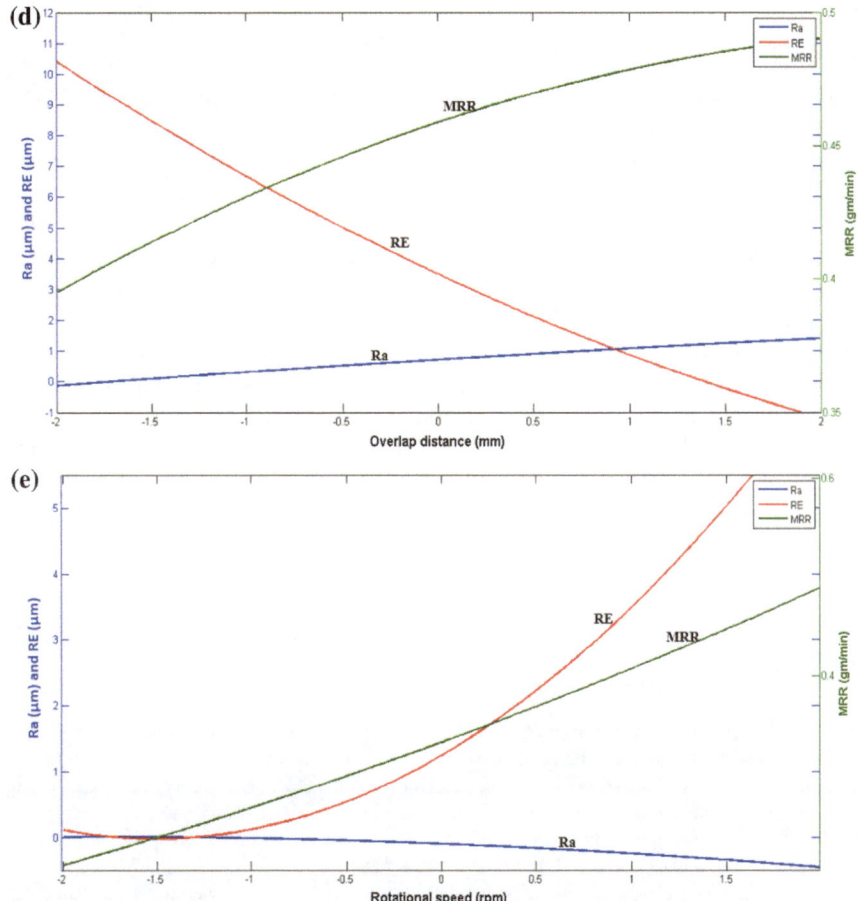

Fig. 7 (continued)

wire diameter (Fig. 7c), it reduces with increase of overlap distance (Fig. 7d), while RE increases with increase of workpiece rotational speed (Fig. 7e). By observing all these facts of the performance parameter with respect to ECT process parameters, it is found that the obtained results of FA-GUI for ECT process are in good with the graph trends.

The considered algorithm in the chapter has found wide range of applications which leads other researchers to work in the other domain of engineering problems. A concept of computer–human interface in the form of FA-based GUI is attempted for solving machining parameter optimization problems. This concept of GUI can be developed for the other domain like wireless Holon network, job-shop scheduling, neural-based computing, image processing, robot path planning, and selection of cutting tools which can be used by the modern industries to serve their purpose.

5   Conclusions

The FA-GUI is successfully attempted for parameter optimization of ECM, EMM, and ECT machining process. The results obtained are compared with other popular population-based metaheuristic to see the effectiveness of the developed FA-GUI. When FA results of GUI are compared with other metaheuristic optimization algorithm, it found that the FA based on GUI outperforms them in terms of obtaining the optimum values for the considered NTM process performance parameters, accuracy, and speed. Thus, the results show that the developed GUI proves its applicability and acceptability in solving the machining optimization problem. The user can use the developed GUI to optimize the process parameters of other NTM processes, i.e., "ultrasonic machining," "electric discharge machining," "laser-beam machining process," etc. From the above results, it demonstrates that FA-GUI can be utilized as a significant tool for optimizing manufacturing optimization problems in the manufacturing industry as well as to other engineering optimization problems. The process engineer can now easily set the optimal combination of several process parameters for the diverse machining process.

References

1. Rajurkar KP, Zhu D, Mcgeough JA, De Silva A (1999) New developments in electro-chemical machining. CIRP Ann Manuf Technol 48:567–579
2. Jain NK, Jain VK (2001) Modeling of material removal in mechanical type advanced machining processes: a state-of-art review. Int J Mach Tools Manuf 41:1573–1635
3. Di ZM, Wang L (2012) Optimization of machining processes from the perspective of energy consumption: a case study. J Manuf Syst 31:420–428
4. Suriano S, Wang H, Hu SJ (2012) Sequential monitoring of surface spatial variation in automotive machining processes based on high definition metrology. J Manuf Syst 31:8–14
5. Jain NK, Jain VK (2007) Optimization of electro-chemical machining process parameters using genetic algorithms. Mach Sci Technol 11:235–258
6. Chakradhar D, Gopal AV (2011) Multi-objective optimization of electrochemical machining of EN31 steel by grey relational analysis. Int J Model Opt 1(2):113–117
7. Das S, Chakraborty S (2011) Selection of non-traditional machining processes using analytic network process. J Manuf Syst 30:41–53
8. Das MK, Kumar K, Barman T, Sahoo P (2014) Optimization of surface roughness and MRR in electrochemical machining of EN31 tool steel using grey-Taguchi approach. Procedia Mater Sci 6:729–740
9. Bhattacharyya B, Sorkhel SK (1998) Investigation for controlled electrochemical machining through response surface methodology-based approach. J Mater Process Technol 86:200–207
10. Schulze HP, Schatzing W (2013) Influences of different contaminations on the electro-erosive and the electrochemical micro-machining. Procedia CIRP 6:58–63
11. Senthilkumar C, Ganeshan G, Karthikeyan R (2011) Parametric optimization of electrochemical machining of Al/15% SiCp composites using NSGA-II. Trans Nonferrous Metals Soc 21:2294–2300
12. Hinduja S, Kunieda M (2013) Modelling of ECM and EDM processes. CIRP Ann Manuf Technol 62:775–797

13. Tang L, Yang S (2013) Experimental investigation on the electrochemical machining of 00Cr12Ni9Mo4Cu2 material and multi-objective parameters optimization. Int J Adv Manuf Technol 67:2909–2916
14. Ayyappan S, Sivakumar K (2015) Enhancing the performance of electrochemical machining of 20MnCr5 alloy steel optimization of process parameters by PSO-DF optimizer. Int J Adv Manuf Technol 82:2053–2064
15. Munda J, Bhattacharyya B (2006) Investigation into electrochemical micromachining (EMM) through response surface methodology based approach. Int J Adv Manuf Technol 35:821–832
16. Chae J, Park SS, Freihei T (2006) Investigation of micro-cutting operations. Int J Mach Tools Manuf 46:313–332
17. Dornfeld D, Min S, Takeuchi Y (2006) Recent advances in mechanical micromachining. CIRP Ann Manuf Technol 55:745–768
18. Samanta S, Chakraborty S (2011) Parametric optimization of some non-traditional machining processes using artificial bee colony algorithm. Eng Appl Artif Intell 24:946–957
19. He H, Zeng Y, Qu N (2015) An investigation into wire electrochemical micro machining of pure tungsten. Precis Eng 45:285–291
20. Goud M, Sharma AK, Jawalkar C (2015) A review on material removal mechanism in electrochemical discharge machining (ECDM) possibilities to enhance the material removal rate. Precis Eng 45:1–17
21. Hajian M, Razfar MR, Movahed S (2016) An experimental study on the effect of magnetic field orientations and electrolyte concentrations on ECDM milling performance of glass. Precis Eng 45:322–331
22. Kao PS, Hocheng H (2003) Optimization of electrochemical polishing of stainless steel by grey relational analysis. J Mater Process Technol 140:255–259
23. Fister I, Yang XS, Brest JA (2013) Comprehensive review of firefly algorithms. Swarm Evol Comput 13:34–46
24. Gandomi AH, Yang X, Talatahari S, Alavi AH (2013) Firefly algorithm with chaos. Commun Nonlinear Sci Numer Simul 18:89–98
25. Zaki EM, El-sawy AA (2013) Hybridizing ant colony optimization with firefly algorithm for unconstrained optimization problems. Appl Math Comput 224:473–483
26. Mukherjee R, Chakraborty S (2013) Selection of the optimal electrochemical machining process parameters using biogeography-based optimization algorithm. Int J Adv Manuf Technol 64:781–791
27. El-Taweel TA, Gouda SA (2011) Performance analysis of wire electrochemical turning process—RSM approach. Int J Adv Manuf Technol 53:181–190
28. Malapati M, Bhattacharyya B (2011) Investigation into electrochemical micromachining process during micro-channel generation. Mater Manuf Process 26:1019–1027
29. Yang XS (2008) Nature-inspired metaheuristics algorithm, 2nd edn. Luniver Press, UK
30. Singh D, Shukla RS (2016) Optimisation of electrochemical micromachining and electrochemical discharge machining process parameters using firefly algorithm. Int J Mech Manuf Sys 9(2):137–159
31. Shukla RS, Singh D (2017) Selection of parameters for advanced machining processes using firefly algorithm. Eng Sci Technol Int J 20(1):212–221
32. Chapman SJ (2008) Matab® programming for engineers. Thomson Asia Ltd, Singapore
33. Hahn BH, Valentine DT (2010) Essential Matlab for engineers and scientist. Elsevier Academic Press, Cambridge
34. Tapie L, Mawussi B, Bernard A (2012) Topological model for machining of parts with complex shapes. Comput Ind 63:528–541
35. Arezoo B, Ridgway K, Al-Ahmari AMA (2000) Selection of cutting tools and conditions of machining operations using an expert system. Comput Ind 42:43–58
36. Kumar R, Rajan A, Talukdar FA, Dey N, Santhi V, Balas VE (2017) Optimization of 5.5-GHz CMOS LNA parameters using firefly algorithm. Neural Comput Appl 28(12):3765–3779
37. Jagatheesan K, Anand B, Samanta S, Dey N, Ashour AS, Balas VE (2017) Design of a proportional-integral-derivative controller for an automatic generation control of multi-area power thermal systems using firefly algorithm. IEEE/CAA J Automatica Sin 6(2):503–515

38. Li J, Li T, Yu Y, Zhang Z, Pardalos PM, Zhang Y (2019) Discrete firefly algorithm with compound neighborhoods for asymmetric multi-depot vehicle routing problem in the maintenance of farm machinery. Appl Soft Comput J 81:105–119
39. Kumar A, Konar A, Bhattacharjee T, Das S (2018) Synergism of firefly algorithm and Q-learning for robot arm path planning. Swarm Evol Comput 43:50–68

Chapter 3
A Firefly Algorithm-Based Approach for Identifying Coalitions of Energy Providers that Best Satisfy the Energy Demand

Cristina Bianca Pop, Viorica Rozina Chifu, Eric Dumea and Ioan Salomie

1 Introduction

Throughout the past years, the electricity market has shifted from a *centralized approach* in which the generation, the transmission, the distribution, and the retail of electricity have been monopolized by the state, to a *decentralized approach*, in which the electricity generation and retail have been liberalized and have become competitive activities between various stakeholders [1]. Currently, electricity is traded in various types of markets according to the trading moment, such as the day-ahead market, the intraday market, the balancing market, and the capacity market. In the day-ahead market, the transactions are made so that to ensure the supply of electricity for the next day. In this market, the electricity providers must submit their selling offers consisting of the provided quantity of electricity and associated price for the 24 h of the next day, while the electricity requesters must submit their purchasing bids, consisting as well of a desired quantity of electricity and price, for the 24 h of the next day. The offers and bids are analyzed and the electricity price as well as the day-ahead schedule is set accordingly. The intraday market aims to adjust the contracts that have been established between electricity providers and requesters in the day-ahead market and to facilitate the integration of renewable energy providers

C. B. Pop (✉) · V. R. Chifu · E. Dumea · I. Salomie
Department of Computer Science, Technical University of Cluj-Napoca,
Baritiu Street, no. 28, Cluj-Napoca, Romania
e-mail: cristina.pop@cs.utcluj.ro

V. R. Chifu
e-mail: viorica.chifu@cs.utcluj.ro

E. Dumea
e-mail: eric.dumea@student.utcluj.ro

I. Salomie
e-mail: ioan.salomie@cs.utcluj.ro

© Springer Nature Singapore Pte Ltd. 2020
N. Dey (ed.), *Applications of Firefly Algorithm and its Variants*,
Springer Tracts in Nature-Inspired Computing,
https://doi.org/10.1007/978-981-15-0306-1_3

53

[1]. The balancing market (i.e., real-time market) is responsible for resolving any real-time deviations of the electricity supply and consumption as established in the day-ahead or intraday markets [1]. The capacity market aims to reserve a particular quantity of electrical energy as backup for critical situations.

Due to the latest technological advancements and also due to the urge of reducing the carbon footprint of electricity production, the model of the electricity grid is shifting from an exclusive brown energy-based grid toward a greener one through the integration of renewable energy sources. The European Union has even defined a *renewable energy directive* which requires the EU member states to contribute to the fulfillment of at least 32% of the energy demand using renewable energy sources by 2030 at the union level [2]. The usage of solar photovoltaic panels, wind turbines, and other behind-the-meter renewable generation devices brings the electricity generation closer to the consumer which besides the beneficial environmental impact can also reduce the losses of electricity transportation and distribution infrastructure costs [3]. These technologies introduce a new type of player in the electricity grid, namely the prosumer, which acts both as an electricity consumer and supplier. In this way, the flow of electricity becomes bidirectional from the grid to the prosumer and vice versa.

Small-scale prosumers do not have the possibility to participate individually in the electricity market due to their unstable provision of electricity that highly depends on weather conditions (e.g., solar photovoltaic panels maximize their production during sunny days, wind turbines maximize their production during windy days, etc.) which cannot guarantee a reliable energy supply such as the one established within an electricity contract [4]. To solve this problem, aggregators have been introduced as entities capable of creating virtual associations of individual prosumers which in this way act as a single entity in the market. In this way, aggregators can trade the energy supplied by such virtual associations in the day-ahead market.

This chapter proposes the application of the state-of-the-art firefly algorithm [5] to solve the problem of aggregating prosumers in virtual associations to be further integrated in the electricity market as possible providers of an amount of demanded electricity on one hour-ahead basis. The aim is to identify a group of heterogeneous individual prosumers that can best satisfy the electricity demand at the lowest cost. In our approach, each individual prosumer has its energy supply forecast for a given time horizon. To apply the firefly algorithm, we have mapped a firefly to a virtual association of energy providers which is evaluated using an objective function that measures the degree by which the forecast energy supply of the providers in an association matches the expected energy demand over the given time horizon. Additionally, we have introduced genetic-based operators to redefine the firefly movement strategy. We have evaluated the proposed firefly-based algorithm on a data set developed in-house consisting of forecast energy supply values for various heterogeneous energy providers (renewable and non-renewable) over a given time horizon.

The rest of this chapter is organized as follows. Section 2 presents an overview of optimization problems and of the firefly algorithm. Section 3 reviews state-of-the-art approaches in the domain of creating virtual associations of renewable energy sources. Section 4 formalizes the problem of creating virtual associations of renewable energy sources as an optimization problem and presents the adapted firefly-based algorithm for identifying virtual associations of renewable energy sources over a given time horizon. Section 5 discusses experimental results. The chapter ends with conclusions.

2   Background

This section overviews theoretical concepts related to optimization problems (see Sect. 2.1) and the state-of-the-art firefly algorithm (see Sect. 2.2) that are relevant for the optimization algorithm we propose for identifying coalitions of energy providers that best satisfy the energy demand for a given time horizon.

2.1   Overview of Optimization Problems

An optimization problem can be defined as follows: "*Given a set of feasible solutions S, identify the solution s that best satisfies an objective function*". According to [6] the optimization, problems can be classified in the following categories:

- Continuous optimization versus discrete optimization
- Unconstrained optimization versus constrained optimization
- Single or multiple objectives
- Deterministic optimization versus stochastic optimization.

In the case of discrete optimization problems, the variables have values in a discrete set, while in the case of continuous optimization problems, the variables can have values in any range.

In the case of unconstrained optimization problems, no constraints are applied on the variables, while in the case of constrained optimization problems, different constraints can be applied on the variables. Depending on the type of constraints, constrained optimization problems can be, in turn, classified in linear optimization, nonlinear optimization, geometric optimization, convex optimization, etc.

Single objective optimization problems have only a single objective function defined, while multiple objectives optimization problems have multiple objective functions defined.

In the case of deterministic optimization problems, all the variables are deterministic, while in the case of stochastic optimization problems, some or all variables are probabilistic.

Because the problem of identifying coalitions of heterogeneous energy providers that provide the demanded energy at a desired cost is a constrained optimization problem, in what follows we will focus on formally defining this type of optimization problem.

A constrained optimization problem can be formally defined as follows:

$$\textbf{optimize } f(x)$$
$$\scriptstyle x \in Sol$$
$$\textbf{subject to } g_i(x) \leq 0, i = 1, ..., q$$
$$h_i(x) = 0, j = i + 1, ..., m$$

where

- *optimize* refers to minimizing or maximizing f, depending on whether it is a minimization or maximization problem;
- f is the objective function that needs to be optimized;
- *Sol* represents the set of all feasible solutions;
- q is a set of inequality constraints;
- h is a set of equality constraints.

Regarding the way in which the constraints are handled in the optimization problem, several approaches have been proposed in the research literature that can be classified in the following categories [7]:

- Methods based on penalty functions: death penalty, static penalties, dynamic penalties, annealing penalties, adaptive penalties, and co-evolutionary penalties;
- Methods based on a search of feasible solutions: repairing unfeasible individuals, superiority of feasible points, and behavioral memory;
- Methods based on preserving feasibility of solutions by searching the boundary of feasible region or homomorphous mapping;
- Hybrid methods.

In the case of penalty-based methods, the constraints are integrated into the objective function as penalty terms are using either the mathematical addition or multiplication operations.

In the case of methods based on a search of feasible solutions, different approaches which make distinction between feasible and unfeasible solutions in the search space are developed. For example, some approaches consider that a feasible solution is better than an unfeasible solution while other approaches apply local repair strategies to transform unfeasible solutions into feasible solutions.

In the case of methods based on preserving feasibility of solutions, either specialized operators are used or searching strategies that identify searching areas that are closer than the boundary of regions with the feasible solutions are applied.

In the case of hybrid methods, different strategies are combined in order to handle the constraints specified for an optimization problems.

2.2   The Firefly Algorithm

The state-of-the-art firefly algorithm [5] is inspired by the mating behavior of fireflies which are attracted to each other based on the intensity of the flashing light they emit. The algorithm works with a population of moving firefly agents, where the position of each firefly is mapped to a solution of the optimization problem being solved. The position of a firefly is updated so that it becomes closer to the position of another firefly having a higher brightness. The brightness of each firefly is computed by considering the value of the objective function of the considered firefly as well as a light absorption coefficient.

The firefly algorithm consists of two main steps, one representing the initialization of the firefly population, and the other one representing the update of fireflies' positions according to their level of attractiveness. The main operations performed in the two steps are presented in the Fig. 1.

According to [5, 8], it has been demonstrated that, in solving NP hard problems, the firefly algorithm is better than the genetic algorithms both in terms of time efficiency and success rate in identifying the global optimal solution. This is due to the fact that the firefly algorithm, in contrast to the other two algorithms, uses the automatic subdivision of the firefly population in subgroups [8] and all subgroups are able to work simultaneously in order to identify the global or a near optimal solution. Another advantage of this algorithm is that its adjustable parameters can be tuned to control the randomness component in order to ensure a faster convergence of the algorithm toward the optimal solution.

Due to its advantages, the firefly algorithm has been widely used in solving various optimization problems, such as job shop scheduling [9], design optimization of a shell and tube heat exchanger [10], emergence evacuation management [11], supply chain optimization [12], ophthalmology imaging [13], optimized non-rigid demons registration [14], autonomous mini unmanned aerial vehicle [15], and power system planning problems [16].

3   Related Work

This section presents some of the most relevant methods in the research literature for the aggregation of the energy resources in order to satisfy the energy demand.

In [17], the authors propose a method for the management of the distributed energy resources in energy hubs. The proposed method integrates an algorithm, namely the DEROP algorithm, which is able to identify the optimal configuration of energy resources so that the energy cost is minimal. Each energy resource has an associated cost that depends on its raw material and time. For example, in the case of photovoltaic panels and wind turbines, the associated cost is represented by the maintenance costs, while in the case of energy storage systems (ESS), the cost is variable and it depends on the charging/discharging profiles of the ESS. The DEROP algorithm starts from

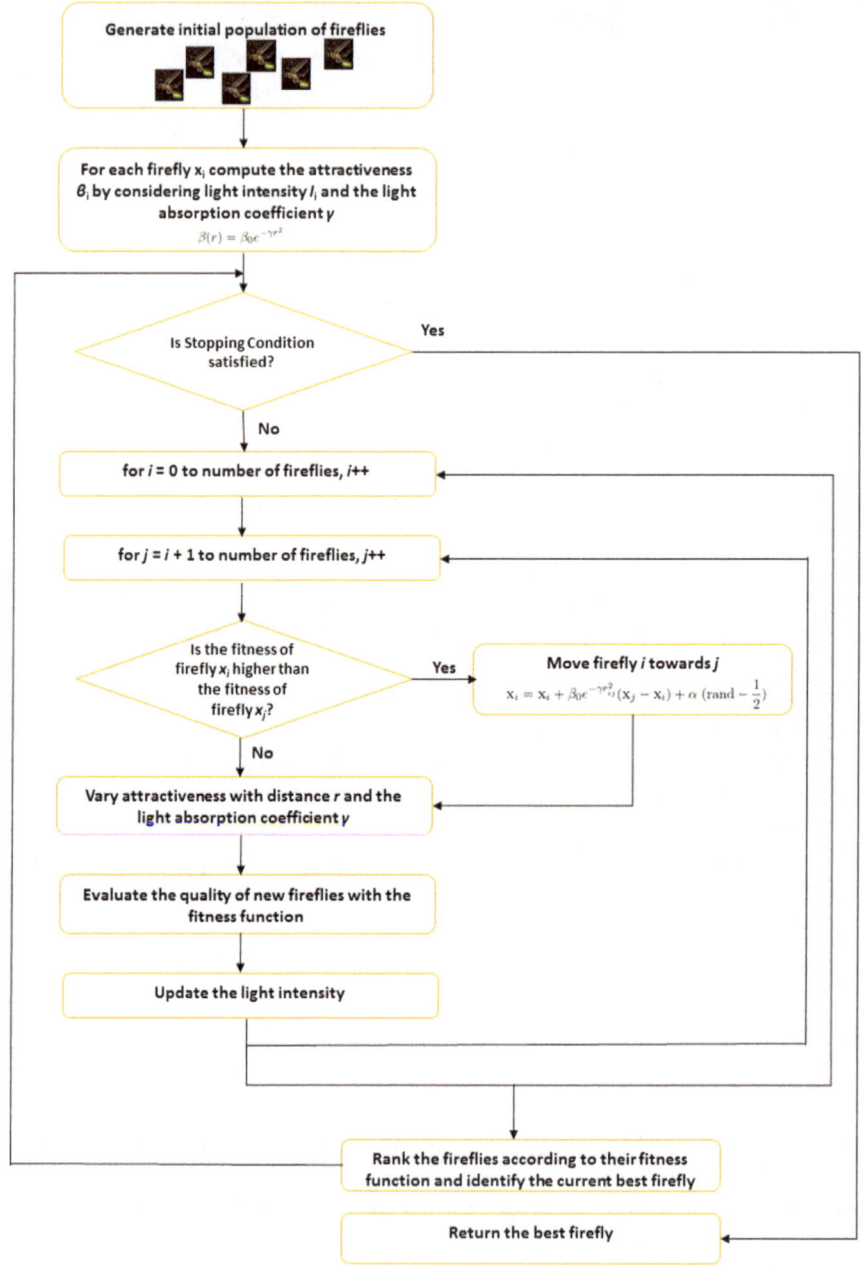

Fig. 1 Flow of the firefly-based algorithm

an initial configuration in which the energy demand is provided by the power grid. Then, the algorithm tries, at every iteration, to minimize the energy cost, by reducing the power provided by the grid during the periods in which the cost is higher. This is achieved by providing energy by power batteries during these periods of time.

In [3], the authors propose a method for the aggregation of renewable energy sources (RESs), which combines a min-max clustering algorithm with different shared distribution strategies. The method is able to create virtual associations (VA) of prosumers, by using a min-max clustering algorithm, so that all the prosumers which are part of a VA are considered as a single entity acting on the energy market. As sharing strategies, the following policies are implemented: (a) Winner Takes All, (b) Equal Volume Sharing, (c) Equal Income Sharing, and (d) Rank-based Fair Sharing. In the case of the Winner Takes All policy, from the VAs obtained by the clustering algorithm, only the one which contains the cheapest prosumers is selected to provide the demanded energy, while the others are eliminated from the market. In the case of the Equal Volume Sharing policy, all the VAs are selected, in equal percentage, to provide the demanded energy. In the case of the Equal Income Sharing policy, each VA is assigned shares which are inversely proportional to their price, while in the case of the Rank-Based Fair Sharing policy, the VAs are ranked based on their competitiveness in the market share distribution, so that the VAs with higher competitiveness have greater shares than the VAs with lower competitiveness.

In [18], the authors propose an approach for creating coalitions of prosumers which are managed by an aggregator entity. The aggregator plays the role of manager of the energy resources provided by the prosumers members of a coalition. The aggregator can be seen either as an entity that has full control over the energy resources or as an entity that could partially control the energy resources only after a planning which is performed by the prosumers. When creating coalitions of prosumers, the following aspects are considered: the energy demand, the production of energy from renewable energy systems, and the market prices. To address these aspects, two stochastic programming approaches are taken. The first approach considers the use of the storage systems together with the balancing market to face the uncertain energy demand. The second approach uses only the balancing market. Both approaches utilize an objective function that is based on a set of weights to give more importance either to the expected profit or to the risk control.

In [19], the authors present three clustering algorithms that are applied for grouping energy prosumers in virtual clusters. A hierarchical architecture is also presented in which an aggregator is defined which act as an intermediate entity between the energy market and the prosumers. The three algorithms applied for clustering are: the spectral clustering technique, the genetic algorithm, and the adaptive algorithm. The goal of each of these clustering algorithms is to create clusters of prosumers with minimum aggregated imbalance and energy penalty costs. The experiments are conducted on a data set of 33 prosumers from Greece. Based on the experimental results, the authors concluded that from the three tested algorithms, the ones that provides the best results in terms of penalty minimization are the genetic algorithm and the adaptive algorithm. However, the main disadvantages of these algorithms in contrast with the spectral clustering algorithm are that they have higher computational cost.

In [20], the authors propose an approach that assists the aggregator of a pro-sumer in defining the demand and the supply bids in a day-ahead energy market. The proposed approach consists of two main steps. In the first step, a K-means clus-tering algorithm is applied to compute the flexibility of aggregated resources such as shiftable loads, electric vehicles, and thermostatically controlled loads, while in the second step, a two-stage stochastic optimization model is defined for optimizing the participation of aggregator in a day-ahead energy market. The main benefits of the defined optimization model are: (i) the energy bids for the day-ahead energy market is defined by considering multiple energy providers and consumers and (ii) the uncertainties of the prosumers is modeled as a set of scenarios.

In [21], the authors propose a decision-making framework that is able to create groups of small energy prosumers that are seen as single entities that participate on the energy market. The prosumers are aggregated based on the presumption profile that best satisfy an energy pattern required by an actor on the energy market. The best groups of prosumers are identified by combining autoregressive forecasting algorithm and a genetic clustering algorithm.

4   Firefly-Based Method for Identifying Coalitions of Energy Providers

This section presents how we have modeled the problem of creating virtual associ-ations of heterogeneous energy providers as an optimization problem as well as the adapted version of the firefly algorithm that will be used for solving our optimization problem.

Figure 2 illustrates how the adapted version of the firefly algorithm can be inte-grated within an energy aggregator component to identify the virtual associations of energy providers satisfying a requested demand. It is assumed that within the energy market, a number of energy consumers/retailers formulate energy demand bids for a particular date and time, consisting of an electricity consumption quantity and a maximum price that they are willing to pay for the desired electricity quantity. Additionally, a set of electricity producers are exposing in the market their portfo-lios consisting of electricity production quantities and the minimum price they are willing to accept for the produced electricity. In order to identify the coalition of elec-tricity price that best matches an electricity demand in terms of desired electricity quantity, price and heterogeneity of the electricity producers, the distribution system operator interacts with an energy aggregator component that implements the firefly algorithm-based method proposed in this chapter.

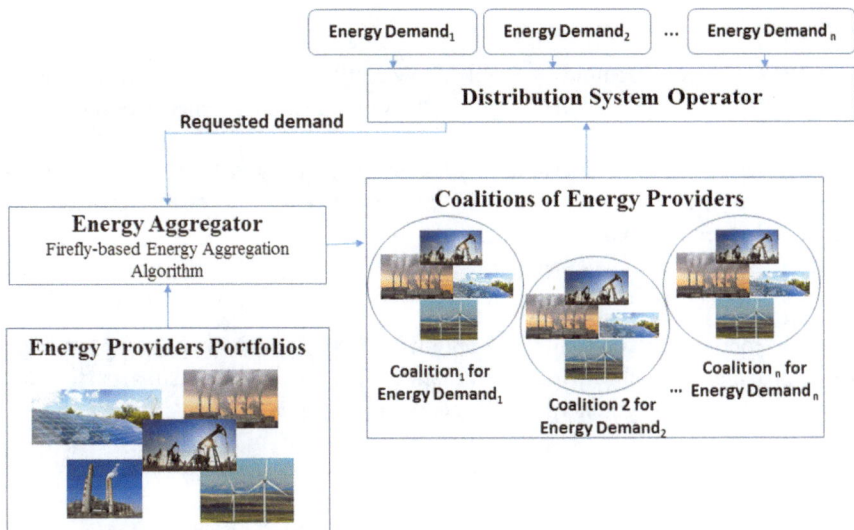

Fig. 2 Integration of the firefly algorithm for creating virtual associations of energy providers

4.1 Problem Formalization

Given a set of heterogeneous energy providers having an estimated energy quantity they can produce for the next day, hourly, and the price for this energy, select a subset of providers which can satisfy the demanded energy, hourly, at a given price, such that the selected subset contains providers of different types as specified by the entity requesting the energy.

Due to the large number of available energy providers, finding the most appropriate combination of energy providers that best satisfies an energy demand, within a given price, can be seen as an optimization problem that in our case will be solved by using the firefly algorithm. Moreover, as the aim is to ensure that each solution contains a desired percentage of several types of energy providers, the optimization problem becomes a constrained optimization problem.

Consequently, our problem of finding the most appropriate combination of energy providers that best satisfies an energy demand quantity, subject to a set of constraints regarding the price and the types of selected energy providers, is defined as follows:

$$\underset{sol}{\textbf{minimize}}\ Energy(sol)$$

$$\textbf{subject to}\ Heterogeneity(sol, type_1) = desired\_percentage_{type_1},$$

$$...$$

$$Heterogeneity(sol, type_{nt}) = desired\_percentage_{type_{nt}}$$

$$Price(sol) \leq maximum\_price$$

where

- *Energy(sol)* (see Formula 5) evaluates whether a subset of providers (i.e., the solution of the optimization problem) satisfies the demanded energy (see Formula 3);
- *Heterogeneity(sol, type$_i$)* evaluates whether the solution *sol* satisfies the heterogeneity constraint with respect to the desired percentage of energy providers of type *type$_i$* (see Formula 6);
- *Price(sol)* evaluates whether the energy provided by the providers part of the solution *sol* is less than the price desired by the demanding entity (see Formula 8).

In our approach, a solution of the optimization problem represents the set of all energy providers available in an energy market, out of which only a subset is selected to provide the demanded energy for a specified *date_time*:

$$sol = \{(provider_1, flag_1), (provider_2, flag_2), \ldots, (provider_n, flag_n)\} \quad (1)$$

where *provider$_i$* is defined using Formula 2, and *flag$_i$* is a flag having the value 0 in case *provider$_i$* has not been selected to supply energy, or 1 in case *provider$_i$* has been selected to supply energy.

The energy provider is formally defined as follows:

$$
\begin{aligned}
provider = (&id, type, \\
&\{date\_time_1, supplied\_energy\_val_1, \\
&date\_time_2, supplied\_energy\_val_2, \\
&\ldots \\
&date\_time_n, supplied\_energy\_val_{24}\})
\end{aligned}
\quad (2)
$$

where:

- *id* represents the identification number of the provider;
- *type* represents the type of the provider (e.g., providers that generate electricity using solar panels, wind turbines, etc.);
- *supplied_energy_val$_i$* represents the quantity of energy forecast to be generated by the provider at the specified date and time (i.e. *date_timei*).

The demanded energy for a specified date and time is represented as:

$$
\begin{aligned}
demanded\_energy = (&demanded\_energy\_val, date, \\
&\{pc_{type_1}, \ldots, pc_{type_{nt}}\}, price)
\end{aligned}
\quad (3)
$$

where *pc$_{t_i}$* represents the desired percentage of energy providers of type *type$_i$*, *nt* represents the total number of types considered for a demand, and *price* is the maximum accepted price for the demanded energy.

We have converted the constraints of the optimization problem in penalty functions which are integrated in the initial objective function, $Energy(sol)$, thus obtaining the new objective function defined below:

$$Fitness(sol) = Energy(sol, demanded\_energy) +$$
$$r_1 * Penalty_h(sol) + \qquad (4)$$
$$r_2 * Penalty_p(sol)$$

where:

- $Penalty_{heterogeneity}$ (see Formula 6) and $Penalty_{price}$ (see Formula 8) represent the penalty components of the objective function that compute the penalty values for a solution of the optimization problem based on the violated constraints regarding the energy providers' heterogeneity and price.
- r_1 and r_2 represent multiplication constants experimentally determined which are used to set the degree by which a solution is penalized according to the violated constraints.

The $Energy$ component of the objective function (see Formula 4) is defined as:

$$Energy(sol) = |demanded\_energy\_value - \sum_{p}^{i=1}(supplied\_energy\_value_i)| \quad (5)$$

where p is the number of energy providers having the flag set to 1 in solution sol, $supplied\_energy\_val_i$ represents the energy forecast to be supplied by the energy provider i.

The $Penalty_h$ component of the objective function (see Formula 4) is defined as:

$$Penalty_h(sol) = \sum_{nt}^{i=1} Heterogeneity(sol, type_i) \qquad (6)$$

where the $Heterogeneity$ component is defined as:

$$Heterogeneity(sol, type_i) = |Percentage(sol, type_i) - perc_{type_i})| \qquad (7)$$

where $Percentage(sol, type_i)$ computes the percentage of energy providers from the solution sol that have the flag equal to 1 and are of type $type_i$, and $perc_{type_i}$ represents the desired percentage as specified in the demand (see Formula 3).

The $Penalty_p$ component of the objective function is defined as:

$$Penalty_p(sol) = max(0, |Price(sol) - price|) \qquad (8)$$

where *Price(sol)* computes the total price for the energy supplied by the energy providers having the flag set to 1 in solution *sol*, and *price* is the desired maximum price.

4.2 Firefly-Based Algorithm for Identifying Coalitions of Energy Providers

The proposed firefly-based algorithm (see Algorithm 1) adapts the state-of-the-art firefly algorithm [5] to solve the problem of identifying coalitions of energy providers.

Algorithm 1: Firefly Algorithm for Identifying Coalitions of Energy Providers

---
1 **Inputs:** $ENERGY\_MARKET$, $demanded\_energy$, noF, r_1, r_2, $maxNoIt$
2 **Output:** sol_{opt}
3 **Begin**
4   $FSOL = \emptyset$
5   **for** $i = 1$ **to** noF **do**
6     $FSOL = FSOL \cup Generate\_Initial\_Solution(ENERGY\_MARKET)$
7   **end for**
8   $iterationNumber = 0$
9   **repeat**
10   **for** $i = 1$ **to** noF **do**
11   **for** $j = i + 1$ **to** noF **do**
12     $fitnessValue_i =$ **Fitness**$(FSOL[i], demanded\_energy, r_1, r_2)$
13     $fitnessValue_j =$ **Fitness**$(FSOL[j], demanded\_energy, r_1, r_2)$
14   **if**$(fitnessValue_i > fitnessValue_j)$ **then**
15       $\Delta =$ **Fitness**$(FSOL[i]) -$ **Fitness**$(FSOL[j])$
16       $nmbC =$ **Nmber_CrossoverP**(Δ)
17       $FSOL[i] =$ **Crossover**$(FSOL[i], FSOL[j], nmbC)$
18       $nmbM =$ **Number_MutationP**$(FSOL[i], demanded\_energy)$
19       $activationFlag =$ **Activation_Flag**$(FSOL[i], demanded\_energy)$
20       $FSOL[i] =$ **Mutation**$(FSOL[i], nmbM, activationFlag)$
21   **end if**
22   **end for**
23   **end for**
24    $sol_{best} =$ **Get_Best_Solution**$(FSOL)$
25    $SOL_{BEST} = SOL_{BEST} \cup sol_{best}$
26    $nmbM =$ **Number_MutationP**$(sol_{best}, demanded\_energy)$
27    $activationFlag =$ **Activation_Flag**$(sol_{best}, demanded\_energy)$
28    $sol_{best} =$ **Mutation**$(sol_{best}, nmbM, activationFlag)$
29    $FSOL =$ **Update**$(FSOL, sol_{best})$
30    $iterationNumber = 0$
31   **until**$(iterationNumber == maxNoIt)$
32   **return Get_Best_Solution**(SOL_{BEST})
33 **End**

In our approach, the position of each firefly represents a solution consisting of the set of all available energy providers (see Eq. 1). The brightness of a firefly is evaluated using the objective function defined in Eq. 4. The update of the firefly position is performed using the genetic crossover and mutation operators. The algorithm takes as input the set of energy providers available in the energy market (*ENERGY_MARKET*), the forecast demanded energy (*demanded_energy*), and some algorithm specific adjustable parameters such as the number of fireflies (*noF*), the weights r_1 and r_2 used for the penalty components in the objective function defined by Formula 4, and the maximum number of iterations.

In the algorithm's initialization step (see lines 4–8), the population of fireflies is generated. In this step, it is important to identify the number of energy providers that should be activated in a solution such that the provided energy is close to the demanded energy; a large number of activated providers might exceed the demanded energy. To address this problem, initially, we have started with solutions having two randomly chosen activated providers; then, if these solutions have an objective function value far from 0, we increase iteratively the number of activated providers for each solution with 1, until we obtain solutions for which the objective function value is closer to 0.

In the algorithm's iterative stage (see lines 9–31), the value of the objective function for each firefly solution *FSOL[i]* is compared to the fitness value of the other fireflies, and if the fitness of firefly solution *FSOL[i]* is higher than the fitness of a firefly solution *FSOL[j]*, then *FSOL[i]* solution is modified according to the distance (i.e., difference between the objective function values) between *FSOL[i]* and *FSOL[j]* through genetic crossover (see lines 16–17) and mutation-based operators (see lines 18–20). The genetic crossover operator is applied between two firefly solutions, *FSOL[i]* and *FSOL[j]*, in a number of randomly chosen crossover points. The number of crossover points, *nmbC*, is computed as the rounded difference between their objective function values. As a result of crossover, two offspring solutions are generated and the best one will replace *FSOL[i]*. Figure 3 illustrates an example of applying the crossover operator. In this example, the crossover is applied in two points randomly selected by exchanging the flags associated to the providers from these points. As a result of applying the crossover operator, two offspring solutions are generated from which the one having the lowest fitness is selected, while the other is discarded.

The mutation operator will modify the resulting *FSOL[i]* in a number of points chosen according to the following strategy: if the fitness value of *FSOL[i]* is positive, then the mutation points will correspond to providers having their flag activated—as a result of mutation, activated providers will be deactivated, while if the fitness value of *FSOL[i]* is negative, then the mutation points will correspond to providers having their flag deactivated—as a result of mutation, deactivated providers will be activated. Figure 4 presents an example of one point mutation operator applied on a solution. It can be noticed that if the fitness of the solution is higher than 0, meaning that the activated providers provide too much energy than it is demanded, the mutation will be

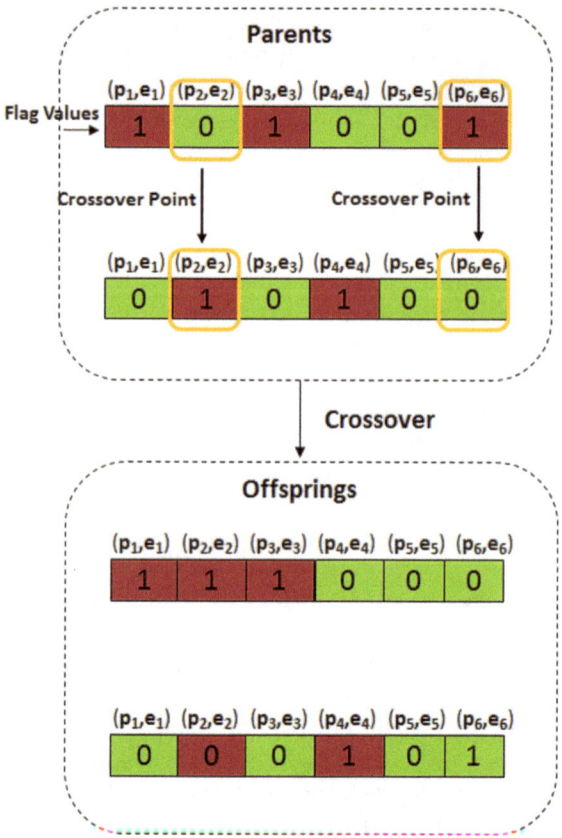

Fig. 3 Example of applying the crossover operator on two firefly solutions with six providers

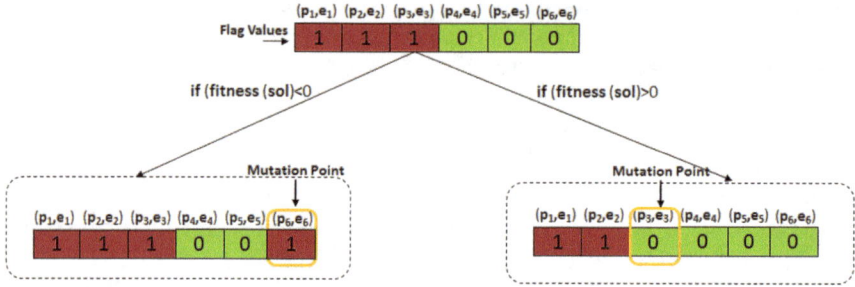

Fig. 4 Example of applying the mutation operator on a firefly solution with six providers

performed on an activated provider randomly selected. Otherwise, if the fitness of the solution is lower than 0, meaning that the activated providers provide less energy than it is demanded, the mutation will be performed on a deactivated provider randomly selected.

After each firefly solution has been compared to the other firefly solutions, the best firefly solution is identified and further updated using the mutation-based operator described previously, to avoid stagnation in a local optimum solution.

The iterative stage ends when the algorithm's number of iterations reaches the maximum number, *maxNoIt*, given as input to the algorithm.

5   Experimental Results

This section presents the experimental results obtained while tuning the values of the adjustable parameters of the firefly algorithm applied for identifying coalitions of energy providers that satisfy an energy demand for one specified hour. The experiments have been performed on a data set developed in-house. Table 1 presents a fragment of the data used in the experiments, where *ID* refers to the identification number of an energy provider, *type* refers to the type of the energy provided by an energy provider (i.e., eolian, solar, etc.), *energy* refers to the forecast energy for a particular hour, and *price* represents the price of the offered energy.

The experimental results that will be further presented in this section have been obtained by applying the firefly algorithm to accommodate the following demand which states that on May 1st, 2019, at hour 13, 10.08 units of electricity are required at a price of maximum 200 EUROS and out of the number of providers, 20% should be providers that generate eolian renewable energy:

$$demanded\_energy = (10.08, Wednesday\ 1st\ May, 2019, 13:00,$$
$$\{0, 0, 20, 0\}, 200)$$

In the first experiments, we have focused on identifying the optimal configuration for the values of the firefly algorithm's adjustable parameters, namely the number of iterations (*maxNoIt*) and the number of fireflies (*noF*). We have varied the value of *maxNoIt* in the interval [50, 950], while the values of *noF* in the interval [10, 240]. Table 2 illustrates a fragment of the experimental results obtained while varying the values of the adjustable parameters. It can be noticed that the best compromise between the fitness function value and execution time have been obtained for $noF = 50$ and $maxNoIt = 100$ (see the highlighted row in the table).

Using the best configuration of the firefly algorithm's adjustable parameters, we have further performed some experiments to analyze how the values of the multiplication constants r_1 and r_2 used to set the degree by which a solution is penalized according to the violated constraints (see Formula 4) influence the fitness function values and algorithm's execution time. The values of r_1 have been varied in the inter-

Table 1 Fragment of the data used in the experiments

ID	Type	Energy	Price
0	Wind	4.800970118	68.31041681
1	Tide	1.396673887	88.01457677
2	Solar	0.676743507	25.19172145
3	Solar	1.497591417	17.93688019
4	Solar	2.48136051	5.133831229
5	Traditional	0.534263429	83.30671202
6	Wind	2.182166562	99.90664902
7	Traditional	0.730821001	29.65040732
8	Solar	3.731143464	92.06405523
9	Tide	0.352630232	52.7896599
10	Solar	1.09760821	65.48904862
11	Wind	2.508687443	82.06160607
13	Traditional	3.139872946	16.25399465
14	Traditional	3.577344426	35.47029621
15	Traditional	1.685347996	83.41238119
16	Wind	1.021026698	12.87006518
17	Traditional	0.734591117	56.09757598
18	Wind	2.296696452	75.23882352
19	Wind	3.769093315	22.19509241
20	Tide	3.731515947	3.390145356
21	Solar	0.84368118	40.50093774
22	Wind	2.783577157	51.76372814
23	Tide	2.330335109	16.55548625
24	Solar	1.220496142	17.16414889
25	Wind	1.579868281	57.25648734

val [0.1, 5], while the values of r_2 have been varied in the interval [0.005, 0.095]. We have considered three cases in these experiments:

- *Case 1*: in this case, the mutation and crossover operators are guided only by the energy component of the defined objective function;
- *Case 2*: in this case, the mutation is guided by all components of the objective function, while the crossover is guided only by the energy component of the objective function;
- *Case 3*: in this case, the mutation and crossover operators are guided by all the three components of the defined objective function, i.e., the energy, the heterogeneity, and the price components.

Table 2 Fragment of experimental results obtained while varying the number of fireflies and the maximum number of iterations

noF	maxNoIt	Fitness value	Execution time (s)
10	750	0.004	0.16
20	950	0.039	0.61
30	850	0.012	1.26
40	350	0.002	0.90
50	100	0.002	0.32
50	200	0.373	0.85
50	250	0.007	1.03
60	50	0.086	0.44
70	150	0.002	1.21
80	900	0.058	9.75
90	850	0.035	11.88
100	300	0.022	5.44
110	600	0.006	12.13
110	950	0.333	20.81
120	150	0.044	3.69
120	50	0.155	1.24
130	600	0.022	17.28
140	350	0.001	11.23
150	700	0.012	25.91
160	600	0.047	27.84
170	750	0.011	40.29
180	550	0.003	32.61
190	700	0.004	44.24
200	250	0.016	17.07
210	450	0.017	36.39
220	100	0.018	8.05
230	550	0.008	47.45
240	400	0.004	39.51

Table 3 presents a fragment of the top 40 experimental results obtained while varying the values of the multiplication constants, when the mutation and crossover operators are guided only by the energy component of the objective function.

Table 4 presents a fragment of the top 40 experimental results obtained while varying the values of the multiplication constants, when the mutation is guided by all components of the objective function, while the crossover is guided only by the energy component of the objective function.

Table 3 Fragment of experimental results obtained while varying the values of the multiplication constants—the mutation and crossover operators are guided only by the energy component

r_1	r_2	Fitness value	Execution time (s)	Obtained heterogeneity (%)	Obtained price
4	0.005	0.012	5.22	20.00	171.58
0.9	0.075	0.037	5.16	15.00	169.99
0.1	0.055	0.053	6.17	26.67	153.33
2.7	0.01	0.055	5.38	12.22	178.04
3	0.01	0.065	4.98	23.33	173.66
0.8	0.03	0.069	5.24	13.33	184.99
3.2	0.01	0.098	5.13	20.00	234.02
0.9	0.005	0.099	5.15	23.33	221.78
3.6	0.01	0.177	5.19	20.00	198.44
2.9	0.005	0.180	5.08	15.00	200.78
3.7	0.05	0.190	5.62	16.67	174.45
1.1	0.075	0.191	4.85	6.67	189.69
3.5	0.025	0.194	5.25	26.11	188.66
1	0.005	0.199	5.04	5.56	203.13
1.8	0.005	0.208	5.15	18.10	191.52
1.9	0.005	0.219	5.05	15.00	145.63
3.6	0.02	0.228	4.97	23.33	169.96
2.7	0.02	0.235	5.08	25.56	209.43
2.4	0.005	0.243	5.11	13.33	169.34
0.1	0.02	0.264	5.41	24.44	159.30
0.4	0.035	0.267	5.17	6.67	173.31
1.5	0.015	0.293	5.40	23.33	183.44
3.8	0.03	0.309	5.23	11.11	125.30
0.1	0.01	0.316	5.14	23.33	212.58
3.1	0.005	0.318	4.81	21.43	224.43
3.2	0.005	0.321	4.80	18.89	243.19
0.4	0.015	0.330	5.06	20.56	193.06
2.2	0.02	0.332	5.61	15.00	199.10
0.3	0.005	0.334	5.55	5.56	239.22
3.1	0.005	0.318	4.81	21.43	224.43
3.2	0.005	0.321	4.80	18.89	243.19
0.4	0.015	0.330	5.06	20.56	193.06
2.2	0.02	0.332	5.61	15.00	199.10
0.3	0.005	0.334	5.55	5.56	239.22
0.2	0.005	0.338	5.56	13.33	195.09
0.6	0.005	0.357	5.02	25.00	189.32
0.5	0.005	0.364	5.16	15.00	207.54
1.8	0.045	0.366	5.72	19.44	153.49
1.5	0.025	0.372	5.60	16.67	194.65
1.2	0.02	0.375	4.92	21.67	190.28

Table 4 Fragment of experimental results obtained while varying the values of the multiplication constants—the mutation is guided by all components of the objective function and the crossover is guided only by the energy component

r_1	r_2	Fitness value	Execution time (s)	Obtained heterogeneity (%)	Obtained price
4.1	0.02	0.02	6.12	16.98	189.86
0.5	0.025	0.04	5.53	26.11	160.10
2.9	0.035	0.04	5.21	25.00	175.65
3.3	0.015	0.04	5.05	20.00	164.19
1.8	0.025	0.05	5.58	8.33	140.26
1.1	0.02	0.08	5.47	20.56	178.68
1.2	0.005	0.08	5.59	4.76	193.41
4.7	0.015	0.09	5.01	20.00	213.27
1.5	0.015	0.10	5.31	13.33	151.94
4.1	0.04	0.10	6.37	21.67	151.52
0.4	0.025	0.12	5.26	20.00	182.53
4.9	0.01	0.13	5.10	21.67	198.97
0.5	0.035	0.15	5.05	27.78	157.64
2.6	0.04	0.16	5.09	23.33	152.94
1	0.005	0.20	4.81	17.78	171.75
2.4	0.005	0.21	4.83	18.10	258.76
1.9	0.035	0.21	5.81	15.00	175.11
0.4	0.005	0.22	4.91	15.00	221.72
0.4	0.01	0.23	4.88	37.22	147.29
1.5	0.005	0.23	5.03	21.67	200.87
3.6	0.035	0.24	5.28	21.67	168.42
0.7	0.035	0.25	5.61	25.00	163.91
0.3	0.055	0.25	5.39	23.33	114.69
2.6	0.085	0.28	4.94	26.67	178.95
0.2	0.005	0.28	5.41	13.33	237.15
2.4	0.02	0.30	5.48	6.67	190.42
1.9	0.09	0.31	5.93	6.67	156.70
2.6	0.005	0.31	5.01	15.00	206.35
2.3	0.03	0.34	5.51	20.00	149.58
0.5	0.01	0.34	4.98	23.33	150.73
1.8	0.04	0.35	5.52	16.67	172.47
2.1	0.025	0.36	5.30	13.33	172.48
4.4	0.005	0.39	6.44	20.56	230.90
2.2	0.025	0.39	4.79	27.78	189.05
0.1	0.015	0.40	5.33	42.22	222.95
2.9	0.005	0.42	4.88	21.75	250.06
0.3	0.005	0.43	4.82	18.89	211.89
4.2	0.005	0.43	4.96	20.56	200.77
0.3	0.02	0.43	5.43	20.56	189.53
2.3	0.035	0.44	5.09	27.78	195.69

Table 5 Fragment of experimental results obtained while varying the values of the multiplication constants—the mutation and crossover operators are guided by all components of the objective function

r_1	r_2	Fitness value	Execution time (s)	Obtained heterogeneity (%)	Obtained price
2.4	0.03	0.020	5.21	20.00	171.31
4.3	0.05	0.020	6.01	18.89	202.59
3.5	0.04	0.047	4.79	8.33	154.92
3.2	0.045	0.055	5.11	8.33	148.60
2	0.005	0.072	5.40	22.22	204.76
1.9	0.015	0.093	5.65	13.10	185.99
0.2	0.005	0.102	6.19	20.00	196.10
1.3	0.035	0.109	6.05	15.00	162.84
3.7	0.03	0.122	5.55	21.67	189.65
0.8	0.095	0.141	5.24	26.11	189.37
0.3	0.005	0.146	5.27	25.56	192.98
0.2	0.01	0.168	5.09	11.11	191.54
2	0.015	0.190	5.02	17.78	182.87
3.1	0.035	0.192	5.43	17.78	152.12
2.8	0.015	0.193	10.47	22.22	148.96
4.9	0.01	0.211	5.16	18.10	184.28
4.5	0.02	0.233	5.54	21.67	185.51
0.1	0.04	0.254	5.93	37.78	189.19
1.4	0.04	0.270	5.34	5.56	162.16
0.5	0.02	0.290	5.39	13.89	203.93
1.5	0.015	0.291	5.31	16.67	186.27
1.7	0.02	0.294	5.17	27.70	165.64
2.9	0.025	0.297	4.86	22.22	174.49
4.2	0.01	0.297	6.02	21.67	145.15
0.5	0.045	0.299	5.07	30.56	166.76
0.7	0.025	0.300	6.29	37.22	168.10
3.8	0.01	0.300	5.45	23.33	156.46
0.1	0.005	0.301	5.02	19.44	247.85
2.1	0.005	0.302	5.52	30.00	237.69
2.9	0.015	0.304	5.13	21.67	200.21
1.2	0.03	0.306	5.32	24.44	194.75
1.5	0.005	0.319	4.72	27.78	137.28
3.4	0.005	0.331	5.12	17.78	200.32
0.7	0.035	0.361	5.30	18.89	180.42
1.1	0.01	0.362	5.12	33.33	192.56
0.9	0.025	0.365	5.40	22.22	182.72
3.5	0.06	0.368	5.89	20.56	199.00
5	0.045	0.378	5.44	21.67	150.61
0.6	0.005	0.387	5.55	16.67	196.56
4.6	0.08	0.389	5.34	26.11	182.28

Table 5 presents a fragment of the top 40 experimental results obtained while varying the values of the multiplication constants, when all the components of the objective function are used to guide the mutation and crossover operators.

By analyzing the results from Tables 3, 4, and 5, it can be noticed that the fitness function values vary in the following intervals:

- Case 1: [0.012–0.375];
- Case 2: [0.02–0.44];
- Case 3: [0.02–0.389].

It can be noticed that the best values are obtained in the first case, while for the other cases, a small deviation is introduced. Therefore, it can be concluded that the usage of the heterogeneity and price components of the objective function to guide the crossover and mutation operators do not introduce significant improvements in the fitness function values.

6   Conclusions

In this chapter, we have proposed an adapted version of the firefly algorithm [5] that is applied for solving the problem of aggregating energy providers in virtual associations. The goal was to identify a group of heterogeneous individual providers that best satisfy a given electricity demand at the lowest cost. The proposed algorithm has been evaluated on a data set developed in-house consisting of forecast energy supply values for various heterogeneous energy providers over a given time horizon. Within the experiments, we have considered three cases to analyze the impact on the fitness function value of using penalty components in applying the mutation and crossover operators. Experimental results demonstrated that the best results have been obtained in the case of using only the energy component in applying the mutation and crossover operators.

Acknowledgements This work has been conducted within the eDREAM project Grant number 774478, co-funded by the European Commission as part of the H2020 Framework Programme (H2020-LCE-2017-SGS).

References

1. Morales JM, Conejo AJ, Madsen H, Pinson P, Zugno M (2014) Integrating renewables in electricity markets—operational problems. Springer, New York, Heidelberg, Dordrecht, London
2. Renewable energy directive, https://ec.europa.eu/energy/en/topics/renewable-energy/renewable-energy-directive
3. Doulamis ND, Doulamis AD, Varvarigos E (2018) Virtual associations of prosumers for smart energy networks under a renewable split market. IEEE Trans Smart Grid 9(6):6069–6083
4. Hrvoje Pandzic, Igor Kuzle, Tomislav Capuder (2013) Virtual power plant mid-term dispatch optimization. Appl Energy 101:134–141

5. Yang X-S (2009) Firefly Algorithms for multimodal optimization. In: International symposium on stochastic algorithms SAGA 2009: stochastic algorithms: foundations and applications, pp 169–178

6. Classification of the optimization problems, https://neos-guide.org/optimization-tree

7. Michalewicz Z, Schoenauer M (1996) Evolutionary algorithms for constrained parameter optimization problems. Evol Comput 4(1):1–32

8. Yang X-S, He X (2013) Firefly Algorithm: recent advances and applications. Int J Swarm Intell 1(1)

9. Udaiyakumar KC, Chandrasekaran M (2014) Application of firefly algorithm in job shop scheduling problem for minimization of makespan. Procedia Eng 97:1798–1807

10. Mohanty DK (2016) Application of firefly algorithm for design optimization of a shell and tube heat exchanger from economic point of view. Int J Therm Sci 102:228–238

11. Hashim NL, Yusof Y, Hussain A (2016) The application of Firefly algorithm in an adaptive emergency evacuation centre management (AEECM) for dynamic relocation of flood victims. In: AIP conference proceedings

12. Mariam E, Zoubida B, Hanaa H, Aouatif A, Youssfi Elkettani (2017) Firefly algorithm for supply chain optimization. J Math 39(3):355–367

13. Dey N, Samanta S, Chakraborty S, Das A, Chaudhuri SS, Suri JS (2014) Firefly algorithm for optimization of scaling factors during embedding of manifold medical information: an application in ophthalmology imaging. J Med Imaging Health Inf 4(3):384–394

14. Chakraborty S, Dey N, Samanta S, Ashour AS, Balas VE (2016) Firefly algorithm for optimized nonrigid demons registration. In: Bio-inspired computation and applications in image processing. Academic Press, pp 221–237

15. Samanta S, Mukherjee A, Ashour AS, Dey N, Tavares JMR, Abdessalem Karâa WB, Hassanien AE (2018) Log transform based optimal image enhancement using firefly algorithm for autonomous mini unmanned aerial vehicle: An application of aerial photography. Int J Image Graph 18(04):1850019

16. Neeraj Gupta, Mahdi Khosravy, Nilesh Patel, Tomonobu Senjyu (2018) A bi-level evolutionary optimization for coordinated transmission expansion planning. IEEE Access 6:48455–48477

17. Roldán-Blay C, Escrivá-Escrivá G, Roldán-Porta C, Álvarez-Bel C (2017) An optimisation algorithm for distributed energy resources management in micro-scale energy hubs. Energy 132:126–135

18. Beraldi P, Violi A, Carrozzino G, Bruni ME (2018) A stochastic programming approach for the optimal management of aggregated distributed energy resources. Comput Oper Res 96:200–212

19. Vergados DJ, Mamounakis I, Makris P, Varvarigos E (2016) Prosumer clustering into virtual microgrids for cost reduction in renewable energy trading markets. Sustain Energy Grids Netw 7:90–103

20. Iriaa J, Soares F (2019) A cluster-based optimization approach to support the participation of an aggregator of a larger number of prosumers in the day-ahead energy market. Electr Power Syst Res 168:324–335

21. Mamounakis I, Vergados DJ, Makris P, Makris P, Doulamis ND (2017) Renewable energy prosumers clustering towards target aggregated prosumption profiles based on recursive predictions. In: 2017 IEEE Manchester powertech conference, pp 1–6

Chapter 4
Structural Damage Identification Using Adaptive Hybrid Evolutionary Firefly Algorithm

Qui X. Lieu, Van Hai Luong and Jaehong Lee

1 Introduction

Over the past decades, owing to the increasing need of discovering structural damage, especially in civil and mechanical engineering, structural health monitoring (SHM) has been extensively developed and drawn a remarkable attention from scholars. Such damage can appear under a variety of scenarios such as ageing, environmental impacts, unexpected accidents and so on. This can cause unavoidable inconveniences in use and may lead to unpredictable damage of a structural part, even an undesired collapse of the whole structure. Therefore, an early damage identification helps to propose appropriate solutions for either repairs and reinforcements or replacement of local damage to lengthen their life.

For the above-discussed crucial reasons, many methodologies have been thus released to detect the damage site and severity of a certain structure. Among them, inverse optimization-based damage detection methods have emerged as one of the most popularly used numerical ones, where metaheuristic algorithms have been widely employed as optimizers. In this regard, Seyedpoor [1] used particle swarm optimization (PSO) algorithm for structural damage detection based on multiple stages. Then, the author also introduced a two-stage damage identification method

Q. X. Lieu · J. Lee (✉)
Department of Architectural Engineering, Sejong University,
209 Neungdong-ro, Gwangjin-gu, Seoul 05006, Republic of Korea
e-mail: jhlee@sejong.ac.kr

Q. X. Lieu
e-mail: lieuxuanqui@hcmut.edu.vn

Q. X. Lieu · V. H. Luong
Faculty of Civil Engineering, Ho Chi Minh City University of Technology (HCMUT)-Vietnam
National University (VNU), 268 Ly Thuong Kiet Street, District 10, Ho Chi Minh City 700000,
Vietnam
e-mail: lvhai@hcmut.edu.vn

© Springer Nature Singapore Pte Ltd. 2020   75
N. Dey (ed.), *Applications of Firefly Algorithm and its Variants*,
Springer Tracts in Nature-Inspired Computing,
https://doi.org/10.1007/978-981-15-0306-1_4

utilizing a modal strain energy-based index (MSEBI) and the PSO [2]. Differential evolution (DE) algorithm was also applied in order to handle optimization-based damage detection problems by Seyedpoor [3] and Kim et al. [4]. Majumdar et al. [5] utilized ant colony optimization (ACO) for damage evaluation of truss structures based on the information of natural frequencies. A micro-genetic algorithm (GA) for handling optimization-based two-stage damage exploration of truss structures was presented by Kim et al. [6] and Guo et al. [7]. Masoumi et al. [8] introduced a generalized flexibility matrix to structural damage detection utilizing imperialist competitive algorithm (ICA). Kaveh and Zolghadr [9] used an improved charged system search (CSS) algorithm for damage estimation of trusses relied upon the data of natural frequencies and eigenvectors. Xu et al. [10] suggested a numerical methodology for structural damage detection employing the cuckoo search (CS) optimization tool. Dinh-Cong et al. [11] used teaching–learning-based optimization (TLBO) approach to assess two-stage damage of truss structures with limited sensors. Moreover, the same TLBO algorithm was also employed for SHM in a recent publication of Khatir et al. [12]. Dinh-Cong et al. [13] dealt with optimization-based damage diagnostic problems using a hybrid target function and Jaya algorithm. It can be found that although a large number of population-based optimization algorithms have been successfully applied to structural damage evaluation, there have been still many issues that need to be further refined and examined, especially for those concerning the computational cost, stability and sensitivity of approaches and investigated problems with regard to various objective functions.

More recently, an adaptive hybrid evolutionary firefly algorithm (AHEFA) was successfully developed for size and shape optimization of truss structures in an authors' previously released work [14]. This method has proven its good efficiency and reliability in yielding a more accurate global solution with a better convergence speed against many other algorithms available in the literature. Then, Lieu and Lee also applied the AHEFA to optimization of functionally graded (FG) plates [15–17] and reliability-based design optimization (RBDO) of FG plates [18]. Nonetheless, its extension to damage detection problems of truss structures has not been yet carried out so far. Accordingly, this study conducted herein aims at investigating the effectiveness and reliability of the AHEFA for such problems for the first time.

In this work, the AHEFA is employed to resolve inverse optimization-based two-stage damage identification problems. In the first step, the MSEBI is utilized to find out the location of mistrustfully weakened elements of a damaged structure. The purpose of the remaining stage is to search the actual extent of those elements by addressing an optimization problem relied on the AHEFA. Since only suspected elements are taken account of design variables, the computational cost and the solution accuracy are enhanced dramatically. Three test examples including a 21-bar planar truss, a 31-bar planar one, and a 25-bar space one are investigated to assess the ability of the present paradigm.

The rest of this chapter is outlined as follows. Section 2 exhibits the theoretical formulation of inverse optimization-based two-stage damage detection problems. Three different algorithms including the DE, FA, and AHEFA are discussed in Sect. 3. Section 4 tests three benchmark numerical examples of truss structures to validate the performance of the proposed paradigm. The final Section closes this chapter with several conclusions.

2   Inverse Optimization-Based Two-Stage Damage Detection

2.1   Modal Strain Energy-Based Index

In the first phase, the modal strain energy-based index (MSEBI) suggested by Seyed-poor [2] is utilized to identify doubtfully damaged elements of a structure. This approach requires compulsory data regarding mode shape vectors that are provided from free vibration analyses. The finite element model for such analyses can be generally stated as follows [19]

$$\left(\mathbf{K} - \omega_i^2 \mathbf{M}\right)\boldsymbol{\phi}_i = \mathbf{0}, \quad i = 1, 2, \ldots, n_{\text{dof}}, \tag{1}$$

where $\mathbf{K}$ and $\mathbf{M}$ symbolize the global stiffness and lump mass matrices, respectively; ω_i and $\boldsymbol{\phi}_i$ denote the ith natural frequency and mode shape vector of the structure, respectively, and n_{dof} represents the total degrees of freedom (DOFs) of the structure.

It can be seen that mode shape vectors resolved from Eq. (1) serve as displacement ones in static analyses, each of all elements therefore exists a strain energy which is the so-called modal strain energy (MSE) in this case. Analogously, the eth MSE corresponding to the ith free vibration mode can be defined as follows

$$\mathcal{U}_i^e = \frac{1}{2}\left(\boldsymbol{\phi}_i^e\right)^{\mathrm{T}} \mathbf{K}^e \boldsymbol{\phi}_i^e, \tag{2}$$

where $\mathbf{K}^e$ stands for the stiffness matrix of the eth element, and $\boldsymbol{\phi}_i^e$ symbolizes the mode shape vector of the eth element with respect to the ith mode.

For the sake of computational convenience, the eth MSE is often normalized by the following formula

$$\overline{\mathcal{U}}_i^e = \frac{\mathcal{U}_i^e}{\sum_{e=1}^n \mathcal{U}_i^e}, \tag{3}$$

where n is the whole number of elements of the structure.

In the case of taking account of the first m mode shapes, the eth normalized MSE given in the above equation is computed by

$$\overline{\mathcal{U}}^e = \frac{\sum_{i=1}^m \overline{\mathcal{U}}_i^e}{m}, \tag{4}$$

It is apparent that as damage can appear at a certain structural element, even at an arbitrary position, the global stiffness of whose system is weakened. Suppose that the global mass matrix is still kept constant. In this case, the mode shape vectors provided from Eq. (1) become larger. This may lead to an increase in the element MSE, especially damaged elements in comparison with that of healthy ones. Therefore, by comparing in pairs of the element normalized MSE given by Eq. (4) for a healthy structure and a corresponding damaged one, i.e. $\overline{\mathcal{U}}^{e,h}$ and $\overline{\mathcal{U}}^{e,d}$, the damage presence of an element can be forecasted by means of the so-called modal strain energy base index (MSEBI), I^e_{MSE}, as follows

$$I^e_{MSE} = \begin{cases} \dfrac{\overline{\mathcal{U}}^{e,d} - \overline{\mathcal{U}}^{e,h}}{\overline{\mathcal{U}}^{e,h}} > 0, \text{ damaged element,} \\[2mm] \dfrac{\overline{\mathcal{U}}^{e,d} - \overline{\mathcal{U}}^{e,h}}{\overline{\mathcal{U}}^{e,h}} \le 0, \text{ healthy element.} \end{cases} \tag{5}$$

It should be noted that $\overline{\mathcal{U}}^{e,h}$ can be directly estimated from simulation models such as analytical or numerical ones, whereas $\overline{\mathcal{U}}^{e,d}$ can be obtained from experimental data.

2.2   Optimization Problem Statement

It is obvious that suspiciously damaged elements can be recognized after the first stage is conducted. In this second step, the damage extent of those elements is evaluated via solving an inverse optimization problem. It is worthwhile to emphasize that selecting different objective functions depends on each specific problem and dramatically affects the predicted outcomes of a damaged structure, especially for the issues relating to the solution accuracy and the computational cost. Several typical functions could be listed as efficient correlation-based index (ECBI) [3], flexibility matrix [11], normalized modal strain energy indicator (nMSEDI) [12], a hybrid objective function [13], multiple damage location assurance criterion (MDLAC) [20], modal flexibility [21], and so on. For further details about those objective functions, interesting readers could refer to the afore-cited materials. In this work, after examinations, the objective function minimized is the ECBI $f_{ECBI}(\mathbf{x})$ that describes the difference of natural frequencies between a healthy structure and a corresponding damage one. Design variables are treated as a relative reduction in the Young's modulus of the suspected elements. Accordingly, this problem model can be mathematically represented as follows

$$\text{Min}: f_{ECBI}(\mathbf{x}) = -\frac{1}{2}\left[\frac{\left|\Delta\omega^T\delta\omega\right|^2}{(\Delta\omega^T\Delta\omega)(\delta\omega^T\delta\omega)} + \frac{1}{p}\sum_{i=1}^{p}\frac{\min(\omega_i(\mathbf{x}), \omega_i^d)}{\max(\omega_i(\mathbf{x}), \omega_i^d)}\right],$$

$$\text{St}: \begin{cases} \left[\mathbf{K}(\mathbf{x}) - \omega_i^2\mathbf{M}\right]\boldsymbol{\phi}_i = \mathbf{0}, \ i = 1, 2, \ldots, p, \ldots, n_{dof}, \\ x_{LB,j} \le x_j \le x_{UB,j}, \quad j = 1, 2, \ldots, d, \end{cases} \tag{6}$$

where $\mathbf{x} = \{x_1, x_2, \ldots, x_j, \ldots, x_d\}$ is the design variable vector with d being the number of variables, and x_j is defined by

$$x_j = \frac{E - E_j}{E_j}, \tag{7}$$

in which E and E_j are the intact and actually damaged Young's moduli, respectively. In addition, $\mathbf{\Delta\omega} = \{\Delta\omega_1, \Delta\omega_2, \ldots, \Delta\omega_i, \ldots, \Delta\omega_p\}^{\mathrm{T}}$ and $\mathbf{\delta\omega} = \{\delta\omega_1, \delta\omega_2, \ldots, \delta\omega_i, \ldots, \delta\omega_p\}^{\mathrm{T}}$ are, respectively, the vectors depicting the relative change of the first p natural frequencies in the damaged and predicted structures against those of the healthy one, where $\Delta\omega_i$ and $\delta\omega_i$ are, respectively, given by

$$\Delta\omega_i = \frac{\omega_i^{\mathrm{h}} - \omega_i^{\mathrm{d}}}{\omega_i^{\mathrm{h}}}, \quad \delta\omega_i = \frac{\omega_i^{\mathrm{h}} - \omega_i(\mathbf{x})}{\omega_i^{\mathrm{h}}}, \quad i = 1, 2, \ldots, p, \tag{8}$$

in which ω_i^{h}, ω_i^{d} and $\omega_i(\mathbf{x})$ symbolize the ith natural frequency of the healthy, damaged and predicted structures, respectively.

3   Optimization Algorithm

3.1   Differential Evolution Algorithm

Differential evolution (DE) method known as one of the population-based optimization ones was first released by Storn and Price [22]. In the first iteration, the DE consists of four major phases, namely initialization, mutation, crossover and selection. It is noted that only the repeatable implementation of the last three phases is done for all succeeding iterations until meeting stopping standards. Below concisely summarizes the way of performed all the above phases.

- *Initialization*

First, a population with np individuals is randomly initialized in a given continuous search space. In which, the ith individual ($i = 1 \div np$) is the target vector containing d design variables, and defined as

$$x_{i,j}^{t=0} = x_{\mathrm{LB},j} + \mathrm{rand}_{i,j}(x_{\mathrm{UB},j} - x_{\mathrm{LB},j}), \quad j = 1, 2, \ldots, d, \tag{9}$$

in which $x_{\mathrm{LB},j}$ and $x_{\mathrm{UB},j}$ denote the lower and upper bounds of x_j; $\mathrm{rand}_{i,j}$ stands for a uniformly disposed random number within $[0, 1]$, and the superscript ($t = 0$) represents $x_{i,j}$ only for the first iteration of the initialization step. In general, the ith individual at the tth iteration can be symbolized as follows

$$\mathbf{x}_i^t = \{x_{i,1}^t, x_{i,2}^t, \ldots, x_{i,j}^t, \ldots, x_{i,d}^t\}. \tag{10}$$

• *Mutation*

In this phase, the mutant vector $\mathbf{v}_i^t$ is created from the target vectors $\mathbf{x}_i^t$ utilizing one of the following mutation strategies

$$\text{rand}/1: \mathbf{v}_i^t = \mathbf{x}_{R_1}^t + F\left(\mathbf{x}_{R_2}^t - \mathbf{x}_{R_3}^t\right), \tag{11}$$

$$\text{best}/1: \mathbf{v}_i^t = \mathbf{x}_{\text{best}}^t + F\left(\mathbf{x}_{R_1}^t - \mathbf{x}_{R_2}^t\right), \tag{12}$$

where three integer numbers R_1, R_2 and R_3 are randomly picked out in $[1, np]$ with $R_1 \neq R_2 \neq R_3 \neq i$; $\mathbf{x}_{\text{best}}^t$ symbolizes the best solution with regard to the best objective value, and the scale factor F within $(0, 1]$ aims to create the difference between two vectors $\mathbf{x}_i^t$. In this study, F is set to be 0.8. For more detailed information on other mutation schemes, interesting readers are suggested to consult the material [23].

Obviously, the ith individual after performing one of the mutant operators may be violated its boundary conditions. In that case, it is rejoined its initial design domain as follows

$$v_{i,j}^t = \begin{cases} 2x_{\text{LB},j} - v_{i,j}^t, & \text{if } v_{i,j}^t < x_{\text{LB},j}, \\ 2x_{\text{UB},j} - v_{i,j}^t, & \text{if } v_{i,j}^t > x_{\text{UB},j}, \\ v_{i,j}^t, & \text{otherwise.} \end{cases} \tag{13}$$

• *Crossover*

Next, in order to heighten the variety of individuals obtained from the foregoing procedure, the most commonly used binomial crossover technique is utilized herein. According to this strategy, the ith trial vector $\mathbf{u}_i^t$ is created by mixing the mutant vector $\mathbf{v}_i^t$ and the target vector $\mathbf{x}_i^t$ based on the following rule

$$u_{i,j}^t = \begin{cases} v_{i,j}^t, & \text{if}(j = K) \text{ or } \left(\text{rand}_{i,j} \leq Cr\right), \\ x_{i,j}^t, & \text{otherwise,} \end{cases} \tag{14}$$

where the integer number K is randomly given within $[1, np]$, and the crossover coefficient Cr is set within $[0.7, 0.9]$ in this work.

• *Selection*

Ultimately, the best individuals are filtered for the next iteration as follows

$$\mathbf{x}_i^{t+1} = \begin{cases} \mathbf{u}_i^t, & \text{if } f\left(\mathbf{u}_i^t\right) \leq f\left(\mathbf{x}_i^t\right), \\ \mathbf{x}_i^t, & \text{otherwise.} \end{cases} \tag{15}$$

3.2   Firefly Algorithm

From the observations of the swarm behaviour and flashing patterns of fireflies in nature, Yang proposed the firefly algorithm (FA) [24, 25]. Several related publications could be found in the literature [26, 27]. In order to establish this methodology, the following suppositions are made:

- A unisex assumption is devoted to all fireflies. This means no any sex influence on the attraction of firefly individuals is available;
- The light intensity represents the attractiveness of a firefly. As a result, fireflies with a lesser light intensity will move to the more attractive others. All of those properties are in direct proportion to their distance;
- The objective function value is utilized to characterize a firefly's light intensity.

Analogously to other metaheuristic optimization algorithms, firstly, an initial population containing np fireflies is randomly initialized using Eq. (9), where each firefly individual includes d design variables. The location of the ith individual is then newly moved by the following formula

$$\mathbf{v}_i^t = \mathbf{x}_i^t + \beta\left(\mathbf{x}_k^t - \mathbf{x}_i^t\right) + \alpha^t \boldsymbol{\varepsilon}_i, \tag{16}$$

where $\beta\left(\mathbf{x}_k^t - \mathbf{x}_i^t\right)$ denotes the firefly's attraction; $\alpha^t \boldsymbol{\varepsilon}_i$ is the random term with α randomly selected within $[0, 1]$; $\boldsymbol{\varepsilon}_i = \left(\mathrm{rand}_{i,j} - 0.5\right)\left|x_{\mathrm{LB},j} - x_{\mathrm{UB},j}\right|$ stands for a random number vector; α^t is monotonously lessened according to the expression $\alpha^{t+1} = 0.98\alpha^t$ with $\alpha^0 = 0.5$ and the attractiveness β is defined as

$$\beta = \beta_0 e^{-\gamma r_{i,k}^2}, \quad k = 1 \div np, \tag{17}$$

in which the subscript k is dedicated to the kth individual which is of the better objective value than the ith one; $\beta_0 = 1$ denotes the initial attractiveness at $r_{i,k} = 0$; γ stands for the light absorption coefficient within $[0, \infty)$ and is taken as 1 in this study, and $r_{i,k}$ is the distance between two fireflies $\mathbf{x}_i^t$ and $\mathbf{x}_k^t$, and given by

$$r_{i,k} = \left\|\mathbf{x}_i^t - \mathbf{x}_k^t\right\| = \sqrt{\sum_{j=1}^d \left(x_{i,j}^t - x_{k,j}^t\right)^2}. \tag{18}$$

Similarly, provided that a new solution is out of its bound constraint, it is settled by Eq. (13) to restore its initial design space.

Apart from the first step, the optimization process is repeatedly done until stopping conditions are fulfilled.

3.3 Adaptive Hybrid Evolutionary Firefly Algorithm

In most of population-based optimization methods, a global solution must be searched in its whole design domain, this performance is thus relatively expensive and time-consuming. In addition, the stochastic finding strategy is also one of the crucial keys that considerably governs the exploration and exploitation capabilities. In which, the exploration one is superior in searching a global solution, yet a lot of computational attempts is required. Whereas the other is surpassing to the convergence rate, but it is usually easy to be fallen in trap of local solutions. In order to gain a better trade-off of two properties as discussed above, below details an adaptive hybrid evolutionary firefly algorithm (AHEFA) [14] created by cross-breeding the DE and the FA.

Substituting β from Eq. (17) into Eq. (16), and replacing $\alpha^t \varepsilon_i$ with $\alpha(\mathbf{x}_{R_2}^t - \mathbf{x}_{R_3}^t)$, then gathering the coefficients of $\mathbf{x}_i^t$, Eq. (16) is now rewritten as follows

$$\mathbf{v}_i^t = \left(1 - \beta_0 e^{-\gamma r_{i,R_1}^2}\right)\mathbf{x}_i^t + \beta_0 e^{-\gamma r_{i,R_1}^2}\mathbf{x}_{R_1}^t + \alpha\left(\mathbf{x}_{R_2}^t - \mathbf{x}_{R_3}^t\right), \tag{19}$$

where R_1, R_2 and R_3 are given those by the mutation scheme 'rand/1' of the DE; R_1 is the index of the best objective $f(R_1)$ which is compared with $f(R_2)$ and $f(R_3)$, and α equals 0.8 in this case.

Provided that the attractiveness of the ith firefly $\mathbf{x}_i^t$ is smaller than that of the R_1th one, it is moved to a new site by Eq. (19), contrariwise, Eq. (11) in the DE is adopted.

Obviously, the term $\alpha(\mathbf{x}_{R_2}^t - \mathbf{x}_{R_3}^t)$ in Eq. (19) leads to a broader search domain for the original FA, and hence the exploration ability in searching a global solution is reinforced. As mentioned before, the algorithm therefore domains a higher effort for such a performance. This is fairly analogous to that of the scheme 'rand/1' of the DE. Nonetheless, in general, the difference of mutant vectors becomes lesser and more stable after doing a number of iterations. As a result, the space for seeking solutions becomes progressively narrow. Consequently, if the local search possibility is fortified in this stage as that of the mutation scheme 'best/1' of the DE, the convergence speed is refined remarkably. It is apparent owing to the fact that the heed of other fireflies is only applied to the seek of the best one, the above drawback can be settled. Based on the above discussions, by excluding the $\alpha(\mathbf{x}_{R_2}^t - \mathbf{x}_{R_3}^t)$ in Eq. (19), and replacing $\mathbf{x}_{R_1}^t$ with $\mathbf{v}_i^t$ given by Eq. (11) in the DE, the new location of fireflies is now updated by

$$\mathbf{v}_i^t = \left(1 - \beta_0 e^{-\gamma r_{i,best}^2}\right)\mathbf{x}_i^t + \beta_0 e^{-\gamma r_{i,best}^2}\left[\mathbf{x}_{best}^t + F\left(\mathbf{x}_{R_1}^t - \mathbf{x}_{R_2}^t\right)\right], \tag{20}$$

where $r_{i,best}$ symbolizes the distance of the ith individual and the best one.

In order to use the afore-derived mutation operators for a flexible control on the exploration and exploitation features, the following automatically adapted parameter δ is employed

$$\delta = |f_{mean}/f_{best} - 1|, \tag{21}$$

where f_{best} and f_{mean} denote the best and mean value of the objective function estimated in the previous iteration, respectively.

It is noted that the optimization process usually yields a smaller and more stable value for δ in many later iterations. This issue happens thanks to a lesser variety of individuals in the whole population. In other words, the swarm behaviour of the evolutionary history becomes better. From the above remarks, either Eq. (19) or Eq. (11) is applied when δ is greater than a threshold ε, on the contrary, Eq. (20) is used. It should be worthwhile stressing that provided that ε is too large, the algorithm converges more rapidly owing to essentially utilizing the exploitation ability, yet local solutions may appear. Otherwise, the computational cost of the method becomes more expensive because it focuses only on the exploration capacity for finding a global solution. In this work, the proper value of ε is 10^{-4}.

After the above step, the way of testing the boundary violation of design variables as well as executing the crossover procedure is completely similar to those of the DE.

Ultimately, an elitist selection strategy [28] is applied to extract np best individuals for the next iteration. Firstly, the target individuals $\mathbf{x}_i^t$ and the trial individuals $\mathbf{u}_i^t$ are combined to create a new population with $2np$ individuals. Only np best individuals are then filtered. Clearly, this new population includes all the most potential candidates against with those of the original selection phase. This also helps to significantly increase the solution accuracy and convergence speed of the presented algorithm. Below describes the main procedure of the AHEFA.

The main procedure of the AHEFA

1. *Initialization phase*: Create a initial population np
2. Estimate the objective function
3. Calculate δ
4. **while** *stopping critera* are not satisfied **do**
5.   **Mutation phase**
6.   **for** $i=1\div np$ **do**
7.   **if** $\delta > \varepsilon$ **then**
8.   Randomly select R_1, R_2 and $R_3 \neq i$
9.   Determine R_1 with $f(R_1^t)$ is the best
10.   **if** $f(\mathbf{x}_i^t) > f(R_1^t)$ **then**
11.   Calculate $\mathbf{v}_i^t$ by Eq. (4.11)
12.   **else**
13.   Calculate $\mathbf{v}_i^t$ by Eq. (4.19)
14.   **end if**
15.   **else**
16.   Calculate $\mathbf{v}_i^t$ by Eq. (4.20)
17.   **end if**
18.   **end for**
19.   **Crossover phase**
20.   **Elitist selection phase**
21.   Calculate δ
22. **end while**

4 Numerical Examples

In this section, three test examples are taken into consideration to prove the possibility of the AHEFA for multidamage identification of truss structures. Results gained by the presented methodology are checked with those by the DE and FA for comparison. In all illustrated examples, the first five mode shapes ($m = 5$) are utilized for the performance of the first stage, while the first ten natural frequencies ($p = 10$) are considered for the objective evaluation of the second stage. The population size np of all the above algorithms is set to be 20. The iterative circle is ended when either $|f_{mean} - f_{best}| \leq 10^{-6}$ or 1000 iterations are done. The linear finite element method of two-node truss bar [19] is adopted for free vibration analyses. Code structures are programmed using Python 3.6 on a laptop computer with Core™ i7-2670QM CPU@2.20 GHz, 8.00 GB RAM and Windows 7® Professional (64-bit operating system). Owing to the stochastic feature of metaheuristic algorithms, each investigated case runs ten independent times. Only the best value of design variables (Best) and the corresponding number of finite element analyses (No. FEAs) of the best instance are displayed for comparison. The average time of one run in terms of second (s) is also reported. Moreover, the statistical result, i.e., standard deviation (SD) is also further presented in Tables to prove the good performance of the AHEFA in accurately determining the damage location and extent.

4.1   21-Bar Planar Truss

A planar truss with 21 bars depicted in Fig. 1 is tested as the first examination. In which, all bars are of the same mass density $\rho = 7800\,\text{kg/m}^3$ and the identical Young's modulus $E = 200$ GPa. Three cross-sectional area groups are utilized as follows: (i) $A_1 = \cdots = A_6 = 15 \times 10^{-4}\,\text{m}^2$; (ii) $A_7 = \cdots = A_{17} = 9 \times 10^{-4}\,\text{m}^2$ and (iii) $A_{18} = \cdots = A_{21} = 12 \times 10^{-4}\,\text{m}^2$. This study has been formerly investigated by several approaches such as the ICA [8] and Jaya [13], etc. Table 1 summarizes two different damage scenarios induced for this example.

In the first stage, the most distrustfully weakened positions are determined when their MSE indexes are assumed to be larger than or equal to 0.05. Accordingly, only the element 8 for case 1, the elements 1, 4, 9, 11, 12 and 18 for case 2 are suspected instead of all 21 elements. The number of design variables in cases 1 and 2 is hence lessened from 21 ones to 1 and 6 ones, respectively. This helps to considerably decrease the computational cost of the optimization process, as well as to achieve a more accurate solution.

A comparison of outcomes obtained by various methods is collected in Table 2. As predicted, the outcomes point out that the DE and the FA demand far more number of FEAs and the computational time than the AHEFA, yet the same damage extent results are diagnosed. It should be noted that these outcomes agree well with those discussed by Dinh-Cong [13].

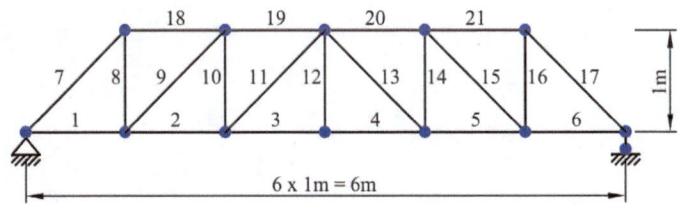

Fig. 1   21-bar planar truss

Table 1   Two different damage cases for the 21-bar planar truss

Case	1	2		
Element	8	1	9	12
Damage ratio	0.25	0.3	0.4	03

Table 2 Damage detection results of the 21-bar planar truss

Case	Ele.	DE				FA				AHEFA			
		Best	SD	No. FEAs	Time (s)	Best	SD	No. FEAs	Time (s)	Best	SD	No. FEAs	Time (s)
1	8	0.25	0	480	2.37	0.25	0	7620	42.2	0.25	0	260	1.5
2	1	0.3	0	2540	14.2	0.3	0	8800	43.5	0.3	0	1420	7.7
	4	0	0			0	0			0	0		
	9	0.4	0			0.4	0			0.4	0		
	11	0	0			0	0			0	0		
	12	0.3	0			0.3	0			0.3	0		
	18	0	0			0	0			0	0		

Fig. 2 Convergence histories for case 1 of the 21-bar planar truss

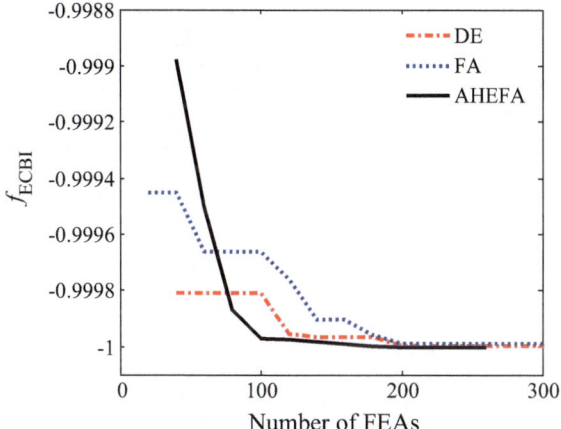

Fig. 3 Convergence histories for case 2 of the 21-bar planar truss

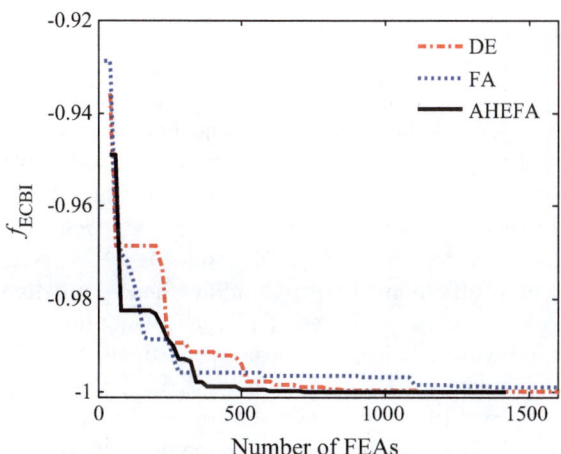

The convergence histories with the best run for cases 1 and 2 are, respectively, plotted in Figs. 2 and 3. It can be found that the AHEFA is of a highly faster convergence than the others. However, for visualization, only a first number of iterations of the convergence histories of the DE and the FA are shown. This implementation is also held for all subsequent examples.

4.2 31-Bar Planar Truss

This example aims at detecting damage of a 31-bar planar truss as described in Fig. 4. In this instance, all elements are taken the same material density $\rho = 2770 \, \text{kg/m}^3$ and Young's modulus $E = 70$ GPa. Moreover, the cross-sectional area is set to be

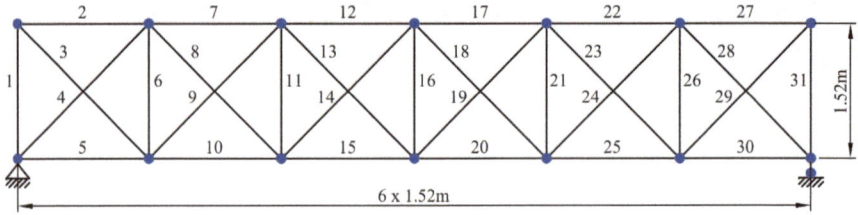

Fig. 4 31-bar planar truss

Table 3 Three different damage cases for the 31-bar planar truss

Case	1		2	3	
Element	11	25	16	1	2
Damage ratio	0.25	0.15	0.3	0.3	0.2

0.01 m². This truss has been previously done by numerous scholars [2–4, 11]. Table 3 exhibits three different damage scenarios.

After the performance of the first stage, only the elements 11, 21, 25 and 26 for case 1, the element 16 for case 2, and the elements 1, 2, 4 and 6 for case 3 are suspected as damaged elements. Table 4 reports the results obtained by various algorithms. It can be found that although the damage values are completely consistent with one another, the number of FEAs and the computational time of the AHEFA are always far lesser than those of the DE and the FA. Moreover, the outcomes obtained by the present algorithm also match well with those previously investigated in Refs. [2, 3].

Figures 5, 6 and 7 present the convergence histories with the best run for cases 1, 2 and 3 without noise. As indicated in the figures, the AHEFA is of a better convergence than the others.

In practice, the measurement error of natural frequencies and mode shapes always exists. Accordingly, to examine the noise influence on the implementation of the current paradigm, an error standard of $\pm 0.15\%$ for all natural frequencies and mode shapes is taken into consideration [3]. In this case, the stopping criterion is set to be $|f_{\text{mean}} - f_{\text{best}}| \leq 10^{-9}$. Note that the most suspiciously damaged elements are determined as those of the foregoing cases based on the MSE index utilized in the first stage.

Table 5 shows the damage detection results when the natural frequencies and corresponding mode shapes are randomly induced utilizing the noise. As found, both the DE and the AHEFA identify the damage site and severity, while the outcomes predicted by the FA are of many false signs. Nonetheless, the AHEFA always outperforms the DE in terms of a fewer number of FEAs and the computational time, yet still achieving a good solution as expected.

The convergence histories with the best run in cases 1, 2 and 3 with noise are drawn in Figs. 8, 9 and 10. Obviously, all good characteristics of the AHEFA drawn from the afore-examined instances are still correct for these ones.

Table 4 Damage detection results of the 31-bar planar truss without noise

Case	Ele.	DE				FA				AHEFA			
		Best	SD	No. FEAs	Time (s)	Best	SD	No. FEAs	Time (s)	Best	SD	No. FEAs	Time (s)
1	11	0.25	0	1560	12.5	0.25	0	7600	58.5	0.25	0	940	7.9
	21	0	0			0	0			0	0		
	25	0.15	0			0.15	0			0.15	0		
	26	0	0			0	0			0	0		
2	16	0.3	0	380	3.0	0.3	0	7220	59.4	0.3	0	240	1.8
3	1	0.3	0	1780	11.9	0.3	0	8080	57.6	0.3	0	860	6.3
	2	0.2	0			0.2	0			0.2	0		
	4	0	0			0	0			0	0		
	6	0	0			0	0			0	0		

Fig. 5 Convergence
histories for case 1 of the
31-bar planar truss without
noise

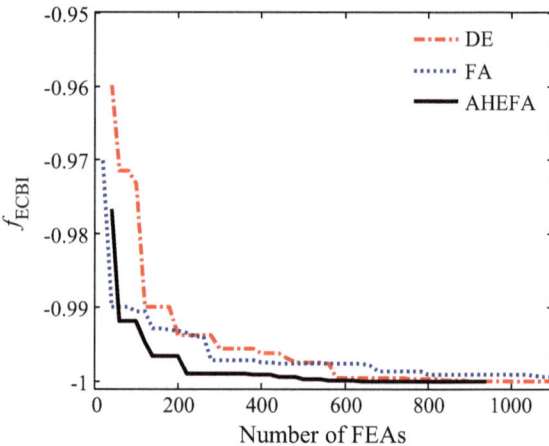

Fig. 6 Convergence
histories for case 2 of the
31-bar planar truss without
noise

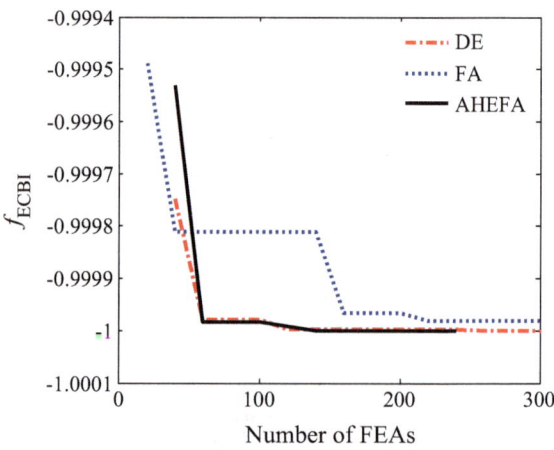

Fig. 7 Convergence
histories for case 3 of the
31-bar planar truss without
noise

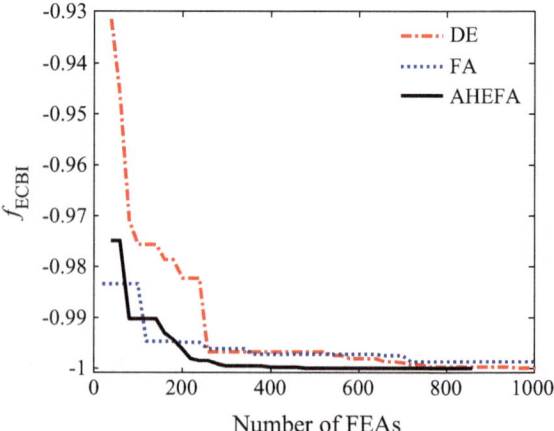

Table 5 Damage detection results of the 31-bar planar truss with noise

Case	Ele.	DE Best	SD	No. FEAs	Time (s)	FA Best	SD	No. FEAs	Time (s)	AHEFA Best	SD	No. FEAs	Time (s)
1	11	0.271	0.04	3060	21.7	0.066	0	14880	107.1	0.240	0	1420	10.7
	21	0.000	0.00			0.000	0			0.000	0		
	25	0.175	0.01			0.080	0			0.159	0		
	26	0.000	0.08			0.169	0			0.000	0		
2	16	0.269	0	460	3.4	0.329	0	7920	61.3	0.298	0	440	3.3
3	1	0.297	0	2720	20.0	0.293	0.002	15,940	117.5	0.301	0	1340	11.3
	2	0.195	0			0.204	0.002			0.193	0		
	4	0.008	0			0.000	0			0.000	0		
	6	0.000	0			0.000	0			0.000	0		

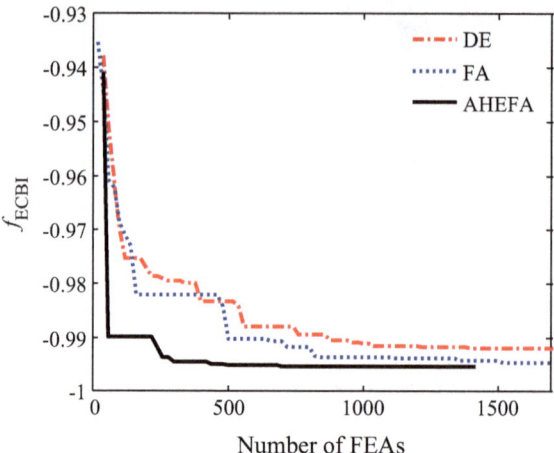

Fig. 8 Convergence histories for case 1 of the 31-bar planar truss with noise

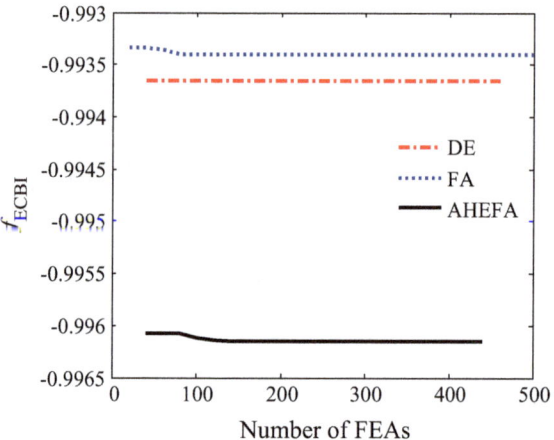

Fig. 9 Convergence histories for case 2 of the 31-bar planar truss with noise

4.3 25-Bar Space Truss

A 25-bar space truss as sketched in Fig. 11 is considered as the last illustration. This structure has ten nodes with four bottom nodes fixed. This means there are only 18 DOFs in total. The material density, Young's modulus and cross-sectional area are

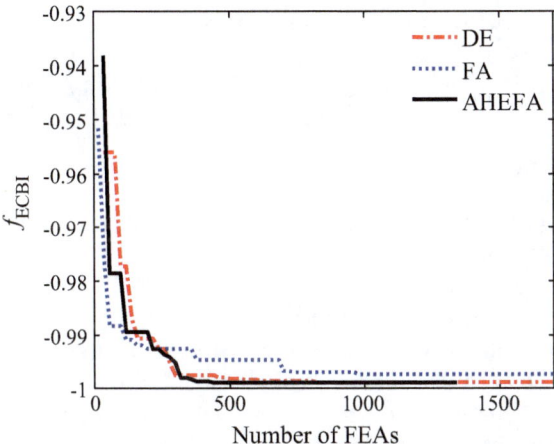

Fig. 10 Convergence histories for case 3 of the 31-bar planar truss with noise

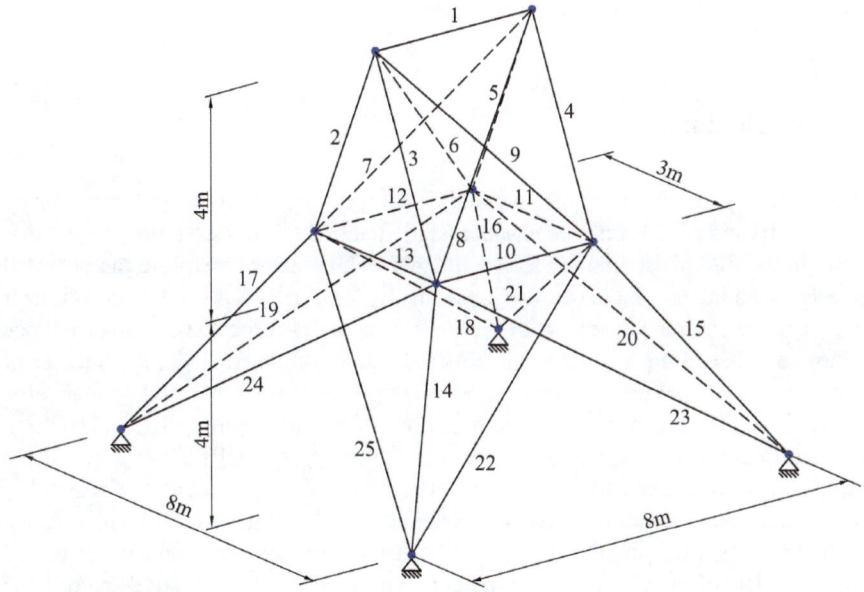

Fig. 11 25-bar space truss

Table 6 Two different damage cases for the 25-bar space truss

Case	1		2		
Element	1	5	1	12	16
Damage ratio	0.383	0.294	0.383	0.457	0.294

$7830\,\text{kg/m}^3$, $210\,\text{GPa}$ and $0.25\,\text{m}^2$, respectively. Two different damage cases reported in Table 6 are examined.

The damage identification results are shown in Table 7. Note that the number of design variables is considerably decreased after the healthy elements are curtailed due to the implementation of the first stage. As observed, the AHEFA has shown a good performance against with the other algorithms. The number of FEAs and the computational time of the current method are always lesser than those of the others as reported in the table. Nonetheless, the present method still results in an accurate damage prediction.

Finally, as observed from Figs. 12 and 13, the AHEFA always stops its process more early that the DE and FA as anticipated.

5 Conclusions

An inverse optimization-driven two-stage damage detection methodology utilizing the MSEBI and AHEFA has been successfully applied to truss structures for the first time. In the first phase, the MSEBI is utilized as a detector to explore the potential damaged candidates of a structure. Consequently, the AHEFA is used as an optimizer to accurately compute the extent of the afore-discovered suspected candidates. Since a large number of healthy elements are eradicated in the first stage, the number of design variables defined in an inverse optimization problem is lessened dramatically. This helps considerably save the computational efforts in searching a global solution when utilizing derivative-free optimization algorithms, especially the AHEFA. Three test examples of planar and space trusses are examined to demonstrate the possibility of the proposed paradigm. The results have revealed that the AHEFA outperforms both the original DE and the FA regarding the solution accuracy and computational attempts. Therefore, the suggested method is promising as an effective and reliable damage detection tool for many types of different structures such as FG plates, shells and so on, even numerous practical problems.

Acknowledgements This research was supported by a grant (NRF-2018R1A2A1A05018287) from NRF (National Research Foundation of Korea) funded by MEST (Ministry of Education and Science Technology) of Korean government.

Table 7 Damage detection results of the 25-bar space truss

Case	Ele.	DE				FA				AHEFA			
		Best	SD	No. FEAs	Time (s)	Best	SD	No. FEAs	Time (s)	Best	SD	No. FEAs	Time (s)
1	1	0.3836	0.0018	2120	12.8	0.3835	0.0234	7600	46.1	0.3820	0.0014	1200	8.3
	5	0.2940	0.0006			0.2940	0.0004			0.2940	0.0009		
	6	0.0000	0.0008			0.0000	0.0012			0.0000	0.0015		
	7	0.0000	0.0002			0.0000	0.0000			0.0000	0.0002		
	12	0.0000	0.0014			0.0000	0.0060			0.0000	0.0025		
2	1	0.3831	0.0957	3080	19.5	0.4552	0.1655	8200	51.3	0.3836	0.1160	1560	11.9
	11	0.0002	0.0266			0.0000	0.1632			0.0000	0.0000		
	12	0.4570	0.0064			0.4555	0.0850			0.4570	0.0435		
	16	0.2940	0.0000			0.2940	0.0030			0.2940	0.0000		
	18	0.0000	0.0000			0.0000	0.0000			0.0000	0.0000		
	21	0.0000	0.0000			0.0000	0.0054			0.0000	0.0035		

Fig. 12 Convergence
histories for case 1 of the
25-bar space truss

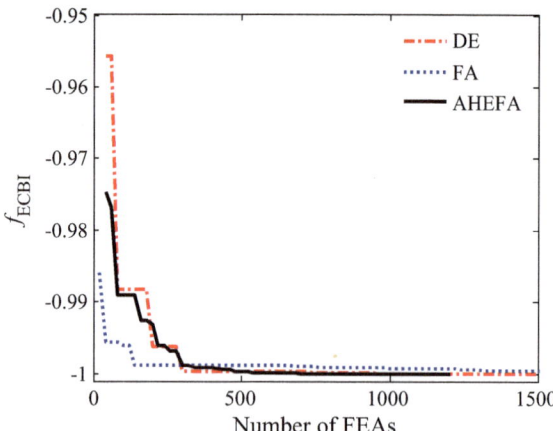

Fig. 13 Convergence
histories for case 2 of the
25-bar space truss

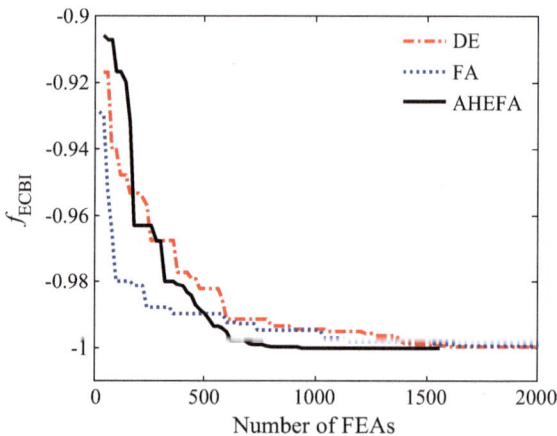

References

1. Seyedpoor SM (2010) Structural damage detection using a multi-stage particle swarm optimization. Adva Struct Eng 14:533–549
2. Seyedpoor SM (2012) A two stage method for structural damage detection using a modal strain energy based index and particle swarm optimization. Int J Non-Linear Mech 47:1–8
3. Seyedpoor SM (2015) An efficient method for structural damage detection using a differential evolution algorithm-based optimisation approach. Civ Eng Environ Syst 32:230–250
4. Kim NI, Kim S, Lee J (2019) Vibration-based damage detection of planar and space trusses using differential evolution algorithm. Appl Acoustic 148:308–321
5. Majumdar A, Maiti DK, Maity D (2012) Damage assessment of truss structures from changes in natural frequencies using ant colony optimization. Appl Math Comput 218:9759–9772
6. Kim NI, Kim H, Lee J (2014) Damage detection of truss structures using two-stage optimization based on micro genetic algorithm. J Mech Sci Tech 28:3687–3695
7. Guo HY, Li ZL (2009) A two-stage method to identify structural damage sites and extents by using evidence theory and micro-search genetic algorithm. Mech Syst Sig Proc 23:769–782

8. Masoumi M, Jamshidi E, Bamdad M (2015) Application of generalized flexibility matrix in damage identification using imperialist competitive algorithm. KSCE J Civ Eng 19:994–1001
9. Kaveh A, Zolghadr A (2015) An improved CSS for damage detection of truss structures using changes in natural frequencies and mode shapes. Adv Eng Soft 80:93–100
10. Xu HJ, Liu JK, Lu ZR (2016) Structural damage identification based on cuckoo search algorithm. Adva Struct Eng 19:849–859
11. Dinh-Cong D, Vo-Duy T, Nguyen-Thoi T (2018) Damage assessment in truss structures with limited sensors using a two-stage method and model reduction. Appl Soft Comput 66:264–277
12. Khatir S, Wahab MA, Boutchicha D, Khatir T (2019) Structural health monitoring using modal strain energy damage indicator coupled with teaching-learning-based optimization algorithm and isogeometric analysis. J Sound Vib 448:230–246
13. Dinh-Cong D, Ho-Huu V, Vo-Duy T, Ngo-Thi HQ, Nguyen-Thoi T (2018) Efficiency of Jaya algorithm for solving the optimization-based structural damage identification problem based on a hybrid objective function. Eng Optim 50:1233–1251
14. Lieu QX, Do DTT, Lee J (2018) An adaptive hybrid evolutionary firefly algorithm for shape and size optimization of truss structures with frequency constraints. Comput Struct 195:99–112
15. Lieu QX, Lee J (2017) Modeling and optimization of functionally graded plates under thermo-mechanical load using isogeometric analysis and adaptive hybrid evolutionary firefly algorithm. Compos Struct 179:89–106
16. Lieu QX, Lee J, Lee D, Lee S, Kim D, Lee J (2018) Shape and size optimization of functionally graded sandwich plates using isogeometric analysis and adaptive hybrid evolutionary firefly algorithm. Thin-Walled Struct 124:588–604
17. Lieu QX, Lee J (2019) An isogeometric multimesh design approach for size and shape optimization of multidirectional functionally graded plates. Comput Methods Appl Mech Eng 343:407–437
18. Lieu QX, Lee J (2019) A reliability-based optimization approach for material and thickness composition of multidirectional functionally graded plates. Compos B Eng 164:599–611
19. Reddy J (2005) An introduction to the finite element method. McGrawHill, New York
20. Messina A, Williams EJ, Contursi T (1998) Structural damage detection by a sensitivity and statistical-based method. J Sound Vib 216:791–808
21. Zhao J, DeWolf JT (1999) Sensitivity study for vibrational parameters used in damage detection. J Struct Eng 125:410–416
22. Storn R, Price K (1997) Differential evolution–a simple and efficient heuristic for global optimization over continuous spaces. J Glob Optim 11:341–359
23. Das S, Mullick SS, Suganthan PN (2016) Recent advances in differential evolution—an updated survey. Swarm Evol Comput 27:1–30
24. Yang XS (2010) Firefly algorithm, stochastic test functions and design optimisation. Int J Bio-Insp Comput 2:78–84
25. Yang XS (2014) Nature-inspired optimization algorithms. Elsevier Science Publishers B.V, The Netherlands
26. Dey N (2017) Advancements in applied metaheuristic computing. IGI Global
27. Jagatheesan K, Anand B, Samanta S, Dey N, Ashour AS, Balas VE (2017) Design of a proportional-integral-derivative controller for an automatic generation control of multi-area power thermal systems using firefly algorithm. IEEE/CAA J Automatica Sinica 6(2):503–515
28. Padhye N, Bhardawaj P, Deb K (2013) Improving differential evolution through a unified approach. J Glob Optim 55:771–799

Chapter 5
An Automated Approach for Developing a Convolutional Neural Network Using a Modified Firefly Algorithm for Image Classification

Ahmed I. Sharaf and El-Sayed F. Radwan

1 Introduction

The conventional neural network (CNN) is a promising deep learning framework that provides a powerful technique to solve real-world problems such as image processing, forecasting, and natural language processing. The CNN depends on many hidden layers to perform deep learning instead of using a simple set of layers as in classical artificial neural networks. Many CNNs have been proposed to perform deep leaning such as AlexNet [1], VGG [2], ResNet [3], and DenseNet [4]. Most of these architectures provide many hidden layers to perform powerful deep learning (e.g., the ResNet provides up to 1202 hidden layers). The fundamental concept behind these architectures is that a deeper CNN has more ability to solve many complicated and large-scale problems. The performance of the CNN depends profoundly on the investigated dataset, which may be not available for the CNN domain expert that may cause a limitation for the design process. Several techniques have been proposed to address this problem in the last decade. These algorithms could be categorized into two main groups according to their learning algorithm.

The first group include algorithms based on evolutionary algorithms [5]. Genetic algorithm (GA) was used to find the optimal network design by providing an encoding technique of network architecture with a fixed-length binary string. Then the GA was used to eliminate weak individuals and then generate more competitive ones. The MNIST and the CIFAR-10 were used for training and benchmarked this

A. I. Sharaf (✉) · E.-S. F. Radwan
Department of Computer Sciences, Faculty of Computers and Information, El-Mansoura University, Mansoura, Egypt
e-mail: ahmed.sharaf.84@gmail.com

E.-S. F. Radwan
e-mail: elsfradwan@gmail.com

Deanship of Scientific Research, Umm Al-Qura University, Mecca, Saudi Arabia

© Springer Nature Singapore Pte Ltd. 2020
N. Dey (ed.), *Applications of Firefly Algorithm and its Variants*,
Springer Tracts in Nature-Inspired Computing,
https://doi.org/10.1007/978-981-15-0306-1_5

methodology [6]. A large-scale neuro-evolutionary network was proposed by [7] for image classification. This method is based on a novel mutation operator to find the best solution. Also, this method concentrated on minimizing human participation in the design process. A hierarchical genetic representation was also proposed to handle this limitation by developing a semi-automated technique based on a modularized design pattern provided by a domain expert in the CNN techniques. This algorithm provided significant results when compared to the ImageNet [8]. The cartesian genetic programming (CGP) has also been used to formulate functional modules such as convolutional blocks to act as the node function in the genetic programming [9]. The genetic programming encoded the structure and the connectively of CNN to maximize the accuracy. This method used the CIFAR-10 for testing and benchmarking and provides high accuracy results. The algorithms in the first category are consuming the resources extensively, and their corresponding accuracies are not satisfactory.

The second group includes many approaches based on reinforcement learning methods to find the optimal structure of CNN [10]. Reinforcement learning has been applied to train a recurrent neural network to generate the model description of a neural network (NAS) and hence maximize the overall accuracy [11]. This method employed the CIFAR-10 and Penn Treebank dataset for training and validation and it also provided a little error rate compared to the state-of-the-art paradigms that employed a related architectural scheme. The MetaQNN is a meta-modeling proposed to generate high performing CNN structure for a specific learning task [12]. This method used Q-learning for the training of the agents to discover the designs with improved performance on the learning task. An efficient architecture search (EAS) [13] used an agent to control the network depth and the width of the layer according to function-preserving transformations. This method used the current weights of the CNN as initial weights to reduce the search space, thus reduce a large amount of computational power. The EAS provided an exploration space for plain CNN only without skip connections or branching. The EAS used the CIFAR-10 dataset for training and validation, and its results were significant when compared to the DenseNet. The learning agent has also been used to generate the fundamental blocks of the CNN, which is trained using the Q-learning to maximize the total accuracy for image classification. This method depends on creating a search space consists of building blocks of the CNN, which reduced the search cost. The method has been trained and validated on the CIFAR-10 and CIFAR-100 datasets. This method provided an error rate of 3.6% on the CIFAR-10 and achieved competitive results when compared to the ImageNet [14]. Because of the high computational cost of reinforcement learning, the second category of algorithms will require more computational resources when compared to the first category.

Swarm Intelligence refers to an artificial intelligence discipline that frequently became widespread during recent years. It has been inspired by the behavior of ants, bees, and fireflies. Although these swarms consist of unsophisticated individuals, they exhibit coordinated function that leads the swarms to their desired objective effectively. Many swarm-intelligent methods have been proposed to simulate the natural world such as ant colony optimization (ACO) [15], particle swarm optimization (PSO) [16], cuckoo-search [17], the bat algorithm [18], and the firefly algorithm

[5]. Firefly optimization is a swarm-intelligent method that has been proposed by Yang [5] to simulate the natural communication behavior of the fireflies. The firefly algorithm is a stochastic meta-heuristic algorithm that can be used to solve any NP problem [19]. The contribution of this chapter is to implement an automated tool to obtain the best CNN structure for an image classification task. The proposed approach should require no previous experience from the end user to design the CNN. The effectiveness and high performance are also required to perform the classification with high accuracy.

In this chapter, an automated tool was proposed to allow non-experts to design their corresponding CNN structure without any prior knowledge in this field. The proposed method encoded the skip connections that represents the basic CNN block to generate a search space of the problem. An enhanced firefly algorithm was developed to find the optimal CNN architecture by navigating the search space. The proposed firefly was developed based on a neighborhood attraction model to reduce the complexity of the algorithm. The CIFAR-10 and CIFAR-100 were used for the training and validation of the proposed algorithm [20]. The proposed algorithm provided significant results and higher accuracy when compared to cutting-edge CNN architecture. The main objectives of this chapter are:

1. Finding the optimal structure of CNN for image classification task without any limit on the CNN depth automatically without any prior knowledge from the operator.
2. Improve the total accuracy of the classification by reducing the gradient vanishing.
3. Generate advanced CNN architectures with deeper depth to perform more complex tasks by using the skip connection as the fundamental building block the of the CNN.

The rest of this work is organized as follows: the preliminary knowledge has been discussed in Sect. 2, and the details of the proposed algorithm were presented in Sect. 3. Section 4 demonstrates the experimental results and discussion. Finally, the conclusion is presented in Sect. 5.

2 Preliminary Knowledge

In this section, the preliminary knowledge of the proposed method is being discussed to inform the reader with the basic knowledge of this chapter. The main components of the proposed algorithms depend on convolutional neural network and firefly algorithm, which are discussed as follows:

2.1   Convolutional Neural Network

Convolutional Neural Network (CNN) is a fundamental deep learning approach which has been applied to many applications because of its significant performance in many areas. The CNN implements filters to perform its convolutional operations on the input image. A two-dimensional array usually represents these filters to perform the essential operation of the input images efficiently. The filter slides in a horizontal line to with predefined step until the end of the image and then it moves vertically for the next horizontal slide. This operation is executed until the entire image is completely scanned. For each input pixel, the filter v was applied to the input image u by multiplying each pixel in the filter with each corresponding pixel in the image, as shown:

$$w = \sum_{j=1}^{m} u(j)v(k - j + 1) \tag{1}$$

where u represents the input image with size m, the filter is denoted by v with a size of n, and the output vector denoted by w with a size of $m + n - 1$. The filter outputs are arranged to form a two-dimensional array named the feature maps, the horizontal and vertical steps size are named the stride width and height, respectively. The filter outputs are arranged to form a two-dimensional array named the features map, the horizontal and vertical steps size are the stride width and height, respectively. The convolutional layer consists of many concurrent filters with an identical size and implementing the same stride to produce a list of features maps, where the count of the features map is considered as an input parameter to design the structure of the network. The count of filters was obtained from the count of the features map and the spatial dimensions of the given image. The convolutional layer can perform two operations, which are the same and the valid operations. The same convolutional performs padding with zeros to the input image when there is no more space available to overlap the filter on the image. On the other hand, the valid operation does not perform any padding. Thus, the required parameters to define the convolutional layer are the count of features map, filter dimensions, stride dimensions, and the convolutional operation. Another component belongs to the architecture of CNN is called the pooling layer. The pooling layer is similar to the convolutional one except for the filter block, which is called the kernel. The kernel type is represented by a clear filter that returns a specific function depends on the input data such as maximum or mean functions. Furthermore, the spatial size of the input is fixed in this kind of layers. Hence, the pooling layer can be defined by the kernel size, stride size, and the pooling type [21]. The skip connection is gate strategy proposed to overcome the problem of gradient vanishing and to improve the training of recurrent neural network with long and short term memories. The regular connections of the CNNs exist among the neurons of two distinct layers, while the skip connection is defined as the connection between the neurons of the non-adjacent layers. The main problem occurs when using the gradient-based algorithms is the gradient vanishing. The reason for this problem is the explosion or the implosion of the gradient value.

Case studies have been implemented by many authors and proved the effectiveness of the skip connection to train the deep neural networks and CNNs [6]. The experiments performed on the skip connection architectures proved a significant performance such as ResNet [3] and CNN-GA [6]. An illustration of the skip layer is presented in Fig. 1, such that the vertical dotted line represents a skip connection from the first layer to the output at the N_{th} layer [6]. The symbol $\oplus$ represents a cumulative addition for each element in the connection. The main reason for using the skip connections is its ability to shorten the path between the layers of the neural network. The main reason for the gradient vanishing is the back-propagation of the error when it iterates through many layers of CNN. The skip connection provided a promising technique to reduce the limitation of the gradient vanishing.

2.2 Firefly Optimization

The firefly algorithm is a promising optimization meta-heuristic proposed to simu-late the behavior of the fireflies swarms [22–24]. The firefly utilizes a flashing light to communicate with the swarm and to attract the mating partner. Each firefly in the swarm makes a step moving toward the highest light source known as the at-tractiveness of the firefly. The firefly algorithm has been applied to many real-world problems such as multi-objective optimization [25], structure design [26], stock mar-ket forecasting [27], and image analysis [28]. Each firefly in the population denotes a candidate a possible solution within the search space. The individuals usually move toward the most attractive firefly to find the optimal solution. The light intensity emit-ted from the candidate determines its attractiveness. The fitness function controls the light intensity of each candidate solution in the search space. Let x_i is a firefly belongs to a population such that $i \in 1, 2, 3, \ldots, N$ where N denotes the population size. The attractiveness between two fireflies is computed as follows:

$$\beta(r_{ij}) = \beta_0 e^{-\gamma r_{ij}^2} \tag{2}$$

$$r_{ij} = \sqrt{\sum_{d=1}^{D} (x_i - x_j)^2} \tag{3}$$

where X_i and X_j are two random fireflies, D is the dimension of the problem such that $d \in 1, 2, 3, \ldots, D$, r is the distance between firefly i and firefly j, β_0 is the initial attractiveness when the distance equals to zero, γ represents the light absorption factor. Each firefly is compared with the whole population, if the x_i is more attractive than firefly x_j then, the firefly x_j will move toward the x_i according to the following formula:

$$x_i(t + 1) = x_i(t) + \beta_0 e^{-\gamma r_{ij}^2} (x_i(t) - x_j(t)) + \alpha \epsilon_i \tag{4}$$

where ϵ_i represents a random variable generated $[-0.5, 0.5]$, and α denotes the step factor $\alpha \in [0, 1]$. The traditional firefly algorithm is shown in Algorithm 1 where $f(.)$ represents the fitness function, N is the population size, and *Max_Loops* represents the maximum number of iterations. The effectiveness and efficiency of the firefly have been proved to solve any problem because of its high coverage rate and its straightforward implementation. However, the traditional algorithm may suffer from high time complexity when many fireflies were generated in the swarm because each firefly could be attracted to the whole population. Moreover, a large number of computations may cause an oscillation in the search space. A novel algorithm based on a neighborhood attraction model (NaFA) was proposed by [29] to solve these problems. In this algorithm, each firefly is attracted by all other brighter fireflies located within its k-nearest neighbors instead of being attracted by the entire population. The NaFA reduced the time complexity of the search process and provided a high coverage rate when compared to the traditional FA. The model suggested that the value of k should be $1 \leq K \leq (N-1)/2$ where N denotes the size of the population.

Algorithm 1 Traditional Firefly Algorithm

1: Initialize the population randomly with N fireflies
2: Compute the fitness function for each firefly
3: **while** (*counter* $\leq$ *Max_Loops*) **do**
4:     **for all** $i = 1 \rightarrow N$ **do**
5:         **for all** $j = 1 \rightarrow N$ **do**
6:             **if** $f(x_i) \geq f(x_j)$ **then**
7:                 Move X_i towards X_j using (4).
8:                 Calculate the $f(X_i)$ using the fitness function
9:                 *counter* $++$
10:            **end if**
11:        **end for**
12:    **end for**
13: **end while**

---

3 Hybrid Firefly Algorithm for CNN Architecture Optimization

In this section, the details of the CNN architecture model optimization were discussed. Many techniques have been proposed by researches to optimize the architecture of such networks. One common approach is to traverse a set of unique solutions where each solution represents specific characteristics of a network layer. The main features of the solution should contain the primary information of this architecture such as filter-size, dilation coefficient, pooling size, and so on. For a typical CNN similar to the architecture of the VGG-16 with dimensions vector of $\mathbb{R}^{5j+2k}$, then the search space is defined as $(2 \times 8^4 \times 16^2 \times 32^2 \times 126)^j \times (2 \times 4096)^k$. When $j = 7$ and $k = 2$ then the search space will create an approximately of $7.1 \times 10^8 7$ potential models. Even though the search space consists of numerous configurations of the architectures, it is incredibly difficult to search for the optimal architecture in this space because of its enormous size. However, hardware constraints, interactions, and conflictions may result in a physically impractical solution. Therefore, this approach is not recommended even when these constraints were added to the optimization process because the model would often prefer to generate the simplest possible unconstrained solution. The proposed method provided a simplified search space by transforming the CNN architecture to its essential components. The search space has been constrained by applying the concept of the sub-sampling width and height of the network with keeping the number of features maps increasing. The overview of the proposed method is presented in Sect. 3.1. Then, the proposed algorithm was discussed in details from Sect. 3.2 up to Sect. 3.5.

3.1 Algorithm Overview

The population is initialized using Algorithm 3 with a predefined size where each firefly contains the encoded CNN blocks. Each firefly was evaluated using the predefined fitness function presented in Algorithm 4. Both of the best and the worst fireflies denoted by x_{best} and x_{worst} were kept into the memory for further usage. For each iteration, while the termination conditions are not met an alternative leader firefly is called S_{best} with a rival fitness function and out of the optimal region. Both the best and the alternatives solutions were implemented to lead the weak fireflies to find the optimal zone using simulated annealing mutation (SAM) algorithm as shown in Algorithm 5. Then a neighborhood attraction variable denoted by k is used to limit the attraction scope of the firefly instead of the full attraction model. The modified attraction model is used to reduce the number of iterations instead of traversing the whole populations of fireflies that also reduce the complexity of the search space. The value of the k is set as $k = \lfloor ((N-1)/2) - 1 \rfloor$ as suggested by [29]. After the attractiveness is computed, if the attractiveness of the firefly j is higher than the firefly i then an offspring is calculated based on the SAM and denoted by $\acute{x}_j$. If the attractiveness of the $\acute{x}_j$ is more significant than firefly j then the firefly j is replaced with the new one. Otherwise, the firefly j is still the same, and then the firefly i is moved toward the optimal firefly. After the comparison among the population of fireflies, then the best and worst solution are obtained again. Finally, the population is ranked according to the attractiveness of each firefly. This algorithm is iterated for the maximum number of repetitions or when a termination condition has occurred. The overall pseudocode is illustrated in Algorithm 2.

Algorithm 2 Optimization algorithm of convolutional neural networks using k-nearest attractive firefly

---

1: **Input:** The decoded CNN skip layers P.
2: **Output:** The best solution x_{best}.
3: Initialize the population using Algorithm 3.
4: Evaluate the fitness function for each firefly using Algorithm 4.
5: measure the fitness function of firefly x_i.
6: Choose the best and the worst solutions x_{best} and x_{worst}.
7: **while** termination conditions are not met **do.**
8:    Define an alternative leader as s_{best}, with a rival fitness and found in a different zone.
9:    Calculate an offspring solution x_{best} using Algorithm 5.
10:   **for all** $((i \to N)$ and $(x_i \neq x_{wosrt}))$ **do.**
11:      **for all** $((j = i - k \to j = i + k)$ and $(x_j \neq x_{wosrt}))$ **do.**
12:         **if** $(f(x_j) > f(x_i))$ **then.**
13:            Modify the firefly j using Algorithm 5 to get a better offspring firefly called x'_j.
14:            Update firefly j with the offspring x'_j.
15:            Perform step movement of firefly j towards the optimal region using (5).
16:         **end if**
17:      **end for**
18:   **end for**
19:   Refresh the worst solution x_{worst}.
20:   **if** $f(x'_{best}) > f(x_{best})$ **then.**
21:      $x_{best} \leftarrow x'_{best}$.
22:   **end if**
23:   Rank the population and modify the best x_{best} and worst x_{worst} solutions.
24: **end while**
25: **return** x_{best}

3.2 Population Initialization

The CNN consists of three main building blocks, which are convolutional, pooling, and fully connected layers. The performance of CNN depends on the depth of its architecture. An encoding strategy was developed to encode the skip connections of the CNN instead of the traditional conventional layers. The skip layer is the fundamental building block of this strategy. A typical example of the skip layer is illustrated as shown in Fig. 1 where this layer consists of two conventional layers and a skip connection. The main features of the convolutional layer are the count of features map, the filter dimensions, and the stride dimensions.

The population was initialized using Algorithm 3 similar to [6]. The first step of this algorithm is to generate the length of the firefly individuals denoted by L, where L represents the depth of the CNN line (5). Then, a dynamic linked list is created with L nodes to simulate the architecture of the convolutional network line (6). The category if the layer is determined by a selection variable set randomly called r. If the value of r is less than a specific threshold, then the node is identified as a skip layer lines (7–23). Otherwise, the node is declared to be a pooling layer. In case of the skip layer, the features maps were generated randomly and assigned to both of the features maps of the node. In the case of the pooling layer, the pooling type is selected

Fig. 1 Illustration of a typical skip layer consists of N conventional layers and one skip connection

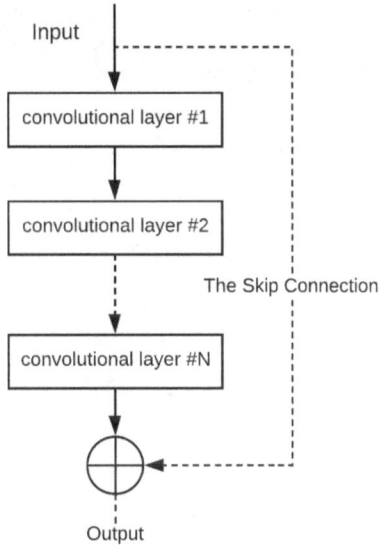

based on the probability of tossing a fair coin. The pooling category $P1$ is set to the maximum when the probability is less than 0.5 while is set to the mean elsewhere. Finally, the linked list is returned as the initial population. The fully connected layers were not included in this strategy of modeling the CNN because of the over-fitting generated from the dense connection of the layer [30]. One technique proposed to overcome this issue was to remove a part of the connection randomly. Nevertheless, each dropout requires one additional parameter to achieve the best performance of the corresponding CNN [31]. This approach generates too many variables obtained from the numbers of the fully connected layer and the neurons in each layer, which is difficult to handle. Consolidating the fully connected layers will substantially enlarge the search-space and will increase the difficulty of the search to find the optimal CNN design. The same convolutional operation was used with 1×1 stride filter to provide the same dimensions of the data. This operation also provided more flexibility for automated design of the CNN. The ResNet [3] is an illustrative example of using two convolutional layers as one skip layer, and its effectiveness has been proved experimentally.

3.3 The Attractiveness Function

The firefly population depends on the attractiveness function of individuals to determine the effectiveness and efficiency of each candidate. The attractiveness function each firefly is discussed in Algorithm 4. First, the CNN was extracted and decoded from the firefly x. The CNN used softmax classifier for the training of the model.

Algorithm 3 Population Initialization

1: **Input:** The population size N
2: **Output:** The initial Population P_0
3: $P_0 \leftarrow \phi$
4: **while** $|P_0| \leq N$ **do**
5:     $L = rand(0, 1)$
6:     $lst \leftarrow$ Create linked list with L nodes
7:     **for all** $nod \in lst$ **do**
8:         $r = rand(0, 1)$
9:         **if** $r \leq \sigma_0$ **then**
10:             $nod.Type =$ 'Skip_Layer'
11:             $nod.F1 = rand(0, 1)$
12:             $nod.F2 = rand(0, 1)$
13:         **else**
14:             $nod.Type =$ 'Pooling_Layer'
15:             $tmp = rand(0, 1)$
16:             **if** $tmp \leq 0.5$ **then**
17:                 $nod.P1 = max$
18:             **else**
19:                 $nod.P2 = mean$
20:             **end if**
21:         **end if**
22:     **end for**
23: **end while**
24: $P_0 \leftarrow P_0 \cup lst$
25: **return** P_0

Moreover, the rectifier activation function and a batch normalization procedure were attached to the output of the model [32, 33]. The CNN was trained using the Stochastic Gradient Descent (SGD) [34] on a training dataset D_{test}, as shown in line (5). The accuracy of the classifier was obtained from CNN in line (6). When the training process is completed, the highest accuracy achieved by the classifier is set as the fitness of the firefly.

Algorithm 4 The attractiveness function

1: **Input:** Firefly x, number of epochs T, training dataset D_{train}, testing dataset D_{test}
2: **Output:** The fitness of firefly $f(x)$
3: $\gamma_{best} \leftarrow 0$
4: **for all** $t \in T$ **do**
5:     $Train(SGD, D_{train})$
6:     $\gamma \leftarrow validate(D_{test})$
7:     **if** $\gamma > \gamma_{best}$ **then**
8:         $\gamma_{best} \leftarrow \gamma$
9:     **end if**
10: **end for**
11: **return** γ_{best}

3.4   Simulated Annealing Mutation

Various methods have been developed to improve the coverage and the offspring solution of evolutionary algorithms based on simulated annealing (SA) [35]. This technique is generic probabilistic evolutionary algorithms used for global optimization. The SA has proved its experimental efficiency in many problems. The proposed algorithm employed SA to generate improved candidates to enhance the search process as possible as shown in Algorithm 5. For each iteration, two fireflies were generated to represents offspring solutions. The first firefly is created from a previous firefly by using a one-step mutation line (5). The second one is produced from by using the traditional simulated annealing heuristics where each firefly has the probability to remains or to be replaced according to predefined acceptance criteria line (6). The fitness function was calculated for the previous solution and the generated one line (8). If the fitness of the new solution is larger than the old one, then it replaces it. Otherwise, a probabilistic variable has the control to change the firefly lines (9–16). The probability depends on the difference between the fitness of the two solutions and a control parameter denoted by *Temp*, which represents the temperature variable in the SA. Applying this criterion provides a minimum chance of the algorithm to be trapped into the local optima. The main advantage of using this method instead of the traditional SM is its ability to generate two neighborhood solutions in each temperature and select one of them according to the probability and the fitness function. This procedure is iterated until it reaches the maximum number of repetitions or when the best neighbor is found.

Algorithm 5 The SAM algorithm to generates offspring firefly

1: **Input:** $iterations\_max, X_{best}, S_{best}, Temp$
2: **Output:** The optimal solution X^*
3: Set $t_i = Temp_{max}, X_{mean} = mean(X_{best}, S_{best}), best = f(X_{mean})$
4: **while** ($i \leq iterations\_max$) **do**
5:     Generate $X_1 \in Neighbors(X_{mean})$
6:     Generate X_2 by one node mutation
7:     **for all** $i = 1 \rightarrow 2$ **do**
8:         $p = \exp((f(X_1) - f(X_2))/t_i)$
9:         **if** $p > rand(0, 1)$ **then**
10:             $X = X_i$
11:         **else**
12:             **if** $f(X_i) > best$ **then**
13:                 $X^* = X_i$
14:                 $best = f(X^*)$
15:             **end if**
16:         **end if**
17:     **end for**
18:     $t_{i+1} = t_i \times 0.99$
19: **end while**
20: **return** X^*

3.5   The Movement Step

After obtaining the global best solution, g_{best}, an alternative leader firefly S_best was defined with a rival fitness but positioned in a different zone. Since both leaders are more likely to discover distinctive search regions, this methodology decreases the possibility of being trapped in the local optima. The generated offspring obtained from the mean position of the leader and the alternative fireflies guides the remaining of the population to find the better zones.

$$x_i = x_i + \beta_0 \, C_k \, (x'_j - x_i) + C_k \, \varepsilon(g'_{best} - x_i) + \alpha' \times sign[rand - 0.5] \qquad (5)$$

$$x'_j = x_j + \sigma_1 \qquad (6)$$

$$g'_{best} = mean\,(g_{best} + S_{best}) + \sigma_2 \qquad (7)$$

where x'_j indicates the offspring firefly with a brighter intensity , x_j determined by the SAM as given in (5) and x'_{best} denotes the obtained candidate using the mean of the leader firefly and the alternative one as expressed in (6). The σ_1 and σ_2 are two stochastic variables initiated using the Gauss normal distribution. The next step of the firefly is defined, as given in (7). Where C_k denotes the chaotic map variable and ε indicates the random vector described in the standard firefly optimization. The variable α' indicates a dynamic step of 0.5 to dominate the diversity of the exploration process.

4   Results and Discussion

The performance of the proposed algorithm was measured during various experiments to study its behavior and prove its efficiency and effectiveness. In the next section, the dataset description was discussed. The performance of the proposed algorithm was measured during various experiments to study its behavior and prove its efficiency and effectiveness. In the next section, the dataset description was discussed. Both of the CIFAR-10 and CIFAR-100 were used as benchmarking datasets for image classification in the experiments [20]. These datasets represent a challenge in terms of the image size and the noise within the image. Moreover, they were generally utilized to evaluate the performance of deep learning, and many authors described their classification results. The CIFAR-10 consists of 10 categories of representing natural objects, birds, animals, and so on. There are approximately 60,000 colorful images with dimensions of 32 × 32. The dataset is divided into two main parts, the first part consists of 50,000 images which were used for training and the remaining 10,000 were used for testing. The CIFAR-100 is much larger than the CIFAR-10 with 100 categories representing the different areas of the whole image. The variations of the CIFAR-10 represent a challenge for image classification algorithms. The training set of images were split into two subsets, and the first one consists of 90% of the training dataset used for training the firefly population. The second one was used for validation of the population. Most of the case studies reported their results about this dataset developed a particular approach to compare their results. In this approach, padding with four pixels was applied to the image in every direction. Next, the images with dimensions of 32 × 32 were randomly cropped. The cropped image was flipped horizontally with a probability equals to 0.5. This approach was suggested by [3, 4]. Thus, the same procedure is implemented in the experiment to provide a fair comparison among the related work. The main objective of this chapter is to design an automated system to discover the optimal CNN architecture with any previous domain experience from the end user. Although the firefly optimization is easy to implement, the system was developed with a built-in setting that does not require the user to gain experience before usage. Thus, the parameters of the algorithm were set based on conventions. The firefly population size was set to 50 fireflies. The light absorption factor was set to 0.1, and the initial attractiveness was set to 0.01. The k nearest attraction factor was set to 8. During each iteration of training, the procedure discussed in [3] such that the learning rate of the SGD was set to 0.1, and the momentum value was set to 0.9 for training 350 epochs as suggested by [36]. Most of the competitors have implemented this training procedure to train their algorithms. After the proposed algorithm terminated, the best firefly was selected to be trained for 350 iterations on the complete dataset. The classification accuracy of the validation dataset was listed in a tabular form to be compared with other related methods. The values of the feature maps were set to $\{2^6, 2^7, 2^8\}$ based on the recommended settings by [4]. The initial values of the firefly parameters were set randomly according to the function $rand(0, 1)$ where the random values obtained from a uniform random distribution model within the interval [0, 1]. A chaotic logistics map set the variable C_k to increase the probability of the randomness in the step movement and hence minimize the probability of being trapped into local optima.

Cross-validation is a strategy to evaluate the accuracy of supervised learning methods by partitioning the original dataset into a training set to train the model, and a test set to evaluate it [19]. In the f-fold cross-validation, the original dataset was divided randomly into k equal-sized partitions. For the k samples, a single partition of data was used for validation and testing, and the remaining $k − 1$ partitions were used for training. This training process is then repeated for k times, where each sample is used only once for the validation. The result was combined, and the average of the results was computed to produce a single estimation. The benefit of this approach is that each observation is used for both training and validation, and each observation is used for validation exactly once. The evaluation and benchmarking of the proposed algorithm are based on a comparison among three categories of CNN and the proposed algorithm. The first category represents the recent proposed CNN that requires a manual inspection to set the parameters of the network. This category consists of ResNet [3], DenseNet [4], VGG [2], Maxout [37], Network in Network [38], Highway Network [39] and All-CNN [40]. It worth to mention that two implementations the ResNet were used with a depth of 101 and 1202, which are named ResNet(101) and ResNet(1202) for convenient. The second category represents the

Table 1 Average accuracy comparison among the cutting-edge techniques and the proposed method on the CIFAR-10 and CIFAR-100

Network title	CIFAR-10%	CIFAR-100%	Parameters count
ResNet(101)	93.57	74.84	1.7×10^6
ResNet(1202)	92.07	72.18	10.2×10^6
DenseNet	94.17	76.58	27.2×10^6
VGG	93.34	71.95	20.04×10^6
Maxout	90.7	61.4	–
Network in network	91.19	64.32	–
Highway network	92.4	67.66	–
All-CNN	92.75	66.29	1.3×10^6
Genetic CNN	92.9	70.97	–
Hierarchical evolution	96.37	–	–
EAS	95.77	–	23.4×10^6
Block-QNN-S	95.62	79.35	6.1×10^6
Large-scale evolution	94.6	–	5.4×10^6
CGP-CNN	94.02	–	1.68×10^6
NAS	93.99	–	2.5×10^6
Meta-QNN	93.08	72.86	
CNN-GA	95.22	–	2.9×10^6
Proposed	96.7	77.75	3.21×10^6

semi-automatic algorithms that require some human interaction to perform its tasks. This category consists of Genetic CNN [41], Hierarchical evolution [8], EAS [11], and Block-QNN-S [14]. Finally, the last category represents the large-scale full-automated algorithms that consist of Large-scale evolution [7], CGP-CNN [9], NAS [9], and MetaQNN [12] (Table 1).

For the first category, the proposed algorithm achieved improved accuracy with percentages of 4.63, 2.53, and 3.36% on the CIFAR-10 dataset when compared to ResNet(1202), DenseNet, and VGG. Moreover, the number of parameters used by the proposed algorithm was reduced by 31.4, 11.8, and 16% for the same algorithms. When compared to the ResNet(101), the proposed algorithm achieved improved with a percentage of 3.13%, but the number of parameters was large than the ResNet(101). When compared to the Maxout, Network in Networks, Highway Network, and the All-CNN, the proposed algorithm improved the accuracy with percentages of 6, 5.51, 4.3, and 3.95%. However, the All-CNN provided the optimal number of parameters rather than the proposed algorithm. For the CIFAR-100 dataset, the proposed algorithm enhanced the accuracy with percentages of 2.91, 5.57, 1.17, 5.8, 16.35, 13.43, 10.09, and 11.46% when compared to ResNet(101), ResNet(1202), DenseNet, VGG, Maxout, Network in Network, Highway Network, and All-CNN. It can be observed from the experiment performed on the manual inspection CNN that the proposed algorithm archived the best classification accuracy on both of the CIFAR-10 and

CIFAR-100. Moreover, the number of the parameters obtained from the proposed algorithm is much less than every architecture in this category excerpt the ResNet(101) and All-CNN.

For the second category, the proposed algorithm achieved improved accuracy with percentages of 3.8% and 0.33% on the CIFAR-10 dataset when compared to genetic CNN and hierarchical evolution, respectively. When compared to the EAS and the Block-QNN-S, the proposed algorithm scored improved accuracy with a percentage of 0.93% and 1.08% respectively. The number of parameters of the proposed algorithm was more signification than the EAS. However, the Block-QNN-S achieved less number of parameters when compared to the proposed algorithm. For the CIFAR-100 dataset, the proposed algorithm achieved the best accuracy, with an enhanced percentage of 6.78% when compared to Genetic CNN. The Block-QNN-S was superior when compared to the proposed algorithm. It can be observed from the experiment performed on the semi-automated category CNN, that the proposed algorithm archived the best classification accuracy on the CIFAR-10. Moreover, the number of the parameters obtained from the proposed algorithm is much less than the EAS. Although the proposed algorithm did not find the best classification accuracy in this category, the algorithms that belong to this category requires an additional user experience when they are employed to solve a real-world problem. In the case of EAS, it requires a hand-crafted CNN on a given dataset. Then the EAS should be adjusted by the network, and if the obtained CNN was designed with significant performance, then the CNN should perform with high accuracy and vice versa. Furthermore, the CNN obtained from the Hierarchical evolution and Block-QNN-S could not be implemented instantly. It must be attached to a more extensive network configured manually. Thus, the significant of the proposed algorithm when compared to the second category of CNNs, relies on its automated behavior.

For the third category, the proposed algorithm achieved improved accuracy with a percentage of 3.62% on the CIFAR-10 dataset when compared to MetaQNN. When compared to the CGP-CNN, NAS, and CNN-GA the proposed algorithm achieved greater accuracy with a percentage of 2.68, 2.71, and 1.48% but the number of parameters was more significant than the CGP-CNN, NAS, and CNN-GA. When compared to the Large-scale evolution, the proposed algorithm improved the accuracy with percentages of 2.1%. Moreover, the number of parameters was reduced by a percentage of 68.22% than the Large-scale evolution. However, the CGP-CNN provided the optimal number of parameters rather than the proposed algorithm. For the CIFAR-100 dataset, the CNN-GA achieved the best value of classification accuracy when compared to each architecture in this category. However, the proposed algorithm placed the second position in this category with a percentage of 77.75%. It can be observed from the experiment performed on the full-automated CNN that the proposed algorithm archived the best classification accuracy on the CIFAR-10 and placed the second order on the CIFAR-100.

The performance analysis of the proposed algorithm for each class is discussed in this section. The performance analysis of classes was compared to the AlexNet [1] and the ResNet [3] because some related architectures did not mention the performance of each class in their corresponding algorithms. The performance analysis

Table 2 Comparison among between the AlexNet, ResNet and the proposed method on the different classes of the CIFAR-10

Class	AlexNet %	ResNet %	Proposed %
Airplane	41.80	90.80	96.80
Automobile	21.80	69.10	97.80
Bird	0.02	72.60	98.50
Cat	0.03	61.90	92.50
Deer	87.60	75.40	94.50
Dog	23.00	82.10	94.80
Frog	24.20	76.60	94.80
Horse	34.70	84.70	96.70
Ship	31.70	83.20	95.20
Truck	95.90	84.60	98.50

of the AlexNet and the ResNet were presented in [42]. As shown in Table 2, the proposed method achieved the best accuracy compared to the AlexNet and ResNet for the CIFAR-10 dataset. The proposed method enhanced the results of the AlexNet and the ResNet with an average accuracy of 59.93% and 17.91%, respectively. The best results were obtained for the birds and trucks classes while the cat class was the worst among the three methods. On the CIFAR-100 dataset, the proposed algorithm achieved higher accuracy with percentages of 38.75% and 18.69% when compared to the AlexNet and the ResNet, respectively. As shown in Fig. 2, a comparison among the AlexNet, ResNet, and the proposed method was performed on the CIFAR-10. The proposed method achieved an average accuracy of 96.7%, lower bound accuracy 93.5%, and upper bound of 98.5%. The ResNet achieved an average accuracy of 78.2%, lower bound of 62%, and upper bound of 91%. The AlexNet achieved an

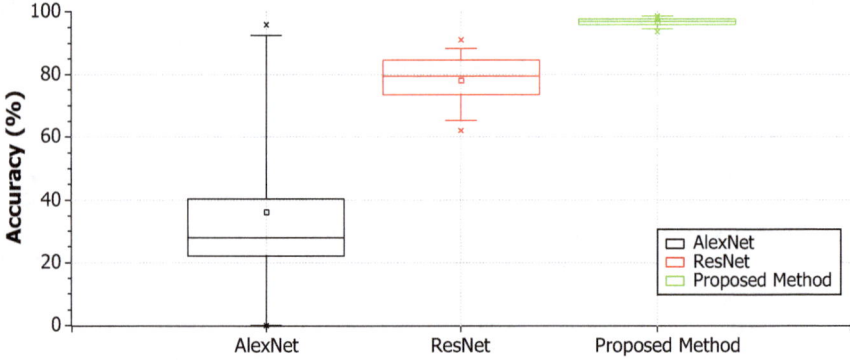

Fig. 2 Classification accuracy obtained by the AlexNet, ResNet, and the proposed method on the CIFAR-10

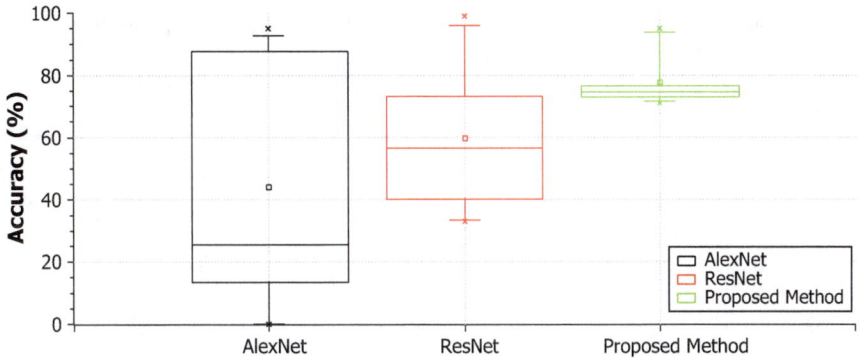

Fig. 3 Classification accuracy obtained by the AlexNet, ResNet, and the proposed method on the CIFAR-100

average accuracy of 36.2% and upper bound of 96%. For the CIFAR-100, the proposed method achieved an average accuracy of 77.7%, lower bound accuracy 71%, and upper bound of 95%. The ResNet achieved an average accuracy of 59.8%, lower bound of 33%, and upper bound of 99%. The AlexNet achieved an average accuracy of 44.1%, upper bound of 95%, and lower bound of 0% as shown in Fig. 3. The firefly optimization provided high accuracy in this system because of its high coverage rate and its ability to escape from local optima. The presented experiments have proved the robustness and effectiveness of the proposed algorithm. There are many factors that let this system to achieve high accuracy and prevent trapping into the local optima. Firstly, the SAM procedure that generated two offsprings to select between them instead of creating only one. The alternative leader firefly also provided a protection mechanism to reduce the probability of trapping into local optima. Secondly, the implementation of the neighborhood attraction model reduced the computational complexity and the time complexity of the firefly algorithm. Determining the value of the variable k improved the search process and reduced the number of comparison of the firefly populations. The neighborhood attraction model enhanced the coverage rate of the firefly population. The main benefits of the proposed system could be shortened as follows:

1. The encoding procedure provided CNN to have any value of layers depth without any limitation because of utilizing implementing the skip layers as a linked list.
2. The implementation of the skip connection reduced the search space by minimizing the behavior of the gradient vanishing.
3. The proposed was developed to be a completely full-automated that can operate without any previous knowledge from the operator, and with high accuracy and minimal complexity.

Although the high accuracy obtained by the proposed method, the Block-QNN-S was the leader technique on the CIFAR-100, which is considered a limitation of this work (Table 3).

Table 3 Comparison among between the AlexNet, ResNet and the proposed method on the different classes of the CIFAR-100

Class	AlexNet %	ResNet %	Proposed %
Bed	0	49.60	73.67
Bicycle	21	55.00	70.75
Bus	84	36.80	73.10
Chair	90	57.60	75.37
Couch	11	76.40	76.57
Motorcycle	95	99.20	95.37
Streetcar	21	63.80	72.81
Table	0	33.40	75.10
Train	30	34.20	72.40
Wardrobe	89	92.20	92.37

5 Conclusion

The primary aspiration of this chapter was to develop an automated tool to design CNN with highest accuracy and performance for image classification. The proposed method was based on the k-nearest neighborhood attraction firefly to address the limitations and obstacles for obtaining the best CNN architecture. The skip connection has been encoded to be the fundamental building block of the CNN architecture. This category of connections provided the proposed method to generate more complicated CNN structure with a deeper depth of layer without any constraint on the value of the depth. The firefly algorithm was utilized based on the k-nearest neighborhood attraction model to navigate the search space with a reduced computational cost. The proposed method was trained and validated on the CIFAR-10 and CIFAR-100 and produced significant results when compared to related approaches. The proposed method achieved the highest accuracy obtained on the CIFAR-10 when compared to three groups of algorithms with a percentage of 96.7%. For the CIFAR-100, the Block-QNN-S was the best method with an accuracy of 79.35% while the proposed method achieved 77.75%.

References

1. Krizhevsky A, Sutskever I, Hinton GE (2012) Imagenet classification with deep convolutional neural networks. In: Advances in neural information processing systems, pp 1097–1105
2. Simonyan K, Zisserman A (2014) Very deep convolutional networks for large-scale image recognition. arXiv preprint arXiv:1409.1556
3. He K, Zhang X, Ren S, Sun J (2016) Deep residual learning for image recognition. In: Proceedings of the IEEE conference on computer vision and pattern recognition, pp 770–778

4. Huang G, Liu Z, Van Der Maaten L, Weinberger KQ (2017) Densely connected convolutional networks. In: Proceedings of the IEEE conference on computer vision and pattern recognition, pp 4700–4708
5. Brownlee J (2011) Clever algorithms: nature-inspired programming recipes. Jason Brownlee
6. Sun Y, Xue B, Zhang M, Yen GG (2018) Automatically designing CNN architectures using genetic algorithm for image classification. arXiv preprint arXiv:1808.03818
7. Real E, Moore S, Selle A, Saxena S, Suematsu YL, Tan J, Le QV, Kurakin A (2017) Large-scale evolution of image classifiers. In: Proceedings of the 34th international conference on machine learning, vol 70, pp 2902–2911. JMLR. org
8. Liu H, Simonyan K, Vinyals O, Fernando C, Kavukcuoglu K (2017) Hierarchical representations for efficient architecture search. arXiv preprint arXiv:1711.00436
9. Suganuma M, Shirakawa S, Nagao T (2017) A genetic programming approach to designing convolutional neural network architectures. In: Proceedings of the genetic and evolutionary computation conference. ACM, New York, pp 497–504
10. Arulkumaran K, Deisenroth MP, Brundage M, Bharath AA (2017) A brief survey of deep reinforcement learning. arXiv preprint arXiv:1708.05866
11. Zoph B, Le QV (2016) Neural architecture search with reinforcement learning. arXiv preprint arXiv:1611.01578
12. Baker B, Gupta O, Naik N, Raskar R (2016) Designing neural network architectures using reinforcement learning. arXiv preprint arXiv:1611.02167
13. Cai H, Chen T, Zhang W, Yu Y, Wang J (2018) Efficient architecture search by network transformation. In: Thirty-Second AAAI conference on artificial intelligence
14. Zhong Z, Yan J, Liu CL (2017) Practical network blocks design with q-learning, 1(2):5. arXiv preprint arXiv:1708.05552
15. Dorigo M, Di Caro G (1999) Ant colony optimization: a new meta-heuristic. In: Proceedings of the 1999 Congress on evolutionary computation-CEC99 (Cat. No. 99TH8406), vol 2. IEEE, New York, pp 1470–1477
16. Kennedy J, Eberhart RC (1999) The particle swarm: social adaptation in information-processing systems. In: New ideas in optimization. McGraw-Hill Ltd., pp 379–388
17. Yang XS, Deb S (2009) Cuckoo search via lévy flights. In: 2009 World Congress on Nature & biologically inspired computing (NaBIC). IEEE, New York, pp 210–214
18. Yang XS (2010) A new metaheuristic bat-inspired algorithm. In: Nature inspired cooperative strategies for optimization (NICSO 2010). Springer, Berlin, pp 65–74
19. Dey N (2017) Advancements in applied metaheuristic computing. IGI Global
20. Krizhevsky A, Hinton G (2009) Learning multiple layers of features from tiny images. Report, Citeseer
21. Khan A, Sohail A, Zahoora U, Qureshi AS (2019) A survey of the recent architectures of deep convolutional neural networks. arXiv preprint arXiv:1901.06032
22. Chakraborty S, Dey N, Samanta S, Ashour AS, Balas VE (2016) Firefly algorithm for optimized nonrigid demons registration. In: Bio-inspired computation and applications in image processing. Elsevier, Amsterdam, pp 221–237
23. Dey N, Samanta S, Chakraborty S, Das A, Chaudhuri SS, Suri JS (2014) Firefly algorithm for optimization of scaling factors during embedding of manifold medical information: an application in ophthalmology imaging. J Med Imaging Health Inf 4(3):384–394
24. Yang XS (2010) Nature-inspired metaheuristic algorithms. Luniver press
25. dos Santos Coelho L, Bora TC, Schauenburg F, Alotto P (2013) A multiobjective firefly approach using beta probability distribution for electromagnetic optimization problems. IEEE Trans Magn 49(5):2085–2088. https://doi.org/10.1109/tmag.2013.2238902
26. Gandomi AH, Yang XS, Alavi AH (2011) Mixed variable structural optimization using firefly algorithm. Comput Struct 89(23–24):2325–2336
27. Kazem A, Sharifi E, Hussain FK, Saberi M, Hussain OK (2013) Support vector regression with chaos-based firefly algorithm for stock market price forecasting. Appl Soft Comput 13(2):947–958

28. Horng MH (2012) Vector quantization using the firefly algorithm for image compression. Exp Syst Appl 39(1):1078–1091. https://doi.org/10.1016/j.eswa.2011.07.108
29. Wang H, Wang W, Zhou X, Sun H, Zhao J, Yu X, Cui Z (2017) Firefly algorithm with neighborhood attraction. Inf Sci 382–383:374–387
30. Hawkins DM (2004) The problem of overfitting. J Chem Inf Comput Sci 44(1):1–12
31. Srivastava N, Hinton G, Krizhevsky A, Sutskever I, Salakhutdinov R (2014) Dropout: a simple way to prevent neural networks from overfitting. J Mach Learn Res 15(1):1929–1958
32. Glorot X, Bordes A, Bengio Y (2011) Deep sparse rectifier neural networks. In: Proceedings of the fourteenth international conference on artificial intelligence and statistics, pp 315–323
33. Ioffe S (2017) Batch renormalization: towards reducing minibatch dependence in batch-normalized models. In: Advances in neural information processing systems, pp 1945–1953
34. Bottou L (2012) Stochastic gradient descent tricks. Springer, Berlin, pp 421–436
35. Abdulal W, Jabas A, Ramachandram S, Jadaan OA (2011) Mutation based simulated annealing algorithm for minimizing makespan in grid computing systems. In: 2011 3rd international conference on electronics computer technology, vol 6. IEEE, New York, pp 90–94. https://doi.org/10.1109/ICECTECH.2011.5942057
36. Sutskever I, Martens J, Dahl G, Hinton G (2013) On the importance of initialization and momentum in deep learning. In: International conference on machine learning, pp 1139–1147
37. Goodfellow IJ, Warde-Farley D, Mirza M, Courville A, Bengio Y (2013) Maxout networks. arXiv preprint arXiv:1302.4389
38. Lin M, Chen Q, Yan S (2013) Network in network. arXiv preprint arXiv:1312.4400
39. Srivastava RK, Greff K, Schmidhuber J (2015) Highway networks. arXiv preprint arXiv:1505.00387
40. Springenberg JT, Dosovitskiy A, Brox T, Riedmiller M (2014) Striving for simplicity: the all convolutional net. arXiv preprint arXiv:1412.6806
41. Xie L, Yuille A (2017) Genetic CNN. In: Proceedings of the IEEE international conference on computer vision, pp 1379–1388
42. Sharma N, Jain V, Mishra A (2018) An analysis of convolutional neural networks for image classification. Proc Comput Sci 132:377–384

Chapter 6
Enhanced Firefly Algorithm for Optimum Steel Construction Design

S. Carbas

1 Introduction

Since the existence of human beings, the rapid increase in the population and the limited places to live and shelter have directed people to different searches. By the developing technology together with these searches, designing the most durable and the most economical structures has been the first important target of structural engineers. The strength of a structure is ensured by the behaviour of the structure under the design loads within the limits determined by the related structural provisions. By the way, the cost is related to the most economical design of the construction by reducing the cross sections of structural members, and/or reducing the material used, without neglecting the safety requirements. Therefore, in order to obtain the most appropriate design as safe and economic in every structural engineering problem encountered, it is necessary to conduct difficult and tiresome researches to find the best one among the acquired solutions. The design, which has the minimum cost among the feasible designs providing the structural safety conditions, can be called the optimum structural design [1]. For this purpose, a number of mathematical-based traditional optimization methods have been developed for producing optimal structural designs [2]. These conventional methods have many disadvantages, such as their specificity to problem type and the need to define the problem with complicated mathematical functions. Above all, in most of these gradient-based mathematical programming techniques, since design variables are considered to be continuous, they are not suitably applied to many engineering problems. Therefore, some scientists have focused their attention on the existing systems in nature and on the events taking place between them in order to develop flexible and high performance methods. These optimization

S. Carbas (✉)
Department of Civil Engineering, Karamanoglu Mehmetbey
University, Karaman, Turkey
e-mail: scarbas@kmu.edu.tr

© Springer Nature Singapore Pte Ltd. 2020
N. Dey (ed.), *Applications of Firefly Algorithm and its Variants*,
Springer Tracts in Nature-Inspired Computing,
https://doi.org/10.1007/978-981-15-0306-1_6

methods based on the existing systems and events in nature are the so-called meta-heuristic methods [3]. The new generation nature-inspired metaheuristic methods are generally developed in order to overcome the above-mentioned inadequacies of classical optimization techniques which are independent of the problem type and model [4–8].

The primary purpose of preliminary design in structural engineering is to find the first stresses and to check the accuracy of the assumptions in this design. The initial assumptions of the preliminary design are almost unlikely to be accurate and economical. If the sections selected in the preliminary design are small, the sections are increased because the stress and displacement conditions are not met, and if the stress and displacement values are much smaller than the section can carry, it is necessary to reduce the sections to make the system more economical. As will be understood, these operations are complex iterative processes.

Some methods have been developed that computer science learns from natural sciences (i.e. biology) and uses it to solve its own problems. These methods are the so-called metaheuristics. Genetic algorithms (GAs) [9, 10] arisen firstly from these trends at the 1970s are influenced by evolution and change in living things, transferring the process of genetic evolution to the computer environment, and instead of improving the learning ability of a single mechanical structure, it is revealed that superior genetic individuals are obtained as a result of genetic processes such as crossover, mutation and selection. As time elapses, it is noticed that GAs have gotten snared in local minimums. In order to overcome this deficiency, a variety of GAs has been developed [11–15]. Afterwards, it has been proven that metaheuristic optimization methods can solve those complicated structural optimization problems as quickly and effectively as possible [9, 16–31]. The common goal of all these state-of-the-art optimization techniques is to make the evaluation and solution process within the algorithm and to optimize the design of the structural problem in terms of strength, rigidity, serviceability and cost. In addition, as mentioned before since these methods include logical approaches inspired from natural phenomenon to solving the problem instead of exact mathematical expressions, they are very easy to apply for design optimization problems of structures. Therefore, it cannot be stated that the solutions obtained with these methods are the precise optimal. In recent years, however, these techniques have been highly enhanced and the results obtained are very close to the global optimum.

Firefly algorithm (FA), suggested by Yang in 2008 [32], remains one of the fashionable stimulating latterly conceived metaheuristic methods and was found on the light blinking features of fireflies. Due to the capability of attaining solutions of either continuous or discrete optimization problems [33–37], FA is highly famous among metaheuristics. The supreme idiocratical behaviour of the FA comes from its mechanism of search by the attitudes of fireflies in their community and making contact with each other through a natural luminosity. In spite of the fact that FA has been appeared to be a viable optimization method in numerous optimization issues, it has a few shortcomings. In FA, all shiny fireflies may allure the others. A lot of alluring can cause hesitations throughout the research procedure, resulting process complication in high time consuming for computations. Namely, it is noticed that the SFA generally presents vital defects when employed to real-sized optimization problems, ending in poor convergence quality and incapable search process. So, the SFA finalizes the solution to the problem with a premature convergence confined in a local optimum solution and cannot skip out of the search space having plenty of local optimal designs. To overcome this allure trouble, a novel FA variant, namely enhanced attractiveness treatment, is implemented to get rid of inefficacy of classical FA, especially in high-sized optimization problems comprised of vast design variables and high discrete sets. By carrying out enhanced attractiveness approach, the alluring of the fireflies by brighter individuals is gradually decreased via dynamic formulation. Furthermore, a fresh reconstruction of arbitrariness parameter is proposed in enhanced firefly algorithm (EFA) by executing it dynamically reduced among the way of optimization course, though it is set to a stable variant in a SFA [38].

Structural members used in steel construction can be classified into two main catalogues. The first and most known kind is the so-called hot-rolled members made up of plates. The latter, which has been gaining popularity in recent times, is brought into being by cold-formed thin-walled steel plates in roll-forming machines and/or by press brake or bending brake operations [39]. In this chapter, the design optimization of both a planar and a spatial steel frame made out of hot-rolled heavy steel profiles is mathematically modelled in conformity with the design provision of American Institute of Steel Construction-Load and Resistance Factor Design (AISC-LRFD) [40]. The proposed EFA chooses the hot-rolled W-shaped sections for the structural member groups in the frames from a steel section list consisting of 272 profiles available in provision so that the weight of the frame can be calculated as minimum while meeting the provision-based structural design constraints, which are mainly strength, displacement, inter- and top-storey drifts, for safety. Besides, a steel space frame built up from cold-formed thin-walled lightweight open profiles is optimally designed according to design constraints practised from American Iron and Steel Institute-Load and Resistance Factor Design (AISI-LRFD) [41, 42]. The strength, displacement, top- and inter-storey drifts, and slenderness ratio are treated as design constraints of the problem. The proposed EFA picks out thin-walled C-shaped profiles produced from cold-formed steel for structural member groups from available profile list in AISI [43]. Moreover, in both frame types made of hot-rolled and cold-formed steel profiles, respectively, some extra constraints are taken into account to meet the realistic design necessities. These requirements can be categorized into three kinds of inequalities. In each beam and column connection at each storey of the frames, (i) the flange wideness of the columns should be equal or wider than the flange wideness of the beams, (ii) the depth of the columns at lower storey should be equal or wider than the depth of the columns at upper storey, and (iii) the mass per metre of the columns of lower storey should be equal or greater than those of the column at the upper storey.

Design optimization of a steel planar frame consisting of two-bay, six-storey and 30 hot-rolled heavy steel structural members, a steel spatial frame comprising of four-storey and 428 hot-rolled heavy steel frame members and the last design example a steel space frame built up using thin-walled open C-shaped cold-formed steel profiles owning two-storey and 379 structural members are accomplished. The enhanced firefly algorithm (EFA) is executed to reach the optimum structural design for all design examples. The optimal designs acquired via both standard and enhanced FAs are evaluated and compared to each other as well as those are scrutinized with optimum reported designs yielded by different metaheuristics, so far.

2   Problem Formulation of Discrete Design Optimization

When designing a structure, the load-bearing system of the construction and the loads acting on the structure should be determined. By analysing the structure under the critical loading conditions, the forces occurring in the structural members and on the nodes should be identified. Then, the structural members are sized according to capacity values determined to satisfy the loads acting on the members and nodes. Afterwards, in the design of the structure, the stresses and displacements occurring in the members should be checked and the serviceability of the structure should be ensured. In addition, the importance of the construction cost of the building should not be neglected during the project design phase. To this context, in steel structures, the lower the structural weight, the lower the cost of the structure. For this reason, the steel structure design should always be performed with the aim of achieving the lowest structural weight to provide all design constraints with provisions. Hence, for all steel frames having predetermined geometries, it is aimed to obtain optimal designs of minimum weight, where all design constraints are fulfilled from AISC-LRFD for steel planar and spatial frames made of hot-rolled steel profiles and from AISI-LRFD for steel space frame built up using cold-formed thin-walled steel profiles. So, the below-mentioned discrete nonlinear problem formulations are derived.

2.1   Objective Function

The minimum weight design problem of steel constructions turned out to be a non-linear programming problem since having a combination of continuous, discrete and integer sets of design variables and is subjected to highly nonlinear, implicit and discontinuous constraints. So, the formulation of such a design problem yields the following nonlinear discrete programming problem. In both of frame types, the objective functions are dealt as minimizing the frame weight.

$$\text{Minimize} \quad W_{\text{f}} = \sum_{g=1}^{\text{ng}} m_{\text{g}} \sum_{\text{fm}=1}^{\text{tm}_{\text{g}}} L_{\text{fm}} \tag{1}$$

Here, the frame weight is described as W_{f}. A steel profile, picked up for member group g from available section lists, has unit weight of m_{g}. The total member number in group g is presented as tm_g. The total group number in the frame is tabulated as ng. The member length of a group g is depicted as L_{fm}.

The objective function is subjected to the following design constraints.

2.1.1 Strength

The steel profiles assigned to member groups of the frame should possess enough amount of strength in order to safely bear the internal forces occurring due to factored external loadings. The strength constraints of each member of both types of frames are defined in Eqs. (2) and (3) executed from Chapter H of AISC-LRFD for steel planar and spatial frames made of hot-rolled steel profiles, and in Eqs. (4) and (5) executed from C5.1 (for combination of axial tensile load and bending), and Eqs. (6) and (7) executed from C5.2 (for combined axial compressive load and bending), of AISI-LRFD for steel space frame built up using cold-formed thin-walled steel profiles.

$$\frac{P_u}{\phi P_n} + \frac{8}{9}\left(\frac{M_{ux}}{\phi_b M_{nx}} + \frac{M_{uy}}{\phi_b M_{ny}}\right) \leq 1.0 \quad \text{for} \quad \frac{P_u}{\varphi P_n} \geq 0.2 \tag{2}$$

$$\frac{P_u}{2\phi P_n} + \left(\frac{M_{ux}}{\phi_b M_{nx}} + \frac{M_{uy}}{\phi_b M_{ny}}\right) \leq 1.0 \quad \text{for} \quad \frac{P_u}{\varphi P_n} < 0.2 \tag{3}$$

Here, the nominal moment capacities of strong and weak axes are M_{nx} and M_{ny}, respectively. The design moments imposed upon a member are M_{ux} and M_{uy}. The capacity of axial load for a member is P_n, the utmost value of which is P_u. These loads could be either tension or compression. Chapter C of AISC-LRFD recommends an option for deriving the moment capacities of frame members taking the geometric nonlinearity into account.

$$\frac{M_{ux}}{\phi_b M_{nxt}} + \frac{M_{uy}}{\phi_b M_{nyt}} + \frac{T_u}{\phi_t T_n} \leq 1.0 \tag{4}$$

$$\frac{M_{ux}}{\phi_b M_{nx}} + \frac{M_{uy}}{\phi_b M_{ny}} - \frac{T_u}{\phi_t T_n} \leq 1.0 \tag{5}$$

Here, M_{ux} and M_{uy} are needful flexural strengths [factored moments] in terms of axes of centroid. ϕ_b is flexural strength [moment resistance] factor which equals to 0.90 or 0.95. The M_{nxt} and M_{nyt} are taken equal to $S_{ft}F_y$ (where S_{ft} is modulus of section for full unreduced section relative to extreme tension fibre in terms of suitable axis and yield stress is F_y). T_u is needful axial tensile strength [factored tension]. The ϕ_t is the factor for tensile axial strength which equals to 0.95. The T_n is nominal tensile axial strength [resistance]. M_{nx} and M_{ny} are nominal flexural strengths [moment resistances] with regard to axes of centroid.

$$\frac{P_u}{\phi_c P_n} + \frac{C_{mx}M_{ux}}{\phi_b M_{nx}\alpha_x} + \frac{C_{my}M_{uy}}{\phi_b M_{ny}\alpha_y} \leq 1.0 \quad \text{for} \quad \frac{P_u}{\phi_c P_n} > 0.15 \qquad (6)$$
$$\frac{P_u}{\phi_c P_{no}} + \frac{M_{ux}}{\phi_b M_{nx}} + \frac{M_{uy}}{\phi_b M_{ny}} \leq 1.0$$

$$\frac{P_u}{\phi_c P_n} + \frac{M_{ux}}{\phi_b M_{nx}} + \frac{M_{uy}}{\phi_b M_{ny}} \leq 1.0 \quad \text{for} \quad \frac{P_u}{\phi_c P_n} \leq 0.15 \qquad (7)$$

Here, P_u is needful axial compressive strength [factored compressive force]. ϕ_c is the factor for axial load equal to 0.85. M_{ux} and M_{uy} are the needful flexural strengths [factored moments] in terms of axes of centroid. ϕ_b is flexural strength [moment resistance] factor equalized to 0.90 or 0.95. M_{nx} and M_{ny} are nominal flexural strengths [moment resistances] in terms of axes of centroid. Also, for cold-formed compression members, the admissible upper bound of the slenderness ratio is set as 200 to prevent the buckling under axial compressive load.

2.1.2 Displacement and Drifts

In both types of frames, because of the serviceability necessities, the displacement occurred at the mid-point of the beam span and the lateral drifts of the beam members are restricted according to the report given in ASCE Ad Hoc Committee [44] [Eqs. (8–10)]. The mid-point deflection of a beam span is limited as $L/360$ (L is beam span length). The permissible bounds of inter- and top-storey drifts are taken as values which are the multiplication of height of the storey and total height of the structure by a value range from 1/500 to 1/200, and 1/750 to 1/250, respectively.

$$g_d(\mathbf{X}) = \frac{\delta_{j,l}}{\delta_{al}} - 10.0 \leq 0 \quad j = 1, 2, \ldots, n_{sm} \quad l = 1, 2 \ldots, n_{lc} \qquad (8)$$

$$g_{td}(\mathbf{X}) = \frac{\Delta_{j,l}^{top}}{\Delta_{al}^{top}} - 1.0 \leq 0 \quad j = 1, 2, \ldots, n_{jtop} \quad l = 1, 2, \ldots, n_{lc} \qquad (9)$$

$$g_{id}(\mathbf{X}) = \frac{\Delta_{jl}^{is}}{\Delta_{al}^{is}} - 1.0 \leq 0 \quad j = 1, 2, \ldots, n_{st} \quad l = 1, 2, \ldots, n_{lc} \qquad (10)$$

Here, the permissible and calculated displacements are $\delta_{j,l}$ and δ_{al}, respectively. The calculated and admissible top-storey drifts of the member j are $\Delta_{j,l}^{top}$ and Δ_{al}^{top}, respectively. The $\Delta_{j,l}^{is}$ and Δ_{al}^{is} are the calculated and permissible inter-storey drifts for the storey j. The n_{sm} is the total critical member number for the deflection. The n_{lc} is the load case number. The n_{jtop} is the node number at top storey, and n_{st} is the total storey number of the frame.

2.1.3 Geometric

In both types of frames, the geometric constraints are described via the following equations [Eqs. (11) and (12)].

$$g_{cc}(\mathbf{X}) = \sum_{i=1}^{n_{ccj}}\left(\frac{D_i^a}{D_i^b} - 1.0\right) + \sum_{i=1}^{n_{ccj}}\left(\frac{m_i^a}{m_i^b} - 1.0\right) \le 0 \qquad (11)$$

$$g_{bc}(\mathbf{X}) = \sum_{i=1}^{n_{j1}}\left(\frac{B_f^{bi}}{D^{ci} - 2t_b^{ci}} - 1.0\right) \le 0 \quad \text{or} \quad \sum_{i=1}^{n_{j2}}\left(\frac{B_f^{bi}}{B_f^{ci}} - 1.0\right) \le 0 \qquad (12)$$

Here, the column-to-column connection number is n_{ccj}. For an ith column-to-column connection, the unit weights of the columns located at above and below storeys are m_i^a and m_i^b, respectively, and the widths of the column flanges of above and below storeys are D_i^a and D_i^b, respectively. The number of beams connected to the column web and the total beam number connected to column flange are n_{j1} and n_{j2}, respectively. The column depth is D^{ci}. The tb^{ci} and B_f^{bi} are the flange thickness and width of a beam connected to column web, respectively. B_f^{ci} and B_f^{bi} are the flange widths of the column and beam connected to flange of a column, respectively.

3 Firefly Algorithm

3.1 Natural Phenomenon of Firefly Algorithm

Fireflies provide a spectacular view thanks to the light beams they emit in tropical and temperate regions, and there are about 2000 different species in nature [45]. The light source emitted by fireflies has two main functions. These are attracting their preys for hunting and wooing to their spouses for mating [46]. Fireflies can produce light through photogenic cells on their body surfaces [47].

All fireflies are single genus, so there is no male and female distinction, and all fireflies can be attracted by other fireflies. The attractiveness of fireflies is directly proportional to their brightness. In other words, the brighter a firefly is, the more attractive it is. The result is that the less bright firefly is attracted by the brighter one. At the same time, the distance decreases the lustre and reduces the charm. If there is a firefly brighter than another one, this firefly moves towards the bright one. If there are no more bright fireflies, they move in random directions. Firefly optimization algorithm is based on these basic principles. The luminosity of fireflies can be considered as the objective function of the metaheuristic firefly algorithm (FA). The FA can be classified under swarm-based nature-inspired optimization algorithms [48].

3.2 Standard Firefly Algorithm (SFA)

The standard firefly algorithm (SFA) was designed and applied by Yang with inspiration from their brightness models and movement directions in nature [1, 3, 48]. The lucency of fireflies is a natural light-emitting pathway for these insects and has the task of mating, attracting a potential prey and protecting from predators. Fireflies communicate with each other according to rhythm, degree and duration of the glitter. The method was first applied to numerical optimization problems [1], then to load distribution problems [49], then to structural optimization problems [50], then to the minimum weighted design of tower structures [51] and to the design optimization of latticed truss structures [52].

There are the following three ideal rules in the FA as mentioned before, briefly:

(i)   All fireflies can communicate with each other without gender discrimination.
(ii) The attractiveness of a firefly is balanced with its luminosity capacity. And so, between two fireflies, more light emits more attention and others move towards it.
(iii) The flashing of the firefly is defined by the configuration of objective function.

There are two principal parameters in the SFA, altering the intensity of light and attraction of firefly (β). As noted above, the firefly attraction is related to its luminousness and the reduction is occurred in attractiveness as the light intensity lessens. On this basis, the attractiveness (β) value for any firefly is computed as follows [Eq. (13)]:

$$\beta(r) = \beta_0 e^{-\gamma r^2}, \ (m \geq 1) \tag{13}$$

Here, γ is light absorption coefficient and β_0 is attractiveness value for $r = 0$. The $r = r_{ij}$ is the distance between the i and j fireflies represented by the design variables x_i and x_j calculated as follows:

$$r_{ij} = \|x_i - x_j\| = \sqrt{\sum_{k=1}^{d} (x_{i,k} - x_{j,k})^2} \tag{14}$$

here, $x_{i,k}$ is the kth ingredient of the x_i coordinate of the ith firefly in space.

The movement of the ith firefly in the direction of the more attractive one is decided by the following equation [Eq. (15)].

$$x_i = x_i + \beta_0 e^{-\gamma r_{ij}^2}(x_j - x_i) + \alpha \left(\text{rand} - \frac{1}{2} \right) \tag{15}$$

Here, the first term to the right of this equation indicates the current location of the firefly. The second term establishes a relationship between the intensity of light seen by neighbouring fireflies and the attractiveness of the current firefly. The last term presents the arbitrary movement to be performed when there is no more attractive firefly near the current firefly. In the last term, α is a parameter of arbitrary selection, and rand is an arbitrary real number generated in the array of [0, 1]. The parameter values in Eq. (15) are usually assigned as $\beta_0 = 1$ and $\alpha \in (0, 1)$. γ can be taken a value between 0 and 100. It delineates the modification of the attractiveness value. This value identifies the convergence velocity and productivity of the FA.

3.3   Enhanced Firefly Algorithm (EFA)

In spite of the fact that the SFA is illustrated to accomplish superior to a few other metaheuristics, it has some stuck and deficiencies while trying to design optimum. The whole parameters of SFA are initially specified and staying stable during the optimization process. It is the fact that they may deal with low-level functional problems in which amount of the design variables and dimension of the problem is limited. Despite this, when the problems turn into a real-sized high-level design optimization, such as steel frame design problems in which the dimensions and the number of the design variables are much, the power and the capacity of initially assigned parameters are not enough to direct the algorithm to the global optimum design and lead the algorithm to get stuck into a local optima [53]. Moreover, SFA lacks in maintaining a proper balance between the exploration and the exploitation capabilities making SFA less efficient. To defeat these shortcomings of SFA, amounts of upgrades are proposed within the literature. Some of those amendments are modified firefly algorithm [54–63], improved firefly algorithm [64–73], adaptive firefly algorithm [53, 74–80] and so forth.

Here, a so-called enhanced attractiveness approach is introduced in the paper to eliminate the observed drawbacks of the SFA. Moreover, a new reformation of randomness parameter is proposed by implementing it dynamically during optimization process, whereas it is set to a static value in a standard algorithm. The detailed explanations of these new strategies are as follows.

In SFA, Eq. (15) and the gamma (γ) ambient light absorption coefficient given in this equation have a very important role. This equation and, therefore, this coefficient play a decisive role in an important task related to the convergence rate and attitude of the firefly algorithm. Theoretically, it can take any value in the range of $[0, \infty]$ and as offered previously it generally is chosen in a range of 0–100 as a constant value 1.0 and/or it is recommended to use a value between 0 and 10. However, the use of light absorption coefficient, which belongs to the environment in which the light is emitted, as a fixed value in SFA created by inspiration from nature, is not sufficient to obtain the optimal solutions of relatively large-scaled problems in terms of convergence success and is not appropriate to get rid of from a local optimal design. For example, if the environment in which light is emitted is considered as air, the air could be foggy, dusty, clear, clean, cloudy, etc., naturally. That is, for example, the light absorption coefficient (γ) in a foggy or dusty air will be very large; i.e. it is assumed to be infinite $(\gamma \rightarrow \infty)$. Since the position update computed by second part of Eq. (15) converges to the zero $\left(e^{-\gamma r_{ij}^2} \to 0\right)$, the displacement amount will be random due to the third part of the equation ($\alpha(\text{rand}-0.5)$). On the contrary, if the air is open and clean, assuming that the light absorption coefficient is zero ($\gamma \to 0$), it causes $\beta(r) \to \beta_0$. Thus, there is a constant attraction where the distance is not so important [81].

For these reasons, it is suggested that the light absorption coefficient (γ) should be dynamically changeable during optimization process rather than staying constant, e.g. in an environment such as air, sudden variations (foggy, rainy, closed, open, etc.) can be occurred in nature. Therefore, it has been suggested that the light absorption coefficient (γ), which is assigned as a constant value at the beginning of the optimization process and remained unchanged in the SFA, should change dynamically in each cycle as shown in the following equation.

$$\beta(r) = (\beta_0 - \beta_{min}) * e^{\left(-\gamma_{(iter)} r_{normalized}^2\right)} + \beta_{min} \tag{16}$$

Here, $\beta(r)$ is the suggested enhanced strategy for attraction of a firefly. β_0 is firefly attraction at its initial location and assigned as 1.0. β_{min} is the minimal firefly attractiveness and depicted as a relatively small value of 0.1. $r_{normalized}$ is the normalized distance as computed in Eq. (17). $\gamma_{(iter)}$ is the enhanced light absorption coefficient as given in Eq. (18).

$$r_{normalized} = \frac{r_{ij}}{r_{max}}$$

here,

$$r_{ij} = \sqrt{\sum_{k=1}^{ng} \left(Sect_i^k - Sect_j^k\right)^2} \quad \text{and}$$

$$r_{max} = \sqrt{\sum_{k=1}^{ng} (LimitUpSect - LimitLowSect)^2} \tag{17}$$

Here, r_{ij} is the distance between any two fireflies. The i and j are defined in Eq. (14). r_{max} is the ultimate distance between two fireflies. The total group numbers in structural frame are ng. Sect is section steel profile number in the section list. LimitLowSect and LimitUpSect are the bottom and top limits of the steel profiles in the list of profiles. All profiles are ordered in an ascending array according to their unit weight; e.g. LimitLowSect is 1.0 and LimitUpSect is 272 according to ready wide-flange (W) steel section list tabulated in AISC-LRFD, and LimitLowSect is 1.0 and LimitUpSect is 85 according to C-shaped steel profile list tabulated in AISI-LRFD.

$$\gamma_{(iter)} = \gamma_{max} - \left[\left(\sqrt{\frac{iter}{iter_{max}}}\right) * (\gamma_{max} - \gamma_{min})\right] \tag{18}$$

Here, $\gamma_{(\text{iter})}$ is the light absorption coefficient at the present iteration. $\gamma_{\min}$ and $\gamma_{\max}$ are the highest and the lowest light absorption coefficient values. $\text{iter}_{\max}$ is the upper limit of the iteration until which tolerance minimization process proceeds. By means of Eq. (18), it is provided relatively higher light absorption coefficient values initially and as it is dynamically changed during optimization process it will be ended with a lower value shortly before the final cycles of design.

As explained before, SFA optimization method can be faced with trapping in local minimum and early convergence problems according to the problem type, e.g. especially in large-sized high-level optimization problems having a great number of design variables and higher problem dimensions, such as design optimization problems of steel frames. Furthermore, in search space, a better optimum point can be missed due to the stopping criteria which ceases the algorithm without the reaching global optimum point, or because the search space is not scanned and/or searched sufficiently. With the dynamic change path proposed in Eq. (19) to enhance performance of the randomness parameter instead of staying constant during optimization process, the SFA, which is transformed into a minimal search error pattern that brings a better natural randomness to the foreground, will result in gaining more scan capability of the search space to the algorithm, and thus, the diversity of fireflies will increase.

$$\alpha_{(\text{iter})} = \alpha_{\max} - \left[\left(\sqrt{\frac{\text{iter}}{\text{iter}_{\max}}} \right) * (\alpha_{\max} - \alpha_{\min}) \right] \tag{19}$$

Here, $\alpha_{(\text{iter})}$ is the value of randomness parameter at present iteration. $\alpha_{\min}$ and $\alpha_{\max}$ are the highest and the lowest values of randomness parameter. $\text{iter}_{\max}$ is the ultimate number of iteration until which minimization process of the tolerance retains. By means of Eq. (19), it is provided a higher arbitrariness value at the beginning of the algorithm, and α will be ended with a lower value so that an extensive search space could be scanned due to the dynamic scheme of the randomness parameter (α).

4 Design Optimization of Steel Frames via EFA

In this chapter, an EFA proposed for minimum weighted design of steel frame structures built up with hot-rolled and cold-formed steel profiles comprises the following steps:

(1) The number of fireflies (nff), the β_0 and $\beta_{\min}$ values of attraction, the higher and the lower values of light absorption coefficient ($\gamma_{\max}$ and $\gamma_{\min}$), the higher and the lower values of random selection parameter ($\alpha_{\max}$ and $\alpha_{\min}$) and the maximum number of iteration (the maximum number of frame structural analyses) parameters are assigned. The iteration value is set as zero (iter = 0.0).

(2) The population of fireflies, each representing a steel frame structure design, is randomly generated. A design variable of each steel frame structure in this population is randomly generated between the maximum $(x_{k,max})$ and the minimum $(x_{k,min})$ limit values as follows:

$$x_k^j = x_{k,min}^j + \text{rand}(0,\ 1) * \left(x_{k,max}^j - x_{k,min}^j \right)$$

$$\text{here } k = 1, 2, \ldots, \text{ng} \quad \text{and} \quad j = 1, 2, \ldots, \text{nff} \qquad (20)$$

Here, rand (0, 1) is a random number generated arbitrarily in the [0, 1] range. In this way, a frame structure design is obtained by producing design variables equal to member grouping number (ng). This process is repeated until the number of fireflies (the number of frame structures) to create a design population.

(3) For each frame structure design in the population, node displacements and member axial forces are obtained by analysis. Considering the proper design constraints adapt to frame type given by Eqs. (2)–(12), the objective function value of each frame structure indicated by $W(X^1)$, $W(X^2)$, ..., $W(X^{\text{nff}})$ is calculated. The objective function values of frame structures in the design population are sorted in an ascending order $(W(X^1) < W(X^2) < \cdots < W(X^{\text{nff}}))$. Accordingly, the first design of the population is defined as the best design $W(X^1) = W(X^{\text{best}})$, and the last design as $W(X^{\text{nff}}) = W(X^{\text{worst}})$.

(4) The number of iteration is increased; iter = iter + 1.

(5) Any ith firefly (ith frame structure design, X_i) is moved by Eq. (16) towards the jth firefly (jth frame structure design, X_j), which is more attractive, and a new design is obtained (X_{new}).

(6) For this new design, node displacements and member axial forces are obtained by analysing. Taking into account the proper design constraints adapt to frame type given by Eqs. (2)–(12), the objective function represented by $W(X^{\text{new}})$ is calculated.

(7) If the objective function belonging to new design is better than the worst design $(W(X^{\text{new}}) < W(X^{\text{worst}}))$ in the population, the worst design is eliminated, and the new design is included in the population. The designs in the population are reordered according to the objective function values, and the design with the best objective function value is assigned to the current optimum.

(8) If the number of iterations is smaller than the maximum number of iterations (iter < iter$_{\text{max}}$), go to step 4. If not, go to step 7.

(9) Once the maximum iteration number is attained, the optimization process is terminated, the current optimum design being considered as the ultimate optimum design. The weight of this design is the optimum steel frame weight.

5 Design Examples

In this chapter, two separate optimum solution algorithms for optimum design of steel frame structures through SFA and EFA have been developed under design loads and constraints of provisions according to AISC-LRFD for steel planar and spatial frames built up using hot-rolled steel profiles and to AISI-LRFD for steel space frames built up utilizing thin-walled cold-formed steel profiles. These steel frame structures are a steel planar frame which has two-bay, six-storey and 30 hot-rolled steel structural members [82, 83], a steel spatial frame comprising of four-storey and 428 hot-rolled steel frame members [84–87], and a steel space frame built up using thin-walled open C-shaped cold-formed steel profiles owning two-storey and 379 structural members [88]. The selected parameter values needed at the beginning of SFA and EFA for all design examples are given in Table 1.

The technical descriptions and the evaluations for obtained optimum designs using SFA and EFA of the example problems are as follows.

5.1 Two-Bay, Six-Storey, 30 Hot-Rolled Steel Member Planar Frames

A six-storey, two-bay planar steel frame comprising of 30 hot-rolled structural steel members which are gathered in eight groups as presented in Fig. 1 is treated as the first design example. Also, the acting external loads on structural members as well as nodes, and the span length of the beams and the heights of the columns are shown in the same figure. The permitted limits of the drift constraints are 1.17 cm for inter-storey drift and 7.17 cm for top-storey drift. The hot-rolled steel ready sections itemized in the wide-flange (W) profile list given in AISC-LRFD are utilized to the structural member sizing. The material properties tabulated in AISC-LRFD provision for hot-rolled steel are modulus of elasticity (E) = 208 GPa (30,167.84 ksi), yield stress (F_y) = 250 MPa (36.26 ksi) and unit weight (ρ) = 7.85 ton/m^3.

Table 1 Parameter sets selected for SFA and EFA

	Two-bay, six-storey, 30 hot-rolled steel member planar frames	Four-storey, 428 hot-rolled steel member spatial frames	Two-storey, 379 thin-walled cold-formed steel member space frames
SFA	Number of fireflies = 20, $\alpha = 0.5$, $\gamma = 1.0$, $\beta = 1.0$	Number of fireflies = 50, $\alpha = 0.5$, $\gamma = 1.0$, $\beta = 1.0$	Number of fireflies = 50, $\alpha = 0.5$, $\gamma = 1.0$, $\beta = 1.0$
EFA	Number of fireflies = 50, $\gamma_{max} = 10$, $\gamma_{min} = 1.0$, $\alpha_{max} = 0.99$, $\alpha_{min} = 0.01$, $\beta_0 = 1.0$, $\beta_{min} = 0.1$, max$_{iter.}$ = 20,000	Number of fireflies = 50, $\gamma_{max} = 10$, $\gamma_{min} = 1.0$, $\alpha_{max} = 0.99$, $\alpha_{min} = 0.01$, $\beta_0 = 1.0$, $\beta_{min} = 0.1$, max$_{iter.}$ = 50,000	Number of fireflies = 50, $\gamma_{max} = 10$, $\gamma_{min} = 1.0$, $\alpha_{max} = 0.99$, $\alpha_{min} = 0.01$, $\beta_0 = 1.0$, $\beta_{min} = 0.1$, max$_{iter.}$ = 75,000

Fig. 1 Two-bay, six-storey, 30 hot-rolled steel member planar frames

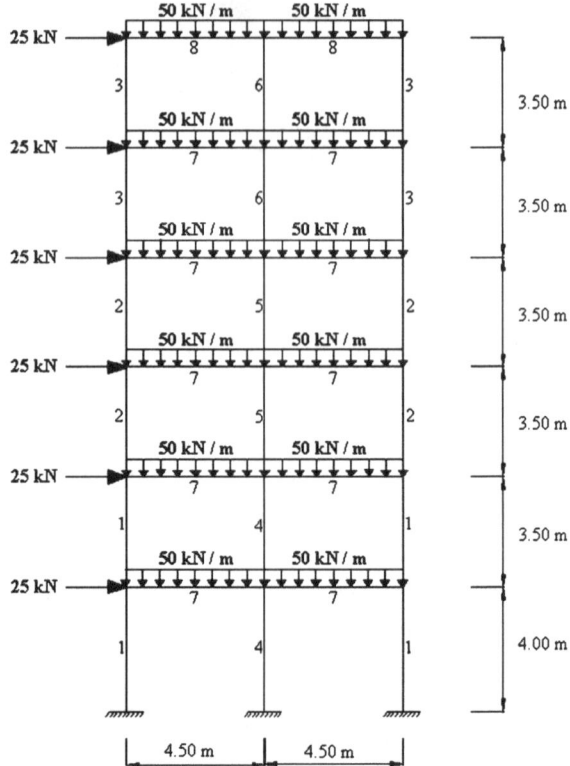

The section designations obtained from SFA and EFA, which are assigned to structural member groups, are represented in Table 2. From this table, it is obvious that the cycle number at which the optimum design is detected in EFA is lower than those of SFA. When the optimal design weights obtained by both algorithms are observed, EFA yielded an overwhelming minimum design weight as 6358.158 kg to those obtained by SFA as 7421.216 kg.

The history search graphs of both algorithms presenting the feasible best designs obtained from structural analyses are shown in Fig. 2. This figure proves the convergence amendment of the SFA after applying the proposed enhanced strategies. The optimal structural frame weight of this problem produced by EFA is 14.33% lighter than the minimum structural frame weight achieved by SFA. Also, the least optimal design weight obtained for this problem so far is reported as 6331.441 kg obtained by biogeography-based optimization (BBO) [83] which is simply 0.42% lighter than the optimal structural design weight yielded by EFA. These results, obviously, show that the proposed enhanced strategies gain distinguishably superiorities to the SFA. It is identified that in the optimally designed frame the inter-storey drift for the third floor is 1.139 cm which is adjacent to its ultimate limit of 1.17 cm while top-storey drift is 5.281 cm whose ultimate bound is 7.17 cm. The biggest strength ratio among

Table 2 Optimum designs for first design example

#	Member type	SFA	EFA
1	Column	W250 × 38.5	W410 × 75
2	Column	W410 × 46.1	W410 × 53
3	Column	W410 × 60	W250 × 32.7
4	Column	W410 × 60	W610 × 113
5	Column	W460 × 74	W360 × 72
6	Column	W460 × 74	W360 × 44
7	Beam	W460 × 89	W460 × 46.1
8	Beam	W460 × 113	W360 × 32.9
Max. inter-storey drift (cm)		1.019	1.139
Max. strength ratio		0.987	0.993
Max. top-storey drift (cm)		4.626	5.281
Minimum weight. kg (kN)		7421.216 (72.777)	6358.158 (62.352)
Maximum iteration		20,000	20,000
No. of structural analysis		19,921	17,189

Fig. 2 Structural analysis history graph of the first design example

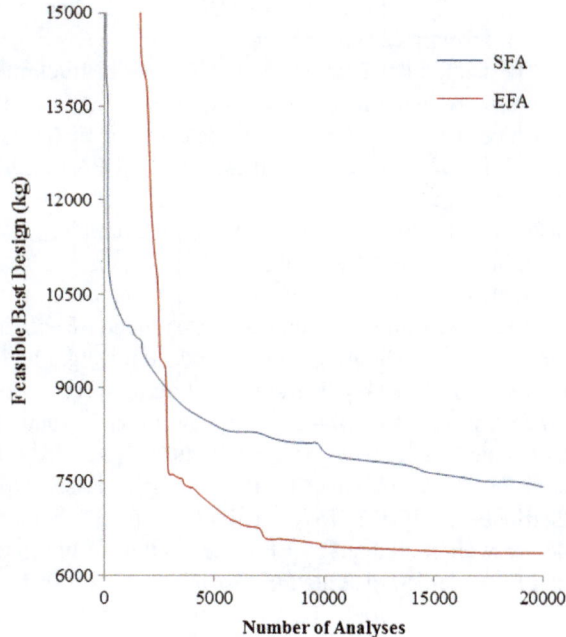

the combined strength constraints is 0.993 against its ultimate bound of 1.0. This specified that the governing design constraint is the strength in the design process.

5.2 Four-Storey, 428 Hot-Rolled Steel Member Spatial Frames

The second design example of this chapter is taken as a four-storey, 428 hot-rolled steel member spatial frames. The geometry of the frame (3D, front and plan shots) as well as the orientations of the columns is presented in Fig. 3. The node and the structural member group numbers are 172 and 20, respectively. Table 3 shows the structural member groups of the frame. The design loads are taken as the lateral and gravity loads calculated to ASCE 7-05 [89]. The live and dead loads for design are stated as 2.39 kN/m^2 and 2.88 kN/m^2, respectively. The wind load is computed with a basic wind speed taken as 85 mph (38 m/s). So, the loading combinations are detected as $1.2D + 1.6L + 0.5S$, $1.2D + 0.5L + 1.6S$, $1.2D + 1.6WX + L + 0.5S$ and $1.2D + 1.6WX + L + 0.5S$. Here, dead load is D, live load is L, snow load is S, and wind loads acting towards global Z and X axes are WZ and WX, respectively. The drift constraints are conducted as 0.875-cm inter-storey drift and 3.5-cm top-storey drift. Displacement of a beam member span is limited as 2.0 cm. Also, the material properties tabulated in AISC-LRFD provision for hot-rolled steel are modulus of elasticity $(E) = 208$ GPa (30,167.84 ksi), yield stress $(F_y) = 250$ MPa (36.26 ksi) and unit weight $(\rho) = 7.85$ ton/m^3.

The section designations assigned to each structural member group via SFA and EFA are presented in Table 4. From this table, it is identified that the cycle number at which the optimum design is detected in EFA is 11,890 which is much lower than those of SFA (31,101). It means that EFA reaches optimum structural spatial frame design 2.61 times quicker than SFA. Figure 4 shows the convergence rates of both algorithms. As previous design example, the EFA algorithm converges to the minimum design weight so rapidly. This means that the enhancement strategies implemented on to SFA demonstrate its efficiency on convergence rate once again.

When the optimal design weights obtained by both algorithms are observed, EFA yielded an overwhelming minimum design weight as 1325.03 kN to those obtained by SFA as 1537.10 kN. The optimal structural frame weight of this problem produced by EFA is 13.79% lighter than the minimum structural frame weight achieved by SFA. The least optimal design weight obtained for this problem so far is reported as 6331.441 kg obtained by Biogeography Based Optimization with Levy Flight distribution (LFBBO) [87] which is only 6.93% lighter than the optimum structural design weight attained by EFA. Also, attained ultimate design constraints via SFA and EFA are demonstrated in that table, as well. When these obtained design constraints are compared with their provision-based pre-calculated upper bounds, it is clearly seen that, at the end of the optimization process while inter-storey drift constraints computed with both algorithms are 0.819 cm (SFA) and 0.816 cm (EFA), the

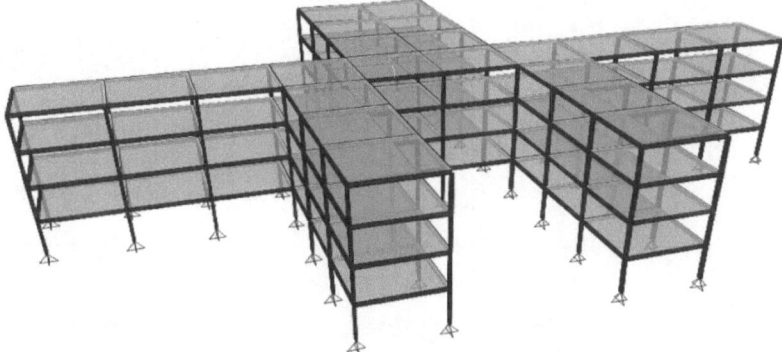

(a) 3D shot of the second design example

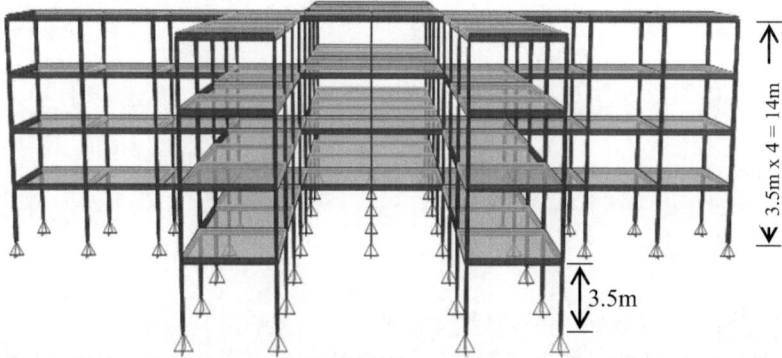

(b) Front shot of the second design example

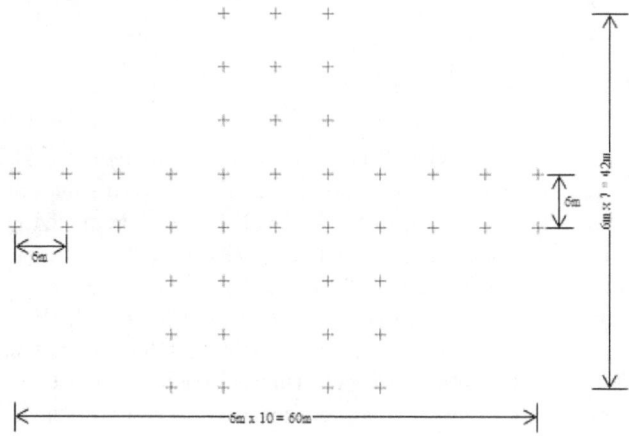

(c) Plan shot of the second design example

Fig. 3 Second design example, **a** 3D shot, **b** front shot, **c** plan shot, **d** column orientation shot

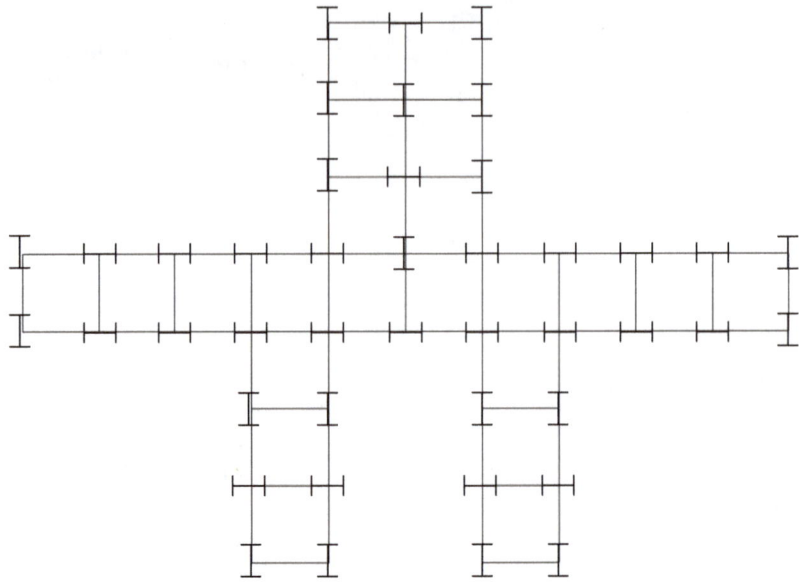

(d) Column orientations shot of the second design example

Fig. 3 (continued)

Table 3 Member grouping of second design example

Storey	Side beam	Inner beam	Corner beam	Side column	Inner column
1	1	2	9	10	11
2	3	4	12	13	14
3	5	6	15	16	17
4	7	8	18	19	20

top-storey drifts yielded via both algorithms are computed as 2.510 cm (SFA) and 2.623 cm (EFA). This result specifies that as the obtained values are closer to their upper bounds, the inter-storey drifts are much dominant than the top-storey drifts on optimization process. In addition to inter-storey drift, one other active design constraint is the strength for both algorithms. The so-called passive constraint for this design example can be pointed out as top-storey drifts for both SFA and EFA. On the basis of optimal design weights, it can easily be said that the proposed enhancement scheme implemented to SFA for performance amendments works very effectively on large and/or real-sized structural design optimization problems.

Table 4 Optimum designs for second design example

#		SFA	EFA
1	Beam	W310 × 28.3	W310 × 28.3
2	Beam	W250 × 25.3	W310 × 32.7
3	Beam	W460 × 60	W360 × 91
4	Beam	W460 × 60	W200 × 52
5	Beam	W460 × 60	W610 × 92
6	Beam	W460 × 60	W310 × 44.5
7	Beam	W460 × 60	W360 × 52
8	Beam	W310 × 60	W410 × 53
9	Column	W310 × 60	W200 × 71
10	Column	W310 × 60	W200 × 46.1
11	Column	W310 × 60	W150 × 37.1
12	Column	W310 × 60	W310 × 97
13	Column	W310 × 60	W200 × 46.1
14	Column	W310 × 60	W250 × 73
15	Column	W410 × 149	W310 × 97
16	Column	W410 × 149	W310 × 97
17	Column	W410 × 149	W310 × 79
18	Column	W530 × 196	W310 × 97
19	Column	W530 × 196	W460 × 97
20	Column	W530 × 196	W360 × 91
Max. strength ratio		0.924	0.868
Top drift (cm)		2.510	2.623
Inter-storey drift (cm)		0.819	0.816
Weight—kN (kg)		1537.10 (156, 741.16)	1325.03 (135, 115.96)
Maximum iteration		50,000	50,000
No. of structural analysis		31,101	11,890

5.3 Two-Storey, 329 Cold-Formed Steel Member Space Frames

The last design example of this chapter is preferred as a two-storey, 379 cold-formed steel member space frames [88]. The 3D, plan and floor shots are shown in Fig. 5. The structural height of the frame is 5.6 m in which each storey has 2.8 m height. The length of the beam spans is 0.6 m. The 239 nodes are connecting totally 379

Fig. 4 Structural analysis history graph of the second design example

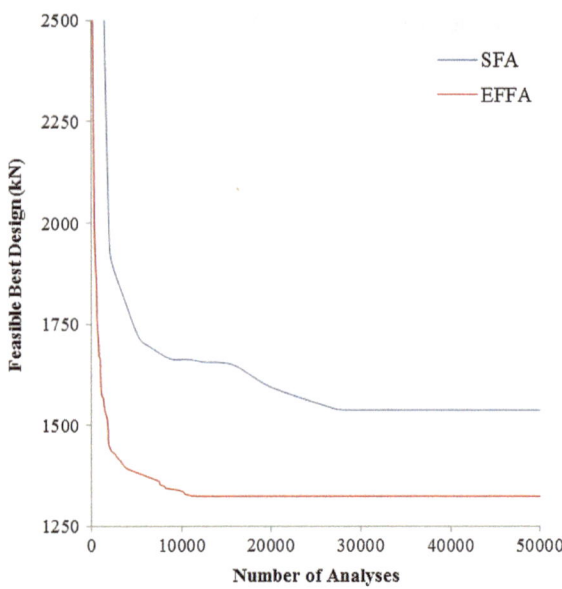

frame members to each other. The dead, live and ground snow loads are imposed on the frame structure according to ASCE 7-05 [89] which are a design dead load of 2.89 kN/m^2, a design live load of 2.39 kN/m^2, a ground snow load of 0.755 kN/m^2, respectively. Also, in Table 5, the unfactored wind load values are depicted. The provision based load cases applied in this design example are Load Case 1: $1.2D + 1.6L + 0.5S$, Load Case 2: $1.2D + 0.5L + 1.6S$ and Load Case 3: $1.2D + 1.6WX + L + 0.5S$ here D is dead, L is live, S is snow loads, and WX is the wind load carried out in the direction of global X-axis. The drift constraints of x and y directions are limited as 1.4 cm for top-storey and 0.7 cm for inter-storey. The thin-walled open C-shaped sections with lips constitute the pool of design variables. The frame members are gathered in ten separate groups. These groupings are represented in Table 6.

The last design example has been optimized using SFA and EFA. Thin-walled open C-shaped sections with lips assigned to member groups are exhibited in Table 7. That is clear from this table that the EFA achieves the optimal design weight as 3763.786 kg which is 4.92% lighter than those achieved by SFA. The lightest optimal design of this cold-formed lightweight frame reported is reached through artificial bee colony (ABC) algorithm [88] which is only 2.31% lighter than attained design weight using EFA. The technicalities of the obtained optimum designs obtained from both algorithms for this structural design problem are exhibited in Table 7. Either the drifts notably top-storey or the strength constraints are operative in this design example. Both of them controlling the optimum designs are attained by the SFA and EFA. From over this table, the ratios of strength of some structural members arrived at the problem solutions using via both algorithms press in upon the ultimate limit of 1.0. The inter-storey drift constraints are not as active as the top-storey drift

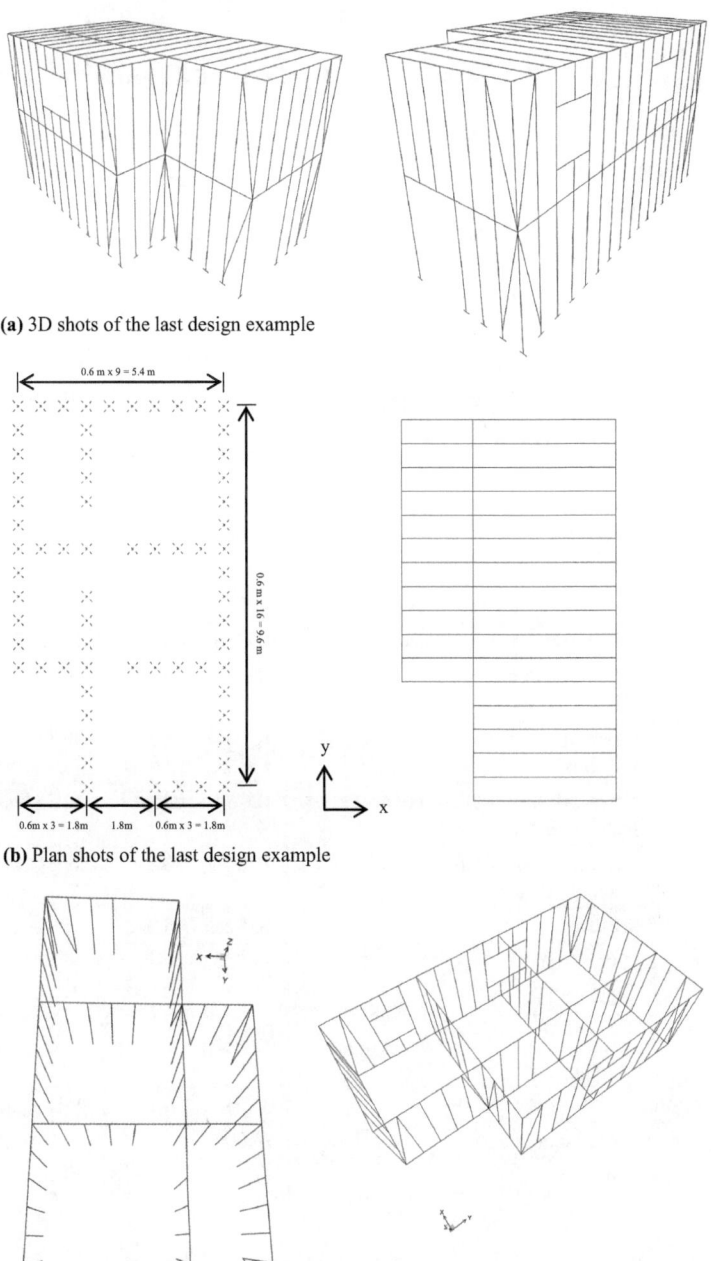

(a) 3D shots of the last design example

(b) Plan shots of the last design example

(c) The first and the second floors top views without slabs

Fig. 5 Last design example; **a** 3D shots, **b** plan shots, **c** the first and the second floor top shots without slabs

Table 5 Wind loads for two-storey 379 cold-formed member space frames

All nodes on windward side of roof	4 kN
All nodes on windward side of first storey	8 kN

Table 6 Member grouping of last design example

# of storey	Y side beams	X side beams	Windows	Columns connected short beams	Columns connected long beams	Braces
1	1	2	–	6	7	10
2	4	5	3	8	9	10

Table 7 Optimum designs for last design example

Group number	Group type	SFA	EFA
1	First floor y side beams	4CS4 × 070	8CS3.5 × 065
2	First floor x side beams	8CS2.5 × 059	8CS2.5 × 059
3	Windows	7CS2.5 × 065	4CS4 × 059
4	Second floor y side beams	4CS4 × 105	7CS4 × 105
5	Second floor x side beams	10CS4 × 085	10CS4 × 070
6	First floor columns connected short beams	10CS4 × 065	8CS3.5 × 059
7	First floor columns connected long beams	4CS4 × 065	4CS4 × 059
8	Second floor columns connected short beams	12CS4 × 085	12CS4 × 085
9	Second floor columns connected long beams	7CS4 × 070	4CS4 × 085
10	Braces	4CS2 × 065	4CS2 × 070
Minimum weight (kN (kg))		38.725 (3948.866)	36.91 (3763.786)
Maximum top-storey drift (mm)		12.54	13.46
Maximum inter-storey drift (mm)		0.304	0.430
Maximum displacement (mm)		2.667	2.556
Maximum strength ratio		0.998	1.0
Maximum number of iterations		75,000	75,000
No. of structural analyses		37,250	33,644

constraints in this example. The maximum values of the displacement obtained are nominal compared to the other design constraints.

The convergence grades to reach the optimum design weights for the design algorithms, namely SFA and EFA, are depicted in Fig. 6. The SFA exhibits a relatively fine convergence, and it started to find favourable design weight very close to optimum weight in the vicinity of 35,000th structural analyses and finalized with the optimum

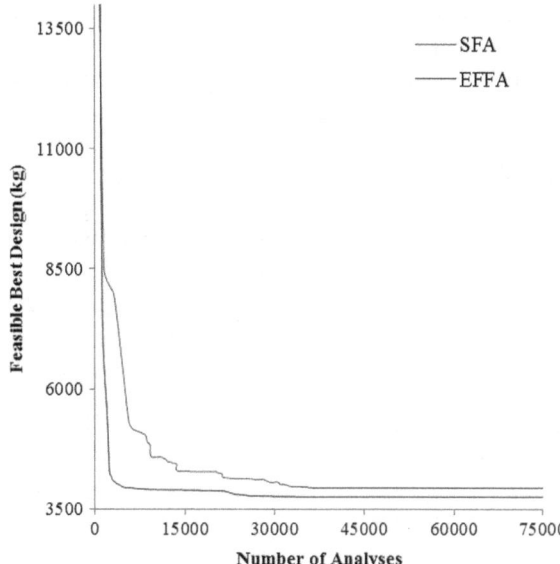

Fig. 6 Structural analysis history graph of the last design example

design at the 37,250th structural analyses. But, as it is apparent from Fig. 6, when the EFA comes to the stage, it displays very smooth convergence fulfilment on the way to the optimum design. It converges to the optimum design weight at very early phases of the optimization process and continues this characterization up to the end of the design optimization process without trapping in a local optimal design. This is due to the productive and operational adaptation of the enhanced scenario to the SFA because of its harmonious working ability.

6   Conclusions

As it is known, metaheuristic algorithms are one of the primary tools that can provide very efficient results in the search for design optimization problem solutions. In the literature, it is possible to find hundreds of metaheuristic algorithms. The main purpose of these algorithms is to always try to get the optimal solutions for the design problems. It is not possible to say that all algorithms, especially in their raw form, provide this aim successfully and satisfy that demand at all times for all kinds of design problems. One of the major reasons for the emergence of new algorithms and/or regenerated versions (enhanced/improved/modified, etc.) of existing algorithms is that a unique algorithm which provides optimum design results for each type of design problem, regardless of the dimension and size of problems, has not yet been discovered. Many recently revitalized, so-called enhanced versions of the metaheuristic algorithms claim to be superior to its raw version. The concept of superiority is expected to reveal new generation algorithms in the area of design optimization as it is a phenomenon that can vary depending upon the type, size and dimension of design problem to be solved.

In this chapter, in order to perform the design optimization of structural steel frames which fall into group of high-level real-sized optimization problems due to having a large amount of design variables and provision-based design constraints to be satisfied, an enhanced version of the firefly algorithm is proposed to improve the exploration capability, avoid immature convergences and move ahead the algorithm away from stuck in a local optimal design. When the optimal results of the design examples are evaluated, it is illustrated that the enhanced firefly optimization algorithm (EFA) supplies more successful designs than the standard firefly algorithm (SFA). The reason for this is that in the enhanced version, it is suggested to use not only a dynamically variable light absorption coefficient but also a state-of-the-art arbitrary movement strategy that also have a dynamic path which takes into account the changes in the environment (i.e. air, etc.) of the fireflies in the standard algorithm created by inspiring the nature. For example, foggy weather in nature has negative effects on light transmittance. Therefore, the change of light absorption coefficient dynamically with a high amount of randomness in each cycle against sudden changes leads to attaining better results in optimization design problems.

When optimum structural weights are compared with those values of optimal designs obtained from other metaheuristic methods reported in the literature, it has been seen that EFA produces very close optimum designs to the best designs attained so far. Compared to the number of structural analysis required to reach the optimum designs for both standard and enhanced versions of firefly algorithm, EFA algorithm has been found to produce much lighter designs than the SFA algorithm by analysing the structure with a very low number of design cycles. Moreover, the reliability of the enhanced algorithm is ensured by not allowing the EFA to make any design constraint violation in any design example.

Where the standard firefly optimization algorithm cannot effectively perform the optimum solution of real-sized large-scale structural steel frame design problems, the enhanced firefly algorithm overcomes convergence troubles and entrapment in local optimal problems, resulting in design of lighter structural weights. As a result, it is the fact that utilizing EFA as an optimization tool is a strong and powerful way for attaining optimal designs of steel frame constructions, whether they are built up with hot-rolled or cold-formed steel profiles, in which a number of provision-based design constraints and a variety of design variables exist.

References

1. Yang XS (2010) Engineering optimization; an introduction with metaheuristic applications. Wiley, London
2. Rozvany GIN (1993) Optimization of large structural sytems. NATO ASI Series, Series E: applied sciences. Springer, Berlin
3. Luke S (2010) Essentials of metaheuristics, 2nd ed. Lulu, http://cs.gmu.edu/~sean/book/metaheuristics/

4. Kochenberger GA, Glover F (2003) Handbook of meta-heuristics. Kluwer Academic Publishers, Dordrecht
5. Blum C, Roli A (2003) Metaheuristics in combinatorial optimization: overview and conceptual comparison. ACM Comput Surv 35(30):268–308
6. De Castro LN, Von Zuben FJ (2005) Recent developments in biologically inspired computing. Idea Group Publishing, Hershey
7. Dreo J, Petrowski A, Siarry P, Taillard E (2006) Meta-heuristics for hard optimization. Springer, Berlin
8. Gonzales TF (2007) Handbook of approximation algorithms and metaheuristics. Chapman & Hall, CRC Press, London
9. Holland JH (1975) Adaptation in natural and artificial systems. University of Michigan Press, Ann Arbor, MI
10. Goldberg DE (1983) Computer-aided pipeline operation using genetic algorithms and rule learning, Ph.D. thesis,. University of Michigan, Ann Arbor, MI
11. Kumar C, Prakash S, Kumar Gupta T, Prasad Sahu D (2014) Variant of genetic algorithm and its applications. Int J Art Neural Net 4(4):8–12
12. Singh B (2014) A survey of the variants of genetic algorithm. Int J Sci Eng Res 5(6):1261–1264
13. Elsayed SM, Sarker RA, Essam DL (2010) A comparative study of different variants of genetic algorithms for constrained optimization. In: Deb K et al (eds) Simulated evolution and learning. SEAL 2010. Lecture notes in computer science, vol 6457. Springer, Berlin, Heidelberg
14. Bineet M, Rakesh Kumar P (2009) Genetic algorithm and its variants: theory and applications. BTech Thesis, National Institute of Technology, Rourkela
15. Affenzeller M (2003) New variants of genetic algorithms applied to problems of combinatorial optimization. In: Trappl R (ed) Cybernetics and systems, vol 1. Austrian Society for Cybernetic Studies
16. Rao RV, Savsani VJ, Vakharia DP (2011) Teaching–learning-based optimisation: a novel method for constrained mechanical design optimisation problems. Comput Aided Des 43:303–315
17. Storn R, Price K (1997) Differential evolution–a simple and efficient heuristic for global optimisation over continuous spaces. J Glob Optim 1:341–359
18. Kirkpatrick S, Gelatt CD Jr, Vecchi MP (1983) Optimization by simulated annealing. Science 220(4598):671–680
19. Erol Osman K, Eksin Ibrahim (2006) A new optimization method: Big Bang-Big Crunch. Adv Eng Soft 37(2):106–111
20. Dorigo M (1992) Optimization, learning and natural algorithms. Ph.D. thesis, Politecnico di Milano
21. Kennedy J, Eberhart R (1995) Particle swarm optimization. In: Proceedings of the ieee international conference on neural networks, Perth, Australia, pp 1942–1948
22. Venkata R (2016) Jaya: A simple and new optimization algorithm for solving constrained and unconstrained optimization problems. Int J Ind Eng Comput 7:19–34
23. Geem ZW, Kim JH, Loganathan GV (2001) A new heuristic optimization algorithm: harmony search. Simulation 76(2):60–68
24. Karaboga D, Basturk B (2007) A powerful and efficient algorithm for numerical function optimisation: artificial bee colony algorithm. J Glob Optim 39(3):459–471
25. Fred G (1989) Tabu search-part I. ORSA J Comput 1(3):190–206
26. Kaveh A, Talatahari S (2010) A novel heuristic optimisation method: charged system search. Acta Mech 213(3–4):267–289
27. Reynolds RG (1994) An introduction to cultural algorithms evolutionary programming. In: Proceeding of 3rd annual conference, World Scientific, River Edge, NJ, USA, pp 131–139
28. Saka MP, Carbas S, Aydogdu I, Akin A, Geem ZW (2015) Comparative study on recent metaheuristic algorithms in design optimization of cold-formed steel structures. In: Lagaros N, Papadrakakis M (eds) Engineering and Applied sciences optimization. Computational methods in applied sciences, vol 38. Springer, Cham

29. Saka MP, Carbas S, Aydogdu I, Akin A (2016) Use of swarm intelligence in structural steel design optimization. In: Yang XS, Bekdaş G, Nigdeli S (eds) Metaheuristics and optimization in civil engineering. Modeling and optimization in science and technologies, vol 7. Springer, Cham
30. Kaveh A (2017) Advances in metaheuristic algorithms for optimal design of structures, 2nd edn. Springer, Cham
31. Ewens MJ (2011) What changes has mathematics made to the Darwinian theory? In: Chalub FACC, Rodrigues JF (eds) The mathematics of Darwin's legacy, mathematics and biosciences in interaction. Springer, Basel
32. Yang XS (2008) Nature-inspired metaheuristic algorithms. Luniver Press, Bristol
33. Yang XS (2013) Multiobjective firefly algorithm for continuous optimization. Eng Comput 29(2):175–184
34. Jati GK, Suyanto (2011) Evolutionary discrete firefly algorithm for travelling salesman problem. In: Bouchachia A (eds) Adaptive and intelligent systems. ICAIS 2011. Lecture notes in computer science, vol 6943. Springer, Berlin, Heidelberg
35. Dey N (2017) Advancements in applied metaheuristic computing. IGI Global
36. Jagatheesan K, Anand B, Samanta S, Dey N, Ashour AS, Balas VE (2019) Design of a proportional-integral-derivative controller for an automatic generation control of multi-area power thermal systems using firefly algorithm. IEEE/CAA J Autom Sinica 6(2):589–594
37. Kumar R, Talukdar FA, Dey N, Balas VE (2018) Quality factor optimization of spiral inductor using firefly algorithm and its application in amplifier. Int J Adv Intel Paradigms 11(3–4):299–314
38. Carbas S (2016) Design optimization of steel frames using an enhanced firefly algorithm. Eng Optim 48(12):2007–2025
39. Yu WW (1973) Cold-formed steel structures; design, analysis, construction. McGraw-Hill Book Company, USA
40. AISC-LRFD (2001) Load and resistance factor design (LRFD), vol 1, Structural members specifications codes, 3rd edn. American Institute of Steel Construction
41. AISI (2002) Cold-formed steel design manual, American Iron and Steel Institute
42. AISI S100-07 (2007) North American specification for the design of cold-formed steel structural members. American Iron and Steel Institute
43. AISI D100-08 (2008) Excerpts-gross section property tables, cold-formed steel design manual, Part I: Dimensions and properties. American Iron and Steel Institute
44. Ad Hoc Committee on Serviceability (1986) Structural serviceability: a critical appraisal and research needs. J Struct Eng ASCE 112(12):2646–2664
45. Lukasik S, Zak S (2009) Firefly algorithm for continuous constrained optimization tasks. In: 1st international conference on computational collective intelligence, semantic web, social networks and multiagent systems, Wrodaw, Poland, pp 97–106
46. Fraga H (2008) Firefly luminescence: a historical perspective and recent developments. J Photochem Photobiol Sci 7:146–158
47. Babu BG, Kannan M (2002) Lightning bugs. Resonance 7(9):49–55
48. Yang XS, He XS (2019) Nature-inspired algorithms. In: Mathematical foundations of nature-inspired algorithms. Springer briefs in optimization. Springer, Cham
49. Yang XS, Hosseini SSS, Gandomi AH (2012) Firefly algorithm for solving non-convex economic dispatch problems with valve loading effect. Appl Soft Comput 12:1180–1186
50. Gandomi AH, Yang XS, Alavi AH (2011) Mixed variable structural optimization using firefly algorithm. Comput Struct 89:2325–2336
51. Talatahari S, Gandomi AH, Yun GJ (2012) Optimum design of tower structures using firefly algorithm. Struct Des Tall Special 23:350–361
52. Degertekin SO, Lamberti L (2013) Sizing optimization of truss structures using the firefly algorithm. In Topping BHV, Iványi P (eds) Proceedings of the fourteenth international conference on civil, structural and environmental engineering computing. Civil-Comp Press, Stirlingshire, UK
53. Yu S, Yang S, Su S (2013) Self-Adaptive step firefly algorithm. J Appl Math 2013:1–8

54. Memari A, Ahmad R, Akbari Jokar MR, Abdul Rahim AR (2019) A new modified firefly algorithm for optimizing a supply chain network problem. Appl Sci 9(1):7:1–13
55. Liu C, Gao F, Jin N (2014) Design and simulation of a modified firefly algorithm. In: Proceedings of seventh international joint conference on computational sciences and optimization. IEEE, Beijing, China
56. Gupta M, Gupta D (2016) A new modified firefly algorithm. Int J Eng Sci 4(2):4006–4011
57. Yelghi A, Kose C (2018) A modified firefly algorithm for global minimum optimization. Appl Soft Comput 62:29–44
58. Tilahun SL, Ong HC (2012) Modified firefly algorithm. J Appl Math 2012:1–12
59. Fister I, Yang XS, Brest J, Fister I Jr (2012) Modified firefly algorithm using quaternion representation. Exp Syst Appl 40:7220–7230
60. Kazemzadeh-Parsi MJ (2014) A modified firefly algorithm for engineering design optimization problems. Trans Mech Eng 38(M2):403–421
61. Garousi-Nejad I, Bozorg-Haddad O, Loáiciga HA (2016) Modified firefly algorithm for solving multireservoir operation in continuous and discrete domains. J Water Resour Plann Manage 142(9):1–15
62. Karuvelam S, Rajaram M (2014) Modified firefly algorithm for selective harmonic elimination in single phase matrix converter. Int J Appl Eng Res 9(23):22325–22336
63. Verma OP, Aggarwal D, Patodi T (2016) Opposition and dimensional based modified firefly algorithm. Exp Syst Appl 44:168–176
64. Celik Y, Kutucu H (2018) Solving the tension/compression spring design problem by an improved firefly algorithm. IDDM 1(2255):1–7
65. Xu H, Yu S, Chen J, Zuo X (2018) An improved firefly algorithm for feature selection in classification. Wireless Pers Commun 102(4):2823
66. Wang GG, Guo L (2014) A new improved firefly algorithm for global numerical optimization. J Comput Theo Nano 11(2):477–485
67. Xiang Q (2015) An improved firefly algorithm for numerical optimization. Int J Comput Sci Mat 6(2):201
68. Wahid F, Ghazali R, Shah H (2018) An improved hybrid firefly algorithm for solving optimization problems. In: Ghazali R, Deris M, Nawi N, Abawajy J (eds) Recent advances on soft computing and data mining. SCDM 2018. Advance International System Computing, vol 700. Springer, Cham
69. Baykasoglu A, Ozsoydan FB (2014) An improved firefly algorithm for solving dynamic multidimensional knapsack problems. Exp Syst Appl 41(8):3712–3725
70. Zhang F, Hui J, Guo Y (2018) An improved firefly algorithm for collaborative manufacturing chain optimization problem. Proc Inst Mech Eng, Part B: J Eng Manuf 233(6):1711–1722 (Sage)
71. Al-Wagih K (2015) Improved firefly algorithm for unconstrained optimization problems. Int J Comput Appl Tech Res 4(1):77–81
72. Kaur K, Salgotra R, Singh U (2017) An improved firefly algorithm for numerical optimization. In: Proceedings of international conference on innovations in information, embedded and communication systems (ICIIECS), Coimbatore, India
73. Nguyen TT, Quynh NV, Le Van Dai LV (2018) Improved firefly algorithm: a novel method for optimal operation of thermal generating units. Complexity 2018:1–23
74. Ranganathan S, Kalavathi MS, Rajan CA (2015) Self-adaptive firefly algorithm based multiobjectives for multi-type FACTS placement. IET Gener Transm Distrib 10(11):2576–2584
75. Fister I, Yang XS, Brest J, Fister Jr I (2013) Memetic self-adaptive firefly algorithm. In: Yang XS, Cui Z, Xiao R, Gandomi AH, Karamanoglu M (eds) Swarm intelligence and bio-inspired computation: theory and applications. Elsevier Inc
76. Baykasoglu A, Ozsoydan FB (2015) Adaptive firefly algorithm with chaos for mechanical design optimization problems. Appl Soft Comput 36:152–164
77. Wang W, Wang H, Zhao J, Lv L (2017) A new adaptive firefly algorithm for solving optimization problems. In: Huang DS, Bevilacqua V, Premaratne P, Gupta P (eds) Intelligent computing theories and application. ICIC 2017. Lecture Notes in Computer Science, vol 10361. Springer, Cham

78. Selvarasu R, Surya Kalavathi M (2014) Self-adaptive firefly algorithm based transmission loss minimization using multi type FACTS devices. In: Proceedings of international conference on circuit, power and computing technologies [ICCPCT], Tamil Nadu, India
79. Cheung NJ, Ding XM, Shen HB (2014) Adaptive firefly algorithm: parameter analysis and its application. PLoS ONE 9(11):e112634
80. Saka MP, Aydogdu I, Akin A (2012) Discrete design optimization of space steel frames using the adaptive firefly algorithm. In: Proceedings of the eleventh international conference on computational structures technology, Dubrovik, Croatia
81. Yang XS (2009) Firefly algorithms for multimodal optimization, stochastic algorithms: foundations and applications, SAGA, Lecture Notes in Computer Science, vol 5792. Springer, Berlin
82. Dogan E, Saka MP (2012) Optimum design of unbraced steel frames to LRFD–AISC using particle swarm optimization. Adv Eng Soft 46(1):27–34
83. Carbas S (2017) Optimum structural design of spatial steel frames via biogeography-based optimization. Neural Comput Appl 28:1525–1539
84. Aydodu I, Akin A (2014) Teaching and learning-based optimization algorithm for optimum design of steel buildings. Comput Civil Build Eng, 2167–2175
85. Akin A, Aydogdu I (2015) Optimum design of steel space frames by hybrid teaching-learning based optimization and harmony search algorithms. Int J Mech Aerosp Indust Mechatron Manuf Eng 9(7):1367–1374
86. Aydodu I, Akin A, Saka MP (2016) Design optimization of real world steel space frames using artificial bee colony algorithm with Levy flight distribution. Adv Eng Software 92:1–14
87. Aydogdu I, Carbas S, Akin A (2017) Effect of Levy Flight on the discrete optimum design of steel skeletal structures using metaheuristics. Steel Comp Struct 24(1):93–112
88. Carbas S, Aydogdu I, Tokdemir T, Saka MP (2014) Design optimization of low-rise cold-formed steel frames with thin-walled sections using the artificial bee colony algorithm. In: Topping BHV, Iványi P (eds) Proceedings of the twelfth international conference on computational structures technology. Civil-Comp Press, Stirlingshire, Scotland
89. ASCE 7-05 (2005) Minimum design loads for buildings and other structures. American Society of Civil Engineers, Reston, VA, USA

Chapter 7
Application of Firefly Algorithm for Face Recognition

Jai Kotia, Rishika Bharti, Adit Kotwal and Ramchandra Mangrulkar

1 Introduction

In all of computer science and associated fields, the recognition that machine learning has gained is unparalleled. Currently, the applications of machine learning are undergoing a rapid growth in commercial products. As the industry starts to adopt novel applications of machine learning models, the requirement of the highest accuracy in these models is mandatory for success. The requirement for high accuracy results in any machine learning tasks comes from the increasing competition in this field. While there are already promising results in several machine learning applications, there is still need for improvement if they are to be integrated into commercial products. A large portion of ongoing research work in this field focuses primarily on improving the performance of machine learning algorithms in applications that are already in use. Face Recognition is one such popular application of machine learning, wherein the model is trained to identify a person's face from an input image or stream. There is a huge scope for improvement as far as computerization of Face Recognition goes. The human brain is highly adept in identifying faces and inter-

Jai Kotia and Rishika Bharti: have contributed equally to this chapter.

J. Kotia · R. Bharti · A. Kotwal (✉) · R. Mangrulkar
Dwarkadas J. Sanghvi College of Engineering, Vile Parle, Mumbai, India
e-mail: aditkotwal29@gmail.com; adit29may@gmail.com
J. Kotia
e-mail: jaikotia10@gmail.com

R. Bharti
e-mail: bhartirishika@gmail.com

R. Mangrulkar
e-mail: ramchandra.mangrulkar@djsce.ac.in

© Springer Nature Singapore Pte Ltd. 2020
N. Dey (ed.), *Applications of Firefly Algorithm and its Variants*,
Springer Tracts in Nature-Inspired Computing,
https://doi.org/10.1007/978-981-15-0306-1_7

preting one of the multitude emotions that a facial expression is trying to convey. For a machine learning algorithm to be able to accurately do the same, is a remarkable feat of achievement.

The basic requirement of identifying faces for different reasons is the cause of why Face Recognition is used in various fields. Face Recognition is a very popular application that has great potential to be used in the fields of security, identification, automation and robotics [1, 2]. The success of this application though will largely depend on how trustworthy is the Face Recognition model at recognizing faces in practical scenarios. More often than not, there is a need for recognizing faces not from stable and high-resolution images but in fact from blurred sources that are susceptible to distortion. For instance, consider a law enforcement and criminal identification system that is expected to identify faces from available closed-circuit television (CCTV) footage and match them with the authority database. Another similar scenario involves identifying speeding drivers on a highway. The challenges faced by in such situations include but are not limited to: (a) varying lighting conditions, (b) poor resolution of input images and (c) failure of model to recognize faces in different angles [3]. Even skilled personnel or people who are acquainted with face under consideration may fail to identify the face correctly under the given circumstances of low-resolution and fluctuating images. Therefore, the demands and expectations of a machine learning algorithm to do the same are very high indeed.

One such advancement to overcome the aforementioned limitations in Face Recognition is the use of the Firefly Algorithm. The Firefly Algorithm is a unique algorithm which draws inspiration from the behavioural patterns of fireflies. This algorithm finds use in the feature selection process of Face Recognition models [4]. The process of reducing the set of features that are fed into the model is known as feature selection. The optimization of the feature selection process greatly benefits the entire process as it requires heavy computations on its own. The features used in Face Recognition are large in number and also heavy on an individual level. Through application of the Firefly Algorithm for Face Recognition, there has been a promising increase in accuracy of these models [4–7]. The Firefly Algorithm not only helps in improving the output of Face Recognition models, but also allows researchers to identify key features for image recognition tasks, which improves speed of the model along with its accuracy.

Along with other nature-inspired algorithms, the Firefly Algorithm has risen to prominence in the last few years, thanks to its applications in a broad diversity of fields. The most salient application of this algorithm is dealing with optimizations of various types such as discrete, chaotic, multi-objective and many more. Optimization in various fields is included in digital image compression, image processing, feature selection, fault detection, scheduling, and many others benefit from the Firefly Algorithm.

Feature selection is a crucial part of the image recognition model [8]. An attempt is made to extract features from an image, such that they can assist the recognition model and identify its target. But that may not be enough as it is still required to narrow down to a limited few features that have a greater influence over the recognition accuracy over the others. Feature selection is both challenging and rewarding. It is difficult to choose an optimal set of features that not only performs well under test conditions, but also in all other possible conditions. If somehow, a selection of a set of features is made, assigning them weights based on their impact towards maximizing the function (here, the Face Recognition accuracy) it can greatly improve the performance of the model [9]. This is where the Firefly Algorithm is introduced. The optimization of the feature selection process greatly benefits the entire process as it requires heavy computations on its own. The features used in Face Recognition are large in number and also heavy on an individual level. The flow of the chapter is organized to present an abstract view of related work followed by the Firefly Algorithm, the optimization technique and the Face Recognition system. The chapter ends with feature selection and performance evaluation of Firefly Algorithm.

2 Related Work

Genetic algorithms (GAs) are an adaptive search heuristic that form a subclass of evolutionary algorithms. They are inspired by a concept borrowed from the Darwinian theory of natural selection which in short is *survival of the fittest*. This concept states that species which adapt best to their environment and are capable of surviving and reproducing in it will be passed onto the next generation. Circa 1970s, John Holland was the first to apply the concepts to design genetic algorithms (GAs) to get a superlative approach for search and optimization problems.

The various steps followed in a genetic algorithm are as follows:

1. A set of individual solutions to a problem are passed through an encoding mechanism to represent them as a sequence of variables of a fixed length. This set of solutions is referred to as the initial population, and the variables represent the genes and their sequences chromosomes. The various encoding mechanisms that may be used are binary encoding, value encoding, permutation encoding and tree encoding.
2. A fitness function assigns a fitness score to every individual solution. The fitness score determines the probability of selection of that individual for reproduction to form the next generation.
3. To create a new generation, the following steps are performed:

Selection (Reproduction): Based on their fitness scores, the individuals are chosen as candidates for producing the next generation of individuals. The new generation will have characteristics of both parent individuals.

Crossover (Recombination): A random crossover point is chosen, and genes are shuffled till crossover point is reached to form the offspring and add it to the population. Singh et al. [10] propose three new crossover methods that aid in improving the quality of the solution obtained and even increase the rate of convergence to the optimal solution. The variant of real-coded genetic algorithms that is mentioned in the paper is derived from the concept of asexual reproduction in which only one parent can produce offspring by itself. Boltzmann's distribution (BD) enables the breaking free from local minima and even allows hill climbing in the algorithm. The metropolis algorithm tests the survival of the offspring before it can be passed on to the next generation.

Mutation: Mutation ensures diversity in a population. It involves the alteration of some of the genes to form a new chromosome. Although the occurrence of a mutation is governed by a low probability, it prevents any premature convergence [11].

4. Termination: For the algorithm to terminate, the new population obtained must be the same as the parent population in terms of its fitness for survival. The algorithm is said to converge, and the remaining final population is taken as the solution.

The Firefly Algorithm has recently gained momentum in its applications in machine learning, and as such the authors explore the works in the field relevant to the topic of interest. In [12], Xin-She Yang has extensively described the working of the Firefly Algorithm in its base form. The algorithm delineates upon the fundamental characteristics of the firefly behaviour, i.e. light intensity and attractiveness. While the basic Firefly Algorithm is efficient in itself, the authors observe several variations to it to further improve its performance. In [13], an in-depth taxonomy of the Firefly Algorithm variants and also a taxonomy of its applications are presented. The application of the Firefly Algorithm for Face Recognition falls in the class of optimization problems. In [14], Dey N. explores the application of the Firefly Algorithm for solving an optimization problem. Along with Xingshe He, Xin-She Yang explores the successful applications of the Firefly Algorithm in various fields [15]. The Firefly Algorithm has outperformed other algorithms such as the particle swarm optimization for optimization in parameter tuning [16] and velocity-smoothing kernels [17]. The optimistic results in several such optimization tasks have led to its application for feature selection. Based on results of Banati and Bajaj [18], the Firefly Algorithm performs better than other feature selection methods in terms of time and optimality. They incorporated the use of random set theory (RST) to find subsets of features. These works encourage the use of the Firefly Algorithm for optimizing the feature space for dimensionality reduction (Table 1).

Agarwal V. and Bhanot S. demonstrate that the use of Firefly Algorithm for Face Recognition has outperformed various other techniques [4]. They evaluate that the algorithm owes its success to its fast convergence within only 20 iterations. These results have been experimentally observed in two benchmarked ORL and Yale Face databases. It achieved a remarkable accuracy of above 94% in the Face Recognition task and also fared well in the dimensionality reduction of the feature space. In [19], they further applied the Firefly Algorithm along with the radial basis function neural network (RBFNN), to obtain centres of the hidden layer neurons for Face Recognition. Sánchez et al. [5] proposed the use of a modular neural network (MNN) designed using granular computing and the Firefly Algorithm for human recognition. Similar to Face Recognition, there have been applications of the Firefly Algorithm for facial emotion [6] and expression [7] recognition. In [7], a variant of the Firefly

Table 1   References and their role in contributing to the topic of interest

References	Contribution role
Yang [12]	Working of the FA
Fistera et al. [13], Yang and He [15]	Applications of FA
Dey [14], Samanta et al. [16], Chakraborty et al. [17]	Performance of FA for optimization problems
Banati and Bajaj [18]	Feature selection using FA
Agarwal and Bhanot [4]	FA for Face Recognition
Sánchez et al. [5]	MNN using FA for human recognition
Zhang et al. [6]	Emotion recognition using FA
Mistry et al. [7]	Expression recognition using FA
Agarwal and Bhanot [19]	RBFNN and FA for Face Recognition

Algorithm is used to get past the drawbacks of the feature selection approach from a global exploration perspective. Mutation operators, namely Gaussian, Cauchy and Levy, have been used to alleviate the consequences of a premature convergence. In each unique application, see a different variant of the Firefly Algorithm used, to cater to specific needs. The facial features extracted are extended to the Firefly Algorithm for dimensionality reduction using different distributions. The distributions such as Gaussian, Cauchy and Levy are applied so that the convergence of the algorithm does not occur too soon, and the search space is further searched to identify the best feature solution.

3   The Firefly Algorithm

Nature-inspired computing (NIC) is an advancing field in modern computer science using which computing techniques are developed by observing and delving into the complex problem-solving approaches used in natural phenomena. The modi operandi used by biological groups have inspired a series of computing algorithms which are used in fields like engineering, physics, economics and even management and given rise to advances in swarm intelligence, artificial immune systems and evolutionary computation among many others. For example, the collective intelligence demonstrated by ant colonies in finding the shortest distance path to a source of food has led researchers to develop ant colony routing (ACR) algorithms to optimize flow of traffic in a network by dynamically re-routing it.

The Firefly Algorithm (FA) an algorithm for mathematical optimization is one such design developed by Xin-She Yang in late 2007 and 2008. Fireflies (Lampyridae) are winged beetles that are known as lighting bugs in the science and classical literature alike for their flashing behaviour. Fireflies produce patterns of flashing light by a process known as bioluminescence. Biochemical reactions take place in tube-like organs in the firefly's abdomen. These light-producing organs, also known as lanterns, can produce regulated glows and discrete flashes in adult male fireflies. The light produced by a firefly can be used to distinguish between the various species and sex, as courting signals to attract female fireflies for mating and also to ward off prospective predators.

The Firefly Algorithm falls under the class of meta-heuristic algorithms. In the field of optimization, all algorithms belong to one of the two types, exact or heuristic. Exact algorithms are constructed in such a way that they are able to definitively find an optimal solution within a finite amount of time. The optimization of the problem set can be achieved in polynomial time as they are simple and easy to optimize. However, as the size of the optimization space increases, exact algorithms become slower and slower. There comes a stage when the exact algorithms become so slow that they are avoided. What is needed is a good optimum which may not be the best but is retrieved faster. A heuristic algorithm searches precisely in this way. It does not explore the entire search space and returns reasonably good optima in a short time. A meta-heuristic approach is not problem-specific or limited. Rather, it is a higher-level concept that provides a framework to create a heuristic optimization of an unforeseen problem.

3.1   Optimization Technique

The study of biological behaviour has inspired the exploration of their applications for optimization problems. This knowledge transferring process from natural occurrences to human applications opens up new methods to approach a problem [20]. Such algorithms, inspired by biological behaviours, have proven to be successful in achieving promising results in optimization tasks [21]. In this section, working of the Firefly Algorithm for optimizing a function will be explained by the authors (Fig. 1).

For use in optimization, three essential rules are borrowed from the firefly behaviour:

(i) The fireflies are unisex, and a given firefly may be attracted to any other firefly.
(ii) The attraction force of a firefly is directly proportional to the intensity of its brightness. The two are, in turn, inversely proportional to distance. A firefly that is relatively less brighter will attempt to move towards a firefly that is relatively more bright. Since the brightest firefly has no other brighter firefly to move towards, its motion will be random.
(iii) The intensity of brightness of a particular firefly is evaluated on the basis of the objective function at a particular point.

To embark on any optimization problem using the Firefly Algorithm, it is requisite to first define the brightness of a firefly as seen by the other fireflies. Based on variable distance, the brightness of the firefly can be represented as:

$$\beta = \beta_0 e^{-\gamma r^2}$$

Fig. 1 Firefly Algorithm flow chart

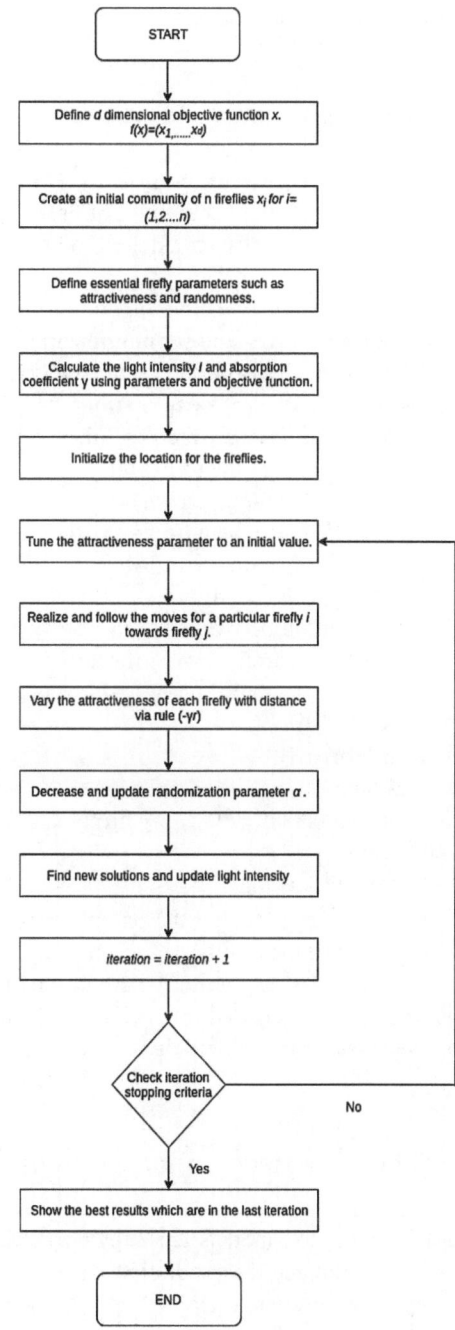

Here, r indicates how far apart the two fireflies are and β_0 indicates the distance when $r = 0$.

The Firefly Algorithm when used for optimization can be described in pseudocode form as shown below:

(i) Define objective function $f(a)$, $x = (a_1, a_2, \ldots a_d)$.

(ii) Create an initial community of fireflies a_i, $(i = 1, 2, 3, 4 \ldots n)$.

(iii) Define the firefly parameters such as randomness, attractiveness and absorption.

(iv) Modulate the objective function to fit light intensity I of fireflies. (For maximization, $f(a) \propto I$ and for minimization $f(a) \propto 1/I$).

(v) Define the absorption coefficient γ for controlling the speed of fireflies and density of search space exploration. A high γ value signifies slower firefly movement in feature space which will thereby result in a denser search for the optima.

(vi) Iterate through the following:

while $(t < MaxGeneration)$ **do**
　　for $i = 1 : n$(for the n fireflies) **do**
　　　　for j = 1 : i (n fireflies) **do**
　　　　　　if $(I_j < I_i)$ **then**
　　　　　　　　Alter attractiveness of the firefly with $e^{-\gamma}r$ as distance r varies.
　　　　　　　　Let the firefly move i towards j.
　　　　　　　　Assess new solutions and revise the light intensity.
　　　　　　end if
　　　　end for
　　end for
　　Associate a rank with every firefly and find the best currently.
end while

(vii) Post-processing the results and visualization.

It is worth to note that the parameter γ is a significant part in maximizing the effectiveness of the Firefly Algorithm and must therefore be selected with great care. A very large γ value could mean that the fireflies are moving too slowly to reach the optima in the desired time. Conversely, a very small γ value could result in the fireflies missing the optima due to their high speed [4].

4　The Face Recognition System

Face Recognition models are known to suffer from challenges like low illumination, occlusion and low resolution. These hurdles intensify when to aim to use Face Recognition in commercial set-ups. As in practical application, the model should be prepared for a lot of distortions and unfavourable noise in the image. The use of Face Recognition is crucial to applications which include surveillance, authentication/control of access and photograph retrieval.

The human brain has an excellent capacity to remember and recognize faces in a lifetime. There are also certain forensic experts that make crucial identity decisions. According to a recent study at the University of Huddersfield [22], it was found that humans and latest algorithms for Face Recognition gave similar accuracy when making decisions for identification. Hence, newer algorithms are constantly being developed that can outperform the human brain. These advanced models for Face Recognition aspire to outperform humans in recognizing faces in real time from low-clarity video sources. Human errors in such situations have often led to serious implications during person identification and fraud detection.

Face Recognition works based on the specific alignment of particular features of an individual relative to other features. The technology exploits the fact that this structure corresponds to a unique ID. Every face has several distinguishable landmarks with distinct crests and troughs that are characteristic facial features of that particular person. The measurement of these facial features makes up the data and is stored as face prints in the database which are used in future for comparison.

4.1 Steps Involved in Face Recognition

The classic steps for performing Face Recognition are [23]:

(i) Face detection comprises extracting the face from the image background. The complexity of this lies in the variation of certain attributes of the human face such as pose, position, expression, presence of glasses or facial hair. There may also be variation in picture attributes such as image resolution, lighting conditions and differences in camera gain.

(ii) Face normalization is a critical step where the extracted features must be normalized so that they are not dependent on position, scale and rotation of the face in the image plane.

(iii) Feature extraction is the process where unique features of the face are extracted and isolated for later comparison. Among all, detecting the position of the eye is crucial because all other features are extracted relative to this.

(iv) Feature selection includes the selection of an optimum subset of the extracted features, which eliminates unnecessary extra features which are unimportant.

(v) Final step is an identification task which encompasses a comparison method. This includes an algorithm for classification and a measure of accuracy (Fig. 2).

For Face Recognition and pattern recognition in general, Multilayered Neurons Systems combined with backpropogation have been widely used. Yet, this system raises two concerns which are: (1) the process as a whole is computationally expensive due to its slow convergence rate and (2) there is no assurance that the global minima will be reached.

Fig. 2 Face Recognition
system design

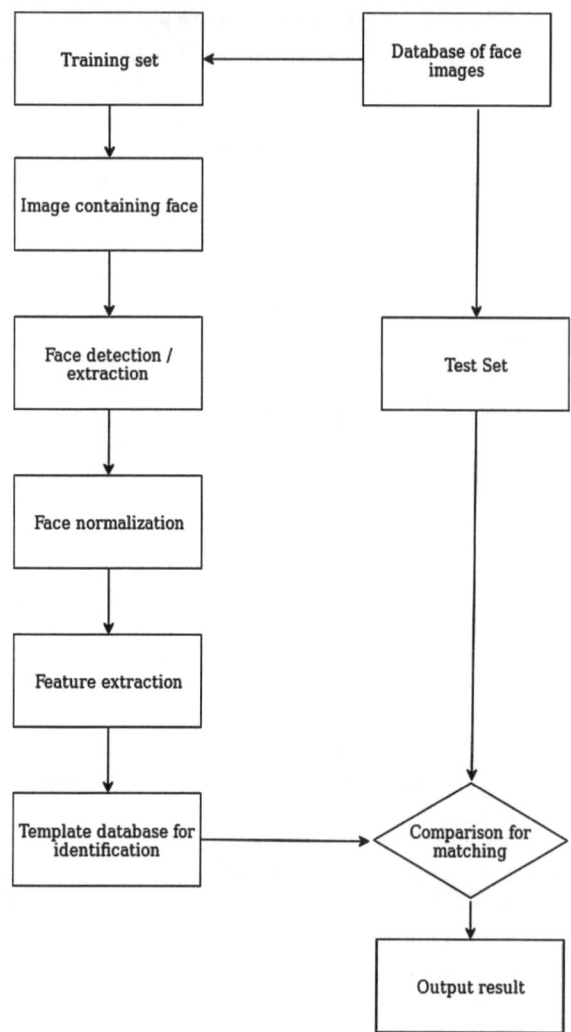

Current research focuses on tackling such challenges and improving the quality of
performance of the Face Recognition systems. In this regard, the Firefly Algorithm
has emerged as a potential solution.

4.2 Radial Basis Function Neural Network for Face Recognition

In the discussed design for the Face Recognition system, the Firefly Algorithm is used on radial basis function neural network (RBFNN) for better results [24]. The RBFNN is an artificial feed-forward neural network that finds application in clustering data points that are nonlinearly separable in smaller dimensions. RBFNN is an improvement of the traditional single-/multi-layer perceptron that is based on Cover's theorem which states that a given nonlinearly separable training set can be made linearly separable with a high probability when modified under a nonlinear transformation to a higher-dimensional space. The RBFNN used in the Face Recognition system has three layers, namely the n-dimensional input vector that is to be classified, one hidden layer which uses the Gaussian activation function and the output layer having nodes for each possible classification. The Gaussian activation function has the following definition:

$$\phi(x) = e^{-||x-C_j{}^2||/2\sigma^2}$$

Here, x is the vector given as input, C_j is a prototype or central measure of the jth RBF unit, and σ_j is the spread of the radial basis function which determines its width.

The Firefly Algorithm is used to create sub-clusters of images having the same identity but subsidiary variations. The fireflies move in the multidimensional hyperspace and help in deciding the number and centres for the aforementioned RBF neurons. RBF neural networks are generally handy in pattern identification and function approximation systems, wherein the input dimensionality is relatively small, width = 12.5cm. The recent yet extensive applications of RBF neural networks within the field of pattern recognition can be attributed to the following properties of these networks:

(i) They are universal approximators. For a wide variety of inputs, the RBF will produce a closely approximated output.
(ii) It possesses the best approximation property.
(iii) Since the neurons are tuned to adapt locally, their learning speed is very fast.
(iv) Their topology is more compact relative to other neural networks.

The deft convergence of the Firefly Algorithm when combined with the learning capabilities of the RBFNN outperforms the other models in high-speed and accurate Face Recognition.

5   Feature Selection and Extraction

In machine learning and data mining, the problem of redundant and noisy data often directly impacts the prediction accuracy of the algorithm. Which is why feature selection and extraction are at the core in this regard. Feature selection involves selecting features that have a relevant contribution in the prediction accuracy without any modifications on the selected features. Feature extraction is a slightly different approach where at the start, initial data is used and then built up to a set that contains the most discriminatory features available. Using feature extraction, one can modify pre-existing features and train the model on derived ones. Both methods are of help in the sense that they reduce over-fitting and training time, thereby improvising upon the accuracy of the model. Usually in the context of facial recognition, the system must work with features that are not only large in number but also difficult to process even individually. Noisy data drastically degrades that quality of performance in Face Recognition. Therefore, reducing the high-dimensionality feature space to a lower dimension becomes a must.

5.1   Conventional Methods for Feature Selection

In the past, feature selection has been performed using various other techniques some of which include the eigenvector-based principal component analysis (PCA), discrete cosine transformation (DCT) [25] and discrete wavelet transformation (DWT) using Gabor and Haar wavelets [4]. These methods are flawed in their own ways when used for facial recognition. For example, PCA-selected features perform poorly when the fed input of the same face is subjected to variation in illumination. In general, DCT introduces block boundaries and this can be an issue with Face Recognition.

5.1.1   Principal Component Analysis

This method aims to find to variables (components) which are responsible for maximum possible variation in the original data set [26]. Eigenface approach makes use of principal component analysis which aids in decreasing the size and dimensionality of a huge data set. Eigenfaces are primarily eigenvectors that attempt to capture variance for an image collection and then utilize the information in order to perform encoding of images of individual faces in a method characterized by a relationship between individual components. The intention behind the encoding and decoding of face images is to retrieve information which highlight the crucial and distinctive image features. These features could be quite random, possibly not relating to facial features as well. They mainly contain relevant data characteristic of a particular face which is then compared with faces having similarly encoded data. Extracting information content of a face corresponds to capturing the variation in collection of face images. Each face image can be approximated using the eigenface that has the greatest eigenvalues, thus accounting for maximum difference between the given data sets of face images.

The downside of using this method is that it is very sensitive to scale and requires an additional amount of normalization. Although this approach is shown to be credible when dealing with expressions and glasses, problems arise when there are significant changes in the pose as well as the expression.

5.1.2   Linear Discriminant Analysis

This is a method similar to principal component analysis which inculcates the process of dimensionality reduction before classification. This method makes use of fact that each image of a face is represented by a large number of pixel values and each new dimension is formed by linearly combining these values which is used as a standard prototype [27]. The linear combinations are called Fisher's faces. The working of this process can be represented as:

(i)   The images in the data set should contain mainly the face region and include different examples of subjects with small alterations in frontal angles and expressions. The test set should also contain an example of the subject.

(ii) Next, a vector is constructed from the initial two-dimensional array of pixel intensity that corresponds to the initial configuration of the face. The set of the faces is represented as a vector of a high dimension.

(iii) Clustering is now performed by representing all specimens of a person's face in one class, while the faces of different subjects (people) are represented by different classes. This is done for all the subjects in the training set. This leads to the formation of a framework, and different classes are subjected to separation analysis in the feature space.

Despite showing good results in recognition accuracy, the LDA projections may not preserve the complex structure needed in the data needed for classification in cases where the information for discrimination is not in the present mean but the variance of the data.

5.1.3   Discrete Cosine Transformation

The main idea behind using this method for feature extraction is to minimize the number of selected coefficients once it has been applied on an image [28]. A coefficient matrix is obtained on applying discrete cosine transformation to a two-dimensional image matrix. The next step is coefficient selection which defines the accuracy of the algorithm. Care needs to be taken while selecting the coefficients and also while finalizing their quantity. A static approach acts as a template while selecting the coefficients, and predefined patterns are followed. This approach has its downsides when it comes to obtaining a higher accuracy. A more superior approach would be to use a data-dependent approach in which the position of the coefficients is chosen based on an acute analysis of the data in the application in which it is being used. Using this approach, the recognition accuracy is highly improved. The eventual goal of data-dependent analysis is to choose coefficients which will be able to differentiate between various individual faces that have been subjected to all kinds of variance. To do this, a specific set of desired properties for the coefficients are defined. Next, the discriminant power of coefficient needs to be processed. This is done using frequency bands, and coefficients are searched locally as opposed to globally because all the coefficients that comprise of the band have an impact on the discriminant value. The bands with large values of the discriminant function are chosen and used to generate a mask which is employed to construct vectors representing features for the training and testing set of the images.

5.1.4 Discrete Wavelet Transformation

Wavelets are primarily functions that, after meeting some mathematical requirements, represent data at different scales or resolutions. The motivation of using wavelet transforms in Face Recognition systems comes from the fact that it represents spatial and frequency domain simultaneously. Since they can capture localized time–frequency information of images, the wavelet coefficients are used as features. In an image, the low-frequency content provides an overview with a few details which is more like a global description. Meanwhile, the content with higher frequency provides a more detail-oriented view with finer characteristics of the image such as edges. The wavelet coefficients can be defined as functions in terms of scale and position. Discrete wavelet transformation involves splitting the signal into new versions of the original wavelet that are scaled and shifted. The next step is to make certain approximations which are low-frequency components of the system which provide maximum variance. Filtering is incorporated to acquire the coefficients where the image is sub-sampled into distinct sub-bands that is estimated using the frequency and spatial components [29]. To establish a threshold value for finalizing the features, focus is shifted towards the sub-band. The low-frequency components that are found in both the vertical and the horizontal orientations are represented by the sub-band.

5.2 Feature Selection Using the Firefly Algorithm

Upon discussing the limitations of the previously used algorithms for feature selection, it can be observed that there is scope for a better and more robust algorithm. In spite of several advances in facial recognition systems in the domain of artificial intelligence, the task of selecting discriminating features that have enough significance to correctly capture the variance in facial expressions remains difficult. Problems also arise when classic feature reduction techniques fail to minimize the features to a significant extent. Many of these might not even contribute to creating a framework for the face and eventually result in over-fitting of the training data. Feature selection problems are represented as a search problem in a d-dimensional hyperspace. Herein, the total number of features that need to be reduced before building the template framework of a face is represented by d. A heuristic approach is adopted to find the best possible combination of features, and the results are evaluated with the help of a fitness function that answers the question of how close the solution is to the optimum solution for the desired problem. Similarly in [30], Kanan H. and Faez K. have made use of another nature-inspired algorithm, ant colony optimization (ACO) for feature selection on Face Recognition system. Downsides of the previously used evolutionary algorithms like the particle swarm optimization (PSO) and the genetic algorithm (GA) are that the convergence rate is slow and it does not guarantee a global optimal solution [4]. Recent work has seen promising results in the form of a flexible algorithm for feature selection which is the evolutionary Firefly Algorithm. The intuition behind Firefly Algorithm is that the firefly will move towards a firefly that has higher intensity of brightness caused by the attractiveness of the latter. The attractiveness of a firefly is dependant on two parameters: its own brightness and the distance between the brighter firefly and itself.

In [13], the Firefly Algorithm is coupled with the rough set-based attribute reduction (RSAR) approach for feature selection. The authors use constructs such as reducts and their approximations in this method, including dynamic reducts, derived from rough set approach [31], in order to perform feature selection. When evaluated against other pairings of the RSAR with algorithms like the particle swarm algorithm, ant colony optimization and others, the firefly-based reduction is seen to outperform the other nature-inspired algorithms. Not only does the Firefly Algorithm produce the minimal feature subset but it also conquers another flip side by generating this subset in the least time among all algorithms without compromising on the quality of feature selection.

Suppose that feature extraction problems consist of n features which are considered for dimensionality reduction, then in mathematical terms, a firefly is represented as an n-dimensional point moving in an n-dimensional search space. During the training phase of face images, a feature vector is created and the most intuitively obvious features of the face are included in it. But of all these, some may not provide enough variance to have a considerable influence on the strength of the algorithm and need to be eliminated. This is where the Firefly Algorithm comes into the picture. The first step is to generate fireflies randomly in the feature space. A firefly in the n-dimensional space is represented as a vector $\langle p_1, p_2, p_3 \ldots, p_n \rangle$. This vector is nothing but the position of the firefly. The value of p_i where $i \in [1, n]$ is generated randomly for every firefly. Every firefly is in association with a value indicating fitness that corresponds to its assessment power in the given domain. A set of features is computed from an n-dimensional firefly and is represented as S_f. These features may or may not have the most discriminating power.

The algorithm commences with a firefly having a lesser fitness value moving towards a firefly having a higher fitness value. This entire process of the fireflies acquiring new positions indicates the algorithm's ability to rank alternative solutions at each branching step based on available information to decide which branch to follow and hence avoid exhaustive search. When fireflies are pushed in direction towards brighter fireflies, they explore the path towards the brighter firefly and are confined to the boundaries of the search space. This movement of fireflies continues and has the potential to scan the search space expeditiously to find multiple optimal solutions. Hence, the convergence rate is superior to other algorithms like the genetic algorithm (GA) or the ant colony optimization (ACO).

Figure 3 illustrates the working of the Firefly Algorithm for feature selection in a Face Recognition system.

The features in the Face Recognition system are defined as S_f. Then, a fitness value G^f for each firefly is computed as:

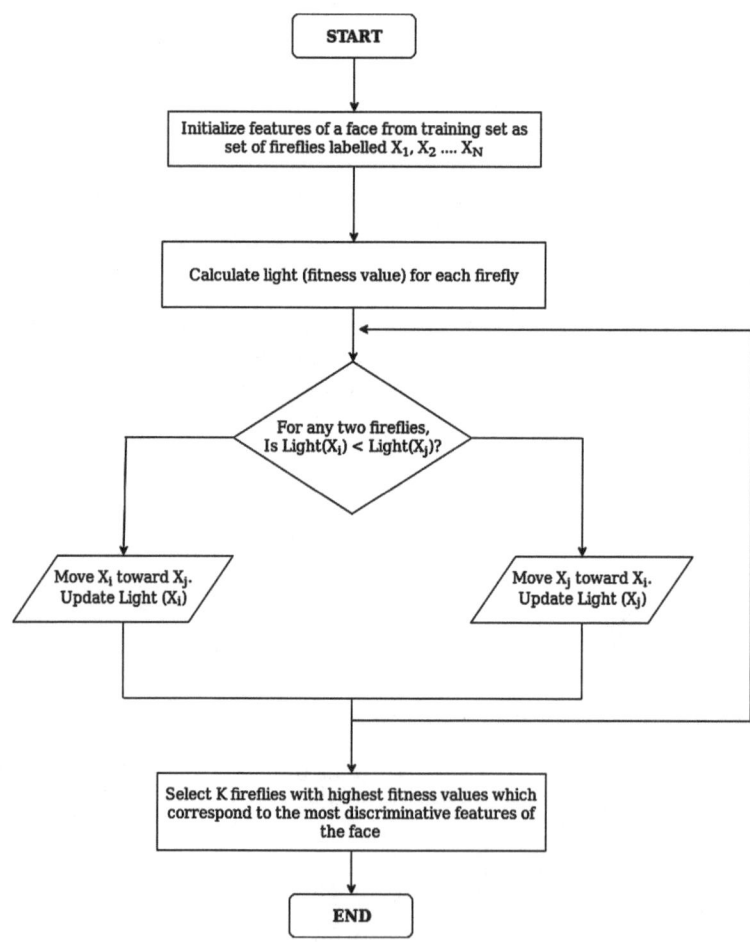

Fig. 3 Firefly Algorithm for Face Recognition

$$G^f = \frac{N_c}{N_t} \times 100$$

where N_c represents the correctly recognized face images and N_t represents the test images.

Higher fitness values correspond to higher recognition accuracy for a set of selected features. Therefore a firefly with higher fitness value is brighter as compared to the ones with lower recognition accuracy.

As illustrated in Fig. 4, the fireflies (the black markers) are initialized by randomly placing them in the feature space (where A, B and C are potential feature clusters). During initialization, the fireflies are not aware of the underlying feature space distribution. This ensures even distribution and lets the firefly converge as it is learning on the training data. In Fig. 5, it is observed how the fireflies have converged to form three distinct clusters. This allows converging to the most distinct features that have maximum impact in identifying the objective target (which here are the facial features). This convergence happens over a number of iterations, which terminates when eventually the fireflies converge to such distinct feature subsets.

To perform feature selection with the help of the Firefly Algorithm, it is requisite to first define a firefly and feature vector as follows:

(i) A firefly is a point with d dimensions in a d-dimensional hyperspace where the total number of features that must undergo dimensionality reduction is given by d. A firefly is mathematically represented as a vector $p = <p_1, p_2 \ldots p_d>$ where $p_i \in [1, d]$. The vector signifies the location of the firefly in the feature space.

(ii) A feature vector $V = <f_1, f_2, \ldots f_d>$ is a vector collection of the relevant features only which have a significant contribution to the prediction accuracy.

Fig. 4 Fireflies randomly initialized in feature space

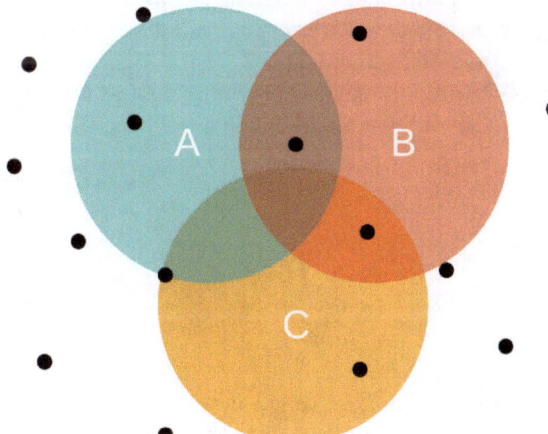

Fig. 5 Convergence of the
fireflies into feature subsets

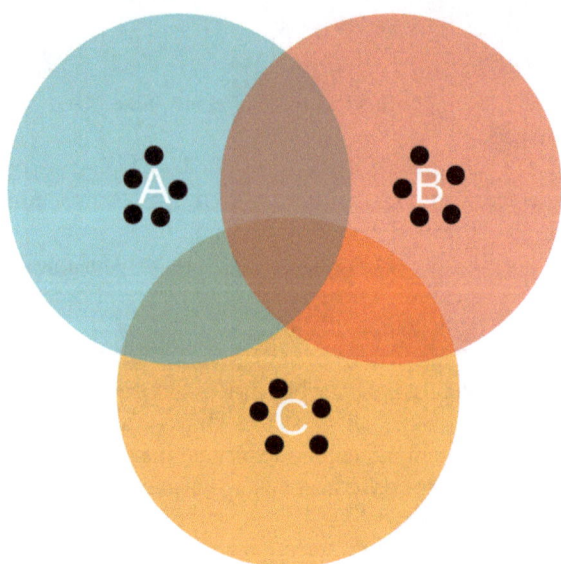

(iii) The features that are selected for the feature vector are evaluated based on the
positions of the fireflies in the feature space which is d-dimensional. Further,
each dimension expands into a range from 1 to d.

(iv) Initially, an ideal number Q of fireflies is chosen to cover the entire feature
space. These fireflies are similar in all respects except their position which is
generated randomly for all fireflies by assigning arbitrary values to each p_i.

(v) Following this, each value of p_i is generated using the formula: $p_i = M(i, 1) +
[M(i, 2) - M(i, 1)] * \gamma$ where γ is an arbitrarily chosen number from the stan-
dard uniform distribution on the open interval $(0,1)$ where $M(i, 1)$ and $M(i, 2)$
represent the smallest and greatest values of $i^t h$ feature spread across the vectors used for training.

(vi) The fireflies move about within the feature space M which can be represented
as a $d \times 2$ matrix described as follows:

$$M = \begin{vmatrix} minf_1 & maxf_1 \\ minf_2 & maxf_2 \\ .. & .. \\ .. & .. \\ .. & .. \\ minf_d & maxf_d \end{vmatrix}$$

(vii) Another attribute called fitness is associated with every firefly in the Face
Recognition system. Fitness is a quantitative measure for the importance of
the firefly in the given problem.

(viii) Now, all the necessary information to compute the set of features, S_k, is present. It is now integrated into the feature vector formula as:

$$S_k = f_q k | q_k = \text{round}(p_k),\ p_k \neq p_m,\ k \neq m,\ p_m \in F$$

The feature set S_k is selected only by a particular firefly, and it may not represent the optimal solution. Using this approach, multiple optima can also be obtained from different fireflies. The migration of the fireflies in realm of the feature space is a very important aspect of the algorithm in feature selection as many measurable quantities like the speed of optimization and rate of convergence depend upon it. The time-consuming comprehensive search of the feature space is avoided due to the fireflies' heuristic movement within the feature space, learning with every new visit. The search space is efficiently minimized. The fireflies participate in two types of movements: (1) heuristic shift and (2) randomized shift.

Heuristic shift depends on the attractiveness of the fireflies under consideration. The firefly which has a lesser intensity of brightness is directed towards the brighter firefly through the heuristic shift as the firefly which is bright is also the more prominent one with a higher fitness value. The path between them must be explored for better results.

The random search allows a firefly to navigate its local neighbourhood and try out nearby positions attempting to optimize the feature selection results. A firefly is restricted to move within the boundary of the feature space by changing its position and brightness. If a firefly ventures out of bounds, its new position is set on the boundary of the feature space.

A metric is required to measure the success demonstrated by the model when it is designed using the features selected by a particular firefly S_k and the accuracy that this model shows. This metric is represented by the fitness function. The fitness function is associated with a firefly F given by G^F, and it is calculated as:

$$G^F = (N_c / N_t) \times 100$$

where N_c is an indication of the total number of correctly classified images while N_t represents the total number of images in the data set.

The proposed algorithm for feature selection (FIFS) as given in [4] is elaborated below:

FIFS(V, Q, γ, β_0, M):
Input $V = (f_1, f_2, \dots f_d)$, γ and $\beta_0 = 1.0$
GenerateRandomFireflies(M,Q,F)
for iteration=1:N **do**
  for for u=1:Q **do**
    Calculate G^{Fu}
    for v=1:Q **do**
      Calculate G^{Fv}
      if $G^{Fu} < G^{Fv}$ **then**

$locVector = |F_u - F_v|$

$r = ||F_u - F_v||$

$\beta = \beta_0(^-\gamma * r^2)$

$D = zeros(d)$

for i=1:d **do**

   $RandomShift = \alpha(i) * (rand - 0.5)$

   $HeuristicShift = \beta * locVec(i)$

   $D(i) = RandomShift + HeuristicShift$

end for

$newLoc = F_u + D$

for k= 1:d **do**

   if $newLoc(k) > M(k, 2)$ **then**

      $newLoc(k) = M(k, 2)$

   end if

   if $newLoc(k) < M(K, 1)$ **then**

      $newLoc(k) = M(1, 2)$

   end if

end for

   $F_u = newLoc$

end if

end for

end for

end for

$F_{opt} = Max(G^{Fi})$

Output = Optimal set of features

As it can be observed, a firefly with a higher fitness function value is the one that best identifies the most discriminatory and concise feature set. As a result, this firefly also has a higher value of brightness and, therefore, even attractiveness when used in the algorithm. This established relationship between the various parameters of the Firefly Algorithm helps the algorithm learn rapidly as compared to its previous counterparts which is also the reason behind its widespread acknowledgement and success.

6 Performance Evaluation

The Firefly Algorithm has consistently outperformed other meta-heuristic algorithms for optimization problems [32]. The performance of the Firefly Algorithm as validated by Agarwal and Banot in [4], using benchmark ORL and Yale databases of images of faces, achieved an average recognition accuracy of 94.375% and 99.16%, respectively. The process of training involves randomly initializing the position of fireflies with a feature set and eventually reaches optima as they evolve over successive generations. It was observed that a small value such as 0.000001, for the coefficient of light absorption (γ), used to calculate the measure of attractiveness of a firefly, produced a maximum average recognition accuracy of 94.375% with a 43.10% decrease in dimensionality. In the training phase, 20 fireflies were used and it was noted that all converged at the end of 20 iterations for both Yale and ORL databases. It was also noted that larger values of gamma (larger than 0.00001) affected the dimensionality reduction percentage in a negative direction.

Another comparative study of nature-inspired algorithms with the rough set-based attribute reduction (RSAR) in [13] shows how the Firefly Algorithm outperforms the RSAR method, entropy-based reduction (EBR), ant colony-based reduction (AntR-SAR) and genetic-based reduction (GenRSR). It performs on a par with particle swarm optimization (PSO-RSAR) and bee colony-based reduction (BeeRSAR). The Firefly Algorithm extended with rough set-based attribute reduction can be used as an efficient solution for problems that are NP-hard. Selection of features was validated on four distinct medical data sets from the UCI repository. In addition to a minimal feature set, the FA-RSAR also takes less time as compared to the algorithms on a par with it in feature set reduction. This is why it outperforms the other algorithms.

6.1 Limitations of the Firefly Algorithm

Although not specific to its applications in Face Recognition, the standard Firefly Algorithm has limitations of its own. Firstly, in [33] it is mentioned that the Firefly Algorithm has a tendency to get trapped in local minima and faces difficulty in avoiding local minima during a local search. As a result of this, it is required to make modifications to the algorithm to ensure that the solution which is returned is optimal and not premature. Also, the parameters of the Firefly Algorithm are fixed before the algorithm starts operating on the data. This can be a downside as a new iteration must be implemented altogether if any change in a parameter is to be accommodated.

Another disadvantage of the Firefly Algorithm as mentioned in [33] is that it does not possess any memory. The previous history of a firefly's movement is not remembered by the algorithm, and due to this the fireflies move independently of the old position even though they might be better. In [34], the Firefly Algorithm is criticized for its performance in high-dimensional and nonlinear problem. In such cases, it claims that the Firefly Algorithm works as a single meta-heuristic algorithm that takes time to attain a solution that is optimal. The Firefly Algorithm meta-heuristic must be combined for it to overcome its drawbacks.

Contrary to the aforesaid drawbacks, in the case of Firefly Algorithm, the limitations do not hamper the popularity of the algorithm. The drawbacks have been recognized and are being worked upon for better results in the various applications.

7 Future Scope

While there have already been several successful applications of the Firefly Algorithm for Face Recognition, there remains further scope for improvements. The Firefly Algorithm can be applied along with various distributions and hybrid combinations. It can even be tuned to target a multi-objective function. This not only allows the algorithm to become more versatile, but also could provide better results across a range of tasks. Further, the research works cited have been performed on databases containing majorly clean and noise-free images. The algorithms need to be tested further with noise images, without which it may fail to perform well in practice. Exploring the variations of the Firefly Algorithm in order to attain an ideal solution should be considered, as it is important to find an algorithm that performs well in all general conditions. Adopting an ensemble of Firefly Algorithms may be taken into consideration, as that would be inclusive of several variations, such that an average best-case output is presented. There could be further research done in finding the optimal set of parameters using fine-tuning.

The noise sensitivity of the technique has not been examined yet, but it should be tested under different levels of signal-to-noise ratio (SNR) in order to effectively test the model for exposure to varying noise levels. This is likely to be an issue for the Face Recognition models as in practical application, there is always going to be some amount of noise as the face image is captured.

Along similar lines, the use of the Firefly Algorithm for other recognition tasks can also be explored. It could be extended to use in hand gesture recognition, pose estimation and object recognition tasks. The optimality with which the Firefly Algorithm derives a set of reduced features can be applied to similar tasks where there is need to converge to a set of determinant features. These tasks depend heavily upon finding a reliable set of condensed features that can have an accurate general inference in identifying the objective target.

Usually, a model is restricted by the available training data and tends to perform relatively poorly on new data, which it may find difficulty in recognizing. This may overcome by augmenting input face images and introducing random noise to distort the images. This adversary can help the model prepare for probable practical conditions where the image may be heavily distorted. Image augmentation can have a significant improvement in the Face Recognition accuracy, as presented in [35]. Hence, the Firefly Algorithm should be trained on augmented images for potentially improving its Face Recognition accuracy.

As far as nature-inspired and evolutionary algorithms go, there is always scope for improvement as there are many biological phenomena present as inspiration for these algorithms. Another state-of-the-art evolutionary optimizer as mentioned in [36] is based on Mendel's theory of evolution which is one of the most prime theories in genetics. Mendel's theory examines how nature optimizes the transfer of hereditary information.

8 Conclusion

In this chapter, the authors have presented an extensive analysis and assessment of the application of the Firefly Algorithm for Face Recognition. The movement of fireflies towards brighter ones has provided an inspiration for solving problems which require an optimal solution. It has been explained how this property of fireflies is exploited and has been used in Face Recognition technology to find the most distinguishing characteristic features of a person's face. It was demonstrated how the Firefly Algorithm aims at finding the most powerful subset of a reduced number of features from a larger subset during the training stage to reduce the time taken during computation in the recognition phase of face detection. The disadvantages of other conventional methods have been discussed, and how the Firefly Algorithm overcomes them has been explored extensively. The Firefly Algorithm converges faster and provides a better output of feature subsets.

Furthermore, the authors have explained the working of the Firefly Algorithm in detail for optimization and feature selection through pseudocode algorithms and flow chart illustrations. Overall, the Firefly Algorithm performed considerably well in the Face Recognition application, providing encouraging results. It even performs better than other nature-inspired algorithms for Face Recognition. While there certainly is scope for improvements in future, the Firefly Algorithm has already proved to be quite useful. The application of the Firefly Algorithm has thus demonstrated itself as a promising solution for improving the accuracy and driving future research work in Face Recognition systems.

References

1. Huang T, Xiong Z, Zhang Z (2011) Face recognition applications. In: Li S, Jain A (eds) Handbook of face recognition. Springer, London
2. Liu JNK, Wang M, Feng B (2005) iBotGuard: an internet-based intelligent robot security system using invariant face recognition against intruder. IEEE Trans Syst Man Cybern Part C (Appl Rev) 35(1):97–105
3. Zhao W, Chellappa R, Phillips PJ, Rosenfeld A (2003) Face recognition: a literature survey. ACM Comput Surv (CSUR) (Surveys Homepage Archive) 35(4):399–458
4. Agarwal V, Bhanot S (2015) Firefly inspired feature selection for face recognition. In: 2015 eighth international conference on contemporary computing (IC3)
5. Sánchez D, Melin P, Castillo O (2017) Optimization of modular granular neural networks using a firefly algorithm for human recognition. Eng Appl Artif Intell 64:172–186
6. Zhang L, Mistry K, Neoh SC, Lim CP (2016) Intelligent facial emotion recognition using moth-firefly optimization. Knowl-Based Syst 111(1):248–267
7. Mistry K, Zhang L, Sexton G, Zeng Y, He M (2017) Facial expression recongnition using firefly-based feature optimization. In: IEEE congress on evolutionary computation (CEC). San Sebastian, pp 1652–1658
8. Guyon I, Gunn S, Nikravesh M, Zadeh LA (2006) Feature extraction, foundations and applications. Series studies in fuzziness and soft computing. Physica-Verlag, Springer
9. Hong Z-Q (1991) Algebraic feature extraction of image for recognition. Patt Recogn 24(3):211–219

10. Singh G, Gupta N, Khosravy M (2015) New crossover operators for real coded genetic algorithm (RCGA). In: 2015 international conference on intelligent informatics and biomedical sciences (ICIIBMS), Okinawa. IEEE, pp 135–140
11. Gupta N, Patel NV, Tiwari BN, Khosravy M Genetic algorithm based on enhanced selection and log-scaled mutation technique. In: Proceedings of the future technologies conference, pp 730–774
12. Yang X-S (2010) Nature-inspired metaheuristic algorithms. Luniver Press
13. Fistera I, Fister I Jr, Yang X-S, Brest J (2013) A comprehensive review of firefly algorithms. Swarm Evol Comput 13:34–46
14. Dey N (2017) Advancements in applied metaheuristic computing. IGI Global
15. Yang X-S, He X (2013) Firefly algorithm: recent advances and applications. Int J Swarm Intell (IJSI) 1(1)
16. Samanta S, Mukherjee A, Ashour AS, Dey N, Tavares J, Abdessalem Karâa WB, Taiar R, Azar A, Hassanien AE (2018) Log transform based optimal image enhancement using firefly algorithm for autonomous mini unmanned aerial vehicle: an application of aerial photography. Int J Image Graph 18:1850019
17. Chakraborty S, Dey N, Samanta S, Ashour AS, Balas VE (2016) Firefly algorithm for optimized nonrigid demons registration. In: Bio-inspired computation and applications in image processing. Academic Press, pp 221–237
18. Banati H, Bajaj M (2011) Fire fly based feature selection approach. (IJCSI) Int J Comput Sci Issues 8(4) (No 2)
19. Agarwal V, Bhanot S (2018) Radial basis function neural network-based face recognition using firefly algorithm. Neural Comput Appl 30(8):2643–2660
20. Zang H, Zhang S, Hapeshi K (2010) A review of nature-inspired algorithms. J Bionic Eng 7(Supplement):S232–S237
21. Yang X-S (2009) Firefly algorithms for multimodal optimization. In: International symposium on stochastic algorithms SAGA 2009: stochastic algorithms: foundations and applications, pp 169–178
22. Phillips PJ et al (2018) Face recognition accuracy of forensic examiners, superrecognizers, and face recognition algorithms. PNAS 115(24):6171–6176 (first published May 29, 2018)
23. Beham MP, Roomi M (2013) A review of face recognition methods. Int J Patt Recogn Artif Intell 27(04):1356005
24. Er MJ, Wu S, Lu J, Toh HL (2002) Face recognition with radial basis function (RBF) neural networks. IEEE Trans Neural Netw 13(3):697–710
25. Zhao W, Krishnaswamy A, Chellappa R, Sts DL, Ng J (1998) Discriminant analysis of principal components for face recognition. In: Chsler H, Phillips PJ, Bruce V, SouliÃl' FF, Huang TS (eds) Face recognition. NATO ASI series (Series F: computer and systems sciences), vol 163. Springer, Berlin
26. Singh A, Kumar S (2012) Face recognition using PCA and eigen face approach
27. Bala M, Singh P, Meena MS (2016) Face recognition using linear discriminant analysis. Int J Electr Electron Res 4(2):96–103. ISSN 2348-6988 (online)
28. Dabbaghchian S, Aghagolzadeh A, Moin MS (2007) Feature extraction using discrete cosine transform for face recognition. In: 9th international symposium on signal processing and its applications
29. Vidya V, Farheena N, Manikantana K, Ramachandran S (2012) Face recognition using threshold based DWT feature extraction and selective illumination enhancement technique. In: 2nd international conference on communication, computing & security (ICCCS-2012)
30. Rashidy H, Faez KK (2008) An improved feature selection method based on ant colony optimization (ACO) evaluated on face recognition system. Appl Mathe Comput 205(2):716–725
31. Swiniarskia RW, Skowronb A (2003) Rough set methods in feature selection and recognition. Patt Recogn Lett 24(6):833–849
32. Gupta D, Gupta M (2016) A new modified firefly algorithm. Int J Recent Contrib Eng Sci & IT (iJES) 4(2)

33. Pal SK, Rai CS, Singh AP (2012) Comparative study of firefly algorithm and particle swarm optimization for noisy nonlinear optimization problems. I.J. Intell Syst Appl 10:50–57
34. Ali N, Othman MA, Husain MN, Misran MH (2014) A review of firefly algorithm. ARPN J Eng Appl Sci 9(10)
35. Lv J-J, Shao X-H, Huang J-S, Zhoub X-D, Zhou X (2017) Data augmentation for face recognition. Neurocomputing 230(22):184–196
36. Gupta N, Khosravy M, Patel N, Sethi I (2018) Evolutionary optimization based on biological evolution in plants. Procedia Comput Sci 126:146–155. https://doi.org/10.1016/j.procs.2018.07.218

Chapter 8
Application of Chaos-Based Firefly Algorithm Optimized Controller for Automatic Generation Control of Two Area Interconnected Power System with Energy Storage Unit and UPFC

K. Jagatheesan, B. Anand, Soumadip Sen and Sourav Samanta

1 Introduction

For keeping power system generated electric power frequency and power flow via tie-line within the predetermined value of power generating system, load frequency control (LFC) plays a crucial role. The changes in system frequency and power flow can be controlled by adjusting the mechanical input power of generator for maintaining stability power system [1–3]. Over the past few decades, varies the level of control strategies have been implemented for load frequency control of the power system by researchers. H∞ theory-based controller and GA tuned proportional–integral (PI) controller are equipped into LFC of the robust decentralized power system in [4]. The adaptive control method is implemented into the interconnected power system for solving LFC issue with variations in power system parameters [5]. The LFC of two areas interconnected power system is introduced with fuzzy gain scheduled PI

K. Jagatheesan (✉)
Department of Electrical and Electronics Engineering, Paavai Engineering College,
Namakkal, Tamil Nadu, India
e-mail: jaga.ksr@gmail.com

B. Anand
Department of Electronics and Instrumentation Engineering,
Hindusthan College of Engineering and Technology, Coimbatore, Tamil Nadu, India
e-mail: b_anand_eee@yahoo.co.in

S. Sen · S. Samanta
Department of Computer Science & Engineering, University Institute of Technology, BU,
Burdwan, West Bengal, India
e-mail: soumadip.95@gmail.com

S. Samanta
e-mail: sourav.uit@gmail.com

© Springer Nature Singapore Pte Ltd. 2020
N. Dey (ed.), *Applications of Firefly Algorithm and its Variants*,
Springer Tracts in Nature-Inspired Computing,
https://doi.org/10.1007/978-981-15-0306-1_8

controller in [6]. The fuzzy logic PI controller is designed by using a tabu search algorithm to overcome the LFC crisis of interconnected two area power system [7]. Hybrid PSO (HPSO) algorithm is implemented for tuning of PI and I (Integral) controllers in LFC of two area power system [8]. The multistage fuzzy controller is designed by using PSO, and it is implemented for LFC of the interconnected power system [9]. PSO technique tuned PID controller is designed and discussed if LFC interconnected power system [10]. The AGC of the interconnected power system is discussed with considering GBD nonlinearity and hybrid neuro-fuzzy approach [11]. Two area interconnected reheat thermal power system is designed by employing the GA optimized PI controller [12]. The automatic LFC (ALFC) of two area interconnected power system is designed by implementing the fuzzy logic controller (FLC) in [13]. Adaptive load saddling scheme is implemented to overcome system frequency decline during the emergency condition in the power system [14]. The PID controller gain values are tuned by applying various heuristic stochastic techniques for Sugeno fuzzy logic controller [15]. Sine–cosine algorithm optimized PIDD^2 controller is implemented in a multi-area multi-source power system to solve the AGC crisis, including geothermal power plant (GTPP) and distribution generation (DG) unit. The result is evident that GTPP unit improves performance of the system with minimal settling time, peak over and undershoot [16] during sudden load disturbance. Sine–cosine algorithm (SCA) tuned fractional order controllers are implemented into interconnected thermal power analyzing the performance of different renewable energy sources and FACTS devices for solving AGC crisis in power system [17]. PID controller performance is analyzed in LFC of power system with static synchronous series compensator and CES unit in multi-area multi-source interconnected power system with HVDC link. During performance analysis of PSO and gravitational search algorithm (GSA) with integral time absolute error (ITAE) cost function is considered. Finally, simulation reveals that GSA yields better performance over PSO and static synchronous series control (SSSC), and capacitive energy storage (CES) improves performance [18]. Five area interconnected thermal power system is analyzed with tilt integral derivative controller with filler is designed by applying differential evolution algorithm considering ITAE cost function and UPFC unit in area 5 alone to get better performance of the system [19].

PID controller gain values are optimized by applying firefly algorithm, and it is equipped to overcome LFC issue of interconnected non-reheat thermal power system. The efficiency of the proposed technique is verified by comparing GA, PSO, differential evolution (DE) algorithm, bacteria foraging optimization algorithm (BFOA), hBFOA-PSO and simulated annealing (SA) technique tuned PID controller response under same criterion [20]. Automatic generation control of multi-area multi-source interconnected power system is analyzed by equipping FA optimized PID controller with the presence of UPFC and superconducting magnetic energy storage (SMES) device in investigated power system and performance is compared with DE, GSA and GA technique tuned controller performance in same analyzed power system [21, 22]. GSA algorithm optimized PID controller is equipped in four areas interconnected power system including doubly fed induction generator (DFIG) wind turbine by varying load disturbance from 0.1 p.u. MW to 0.4 p.u. MW load demand [23]. From the above literature survey, obviously evident that several optimization methods are implemented to solve AGC/LFC of interconnected power system to tune the gain values of controller gain value and system parameters like ant colony optimization technique [24], flower pollination algorithm [25], firefly algorithm [26], stochastic particle swarm optimization technique [27] and bat algorithm [28].

The major aim of this proposed chapter is that CFFA is implemented for tuning of PID controller parameters with considering nonlinearity effect, energy storage unit, and unified power flow controller is connected parallel with tie-line. The performance of the tuning method is verified by comparatively applying GA, PSO and FFA algorithm tuned PID controller performance. During performance analysis, the effect of HAE and UPFC response is analyzed by considering without considering these units during sudden load demand.

The proposed chapter is organized as follows: Sect. 1 gives a literature review related to optimization of controller gain values in LFC/AGC of interconnected power system; in the chapter, Sect. 2 discusses about system under investigation and study; Sect. 3 delivers the detail about the proposed optimization algorithm and optimized gain values for different cases, based on the optimized gain values in Sect. 3 is applied into investigated power system. The performance of the proposed tuning method is verified by comparatively applying GA, PSO and FFA technique that is discussed with different cases in Sect. 4, and finally, conclusion about proposed work is depicted in the conclusion section.

2 System Studied

The transfer function model of two areas interconnected power system is shown in Fig. 1, and it is designed by using MATLAB software for analysis purpose. In this model, area 1 and area 2 are thermal power system. The two thermal areas are interconnected via tie-line. The investigated power system is comprised with the PID controller as a secondary controller to regulate system parameters (frequency and tie-line power flow) during step load perturbation in area 1. Each thermal area comprises reheater, turbine, governor, speed governor and generator unit. During nominal loading condition, each takes care of its own load demand and keeps system parameters which are within the predetermined limit. Sudden load demand happens in any interconnected power system; it will share the power via the tie-line for maintaining system stability. In this analysis error in system frequency and tie-line power flow between connected systems combine together, it is named as an area control error (ACE), and it is given as an input to the secondary controller. In this work, considered secondary PID controller generates the required control signal based on the error input signal. The nominal parameter value of the investigated interconnected thermal power system is given in Table 1 [29, 30].

The power generated by the thermal power plant is controlled within a specified maximum limit by equipping GRC nonlinearity into the investigated power system. The value of GRC 2–5 per min is considered for the thermal system as a nominal

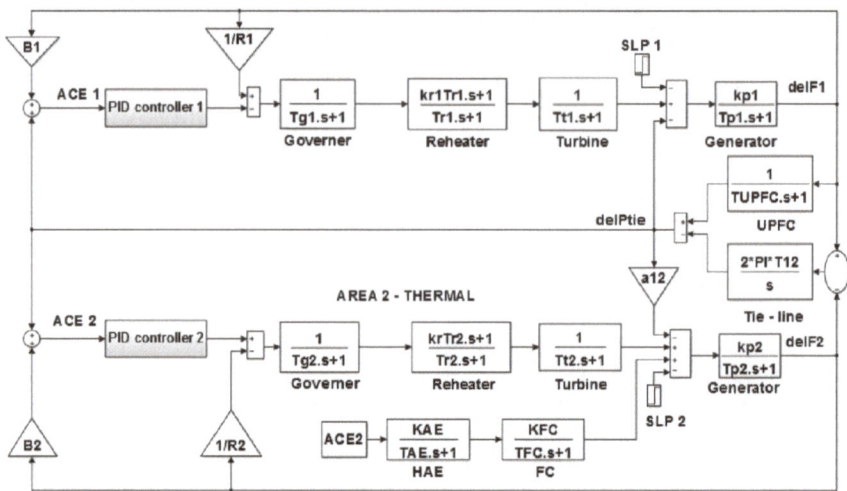

Fig. 1 Transfer function model of investigated power system under the MATLAB/Simulink environment

Table 1 Nominal parameters of the investigated power system [26–30]

System parameters	Nominal values
Generator gain constant of area 1 and 2 ($K_{p1} = K_{p2}$)	120 Hz/p.u. MW
Generator time constant of area 1 and 2 ($T_{p1} = T_{p2}$)	20 s
Speed regulator of area 1 and 2 ($R_1 = R_2$)	2.4 Hz/p.u. MW
Governor time constant of area 1 and 2 ($T_{g1} = T_{g2}$)	0.2 s
Tie-line coefficient (T_{12})	0.0707 p.u.
Area capacity ratio (a_{12})	−1
Reheater constant of area 1 and 2 ($K_{r1} = K_{r2}$)	0.333
Reheater time constant of area 1 and 2 ($T_{r1} = T_{r2}$)	10 s
Frequency bias constant of area 1 and 2 ($B_1 = B_2$)	0.425 p.u. MW/Hz
Turbine time constant of area 1 and 2 ($T_{t1} = T_{t2}$)	0.3 s
Gain constant of aqua electrolyzer (K_{AE})	0.002
Time constant of aqua electrolyzer (T_{AE})	0.5 s
Time constant of UPFC (T_{UPFC})	0.01 s
Gain constant of fuel cell (K_{FC})	0.01
Time constant of fuel cell (T_{FC})	4 s

value. The governor dead band (GDB) nonlinearity is taken into the account of the analyzed power system for analyzing the dynamic performance of an electric power generating system. The GDB nonlinearity is defined as the total amount of a continuous speed change within which there is no change in the position of the valve.

In this work, the apparent steady-state speed regulation of system linearity is increased by 0.06% by adding GDB nonlinearity. In this work, 0.06% is considered as a maximum limit value of GDB nonlinearity. For this present investigation, 3% min and 0.036 Hz value are considered for GRC and GDB nonlinearity nominal values, respectively.

In energy carrier, hydrogen is one of the promising alternative fuels. The major essential components of hydrogen energy storage unit comprise electrolyzer unit with the help of this hydrogen energy conversion system; it converts stored chemical energy in the hydrogen bank which is converted into electric energy by the support of fuel cell stored hydrogen energy which converts into electric energy [29, 30]. The aqua electrolyzer transfer function is given as follows:

$$G_{AE}(s) = \frac{K_{AE}}{1 + sT_{AE}} \tag{1}$$

The energy conversion system converts the chemical energy of the fuel directly into electric energy by the support of fuel cell. The major advantage of FC is pollution free and high efficiency. This is recently playing a major role in the distributed power system [29, 30]. The following simple linear transfer function equation of FC is given by

$$G_{FC}(s) = \frac{K_{FC}}{1 + sT_{FC}} \tag{2}$$

In recent days, power electronics-based FACTS device plays a major role in power system applications [29, 30]. By implementing FACTS family based devices in power system improves system responses and it gives flexible power control. In this work, the UPFC device is considered to control power flow in the transmission line, transient stability, reducing system oscillations and voltage support. In this work, UPFC is connected in series to tie-line, and it reduces damping oscillations in system response. The transfer function for UPFC-based controller for LFC is represented in the following equation:

$$G_{UPFC} = \frac{1}{1 + sT_{UPFC}} \tag{3}$$

The values of ACE and U are given in the following expression [26–28]:

$$ACE_i = B_i.\Delta F_i + \Delta P_{tie\,i,j} \tag{4}$$

and control signal generated by the PID controller in each area is depicted as follows:

$$U_{\text{PID}} = u_i(t) = K_{\text{P}i}.\text{ACE}_i + K_{\text{I}i} \int_0^t \text{ACE}_i dt + K_{\text{D}i} \frac{d\text{ACE}_i}{dt} \tag{5}$$

In this work, integral time absolute error (ITAE) cost function is considered to tune gain values of controller gain values [26]:

$$\text{ITAE} = \int_0^{t_{\text{sim}}} t.|e(t)|dt \tag{6}$$

In the current work, proposed CFFA is implemented to tune PID controller parameters in LFC of interconnected two area thermal power system. The controller gain values are optimized simultaneously. The constraints for optimized controller gain for this analyzed work is subject to as follows:

$$K_{\text{P}i}^{\min} \leq K_{\text{P}i} \leq K_{\text{P}i}^{\max} \tag{7}$$

$$K_{\text{I}i}^{\min} \leq K_{\text{I}i} \leq K_{\text{I}i}^{\max} \tag{8}$$

$$K_{\text{D}i}^{\min} \leq K_{\text{D}i} \leq K_{\text{D}i}^{\max} \tag{9}$$

where, $K_{\text{P}i}$, $K_{\text{I}i}$ and $K_{\text{D}i}$ are proportional, integral and derivative controller gain values, respectively. The minimum and maximum values of controller gain values are considered to be 0 and 1, respectively. The gain values of controllers are optimized by using CFFA depending upon error signal given feedback to the controller. The gain values are optimized by applying step load perturbations in area 1 of interconnected two area thermal power system.

3   Chaos-Based Firefly Algorithm

The behaviors of fireflies are inspired by researchers and developed the firefly algorithm (FFA) [31–34]. FFA is a bio-inspired metaheuristic search algorithm, and it does not declare the global optimal solution that can ever be attained; such global optimality can be generating in many cases. Few forms of stochastic components are used by metaheuristic [35–40]. Over the few millions of years, its powers come from the efforts for following the features of nature and biological systems specially developed by natural selection of the system. In this proposed work, chaos-based firefly algorithm is implemented to tune controller gain values in the interconnected power system [35–37]. Chaos in natural science represents the unpredictable behavior of the deterministic system. Mathematically, it does not show the system with a complete absence of order. The random manner-based optimization algorithm is used as chaotic variables on behalf of random manner-based variables. So, it is called a chaos-based optimization algorithm.

Concept of chaos can be included in normal FFA in two ways, such as in one way replacing some random distributive parameters with a chaotic map to improve the performance of the system. In another method, structure of intrinsic firefly is implemented to the parameters by utilizing a chaotic map. There are three parameters to control the behavior of FFA, such as randomized move of step size (α), attractiveness (β) and absorption coefficient (λ). The implementation details of CFFA for this work is given in the following section.

3.1 Implementation of Chaos-Based Firefly Algorithm in the Proposed Work

With the increasing drawback of obtaining local optima, there has been a shift toward using global optimization algorithms to effectively tackle and deal with the problem arising from local optima [41]. As a result, metaheuristic procedures, which are efficient global optimization methods, became widespread [41]. Firefly algorithm is a popular nature-inspired, metaheuristic optimization algorithm, developed by Yang in 2008 [31, 32] by mimicking the natural characteristic and social behavior of fireflies. This algorithm, which falls under the category of swarm intelligence, has shown stunning performance concerning optimization problems [33, 34, 42, 43].

Over the years, this popular algorithm has obtained various variants of itself which are equally efficient and effective in solving complex optimization problems from diverse areas [34]. Chaotic firefly algorithm can be considered as an enhanced upgraded version of the firefly algorithm after augmenting the concept of chaos into it for better result [41, 44]. In general, chaos refers to randomness or unpredictability [44]. Therefore, when a system behaves in an unpredicted manner due to the incorporation of randomness into it, then we can connect it to chaos. Chaotic algorithms generally use chaotic variables instead of random variables while dealing with optimization problems. The introduction of chaos or randomness is quite vital since they create sufficient impact on exploration and exploitation in search procedures [44], which is also known as diversification and intensification. They play a prime role in determining the search process, thereby exploring the possible set of new solutions. This helps them to speed up the search procedure, thereby gaining the upper hand over the stochastic methods that mostly depends on probabilities [41, 45]. Chaotic algorithms use chaotic sets or maps which projects or maps the set of input values to the set of output values, to improve the overall performance [44]. These algorithms are highly adaptive since they can quickly adapt themselves to the various fitness landscapes of diverse problems [44].

This proposed work is aimed at implementing chaos-based FFA to optimize parameters of PID controller in two area interconnected thermal power systems with nonlinearity and energy storage unit. The essential steps of the proposed CFFA [31–34] can be summarized by using subsequent pseudo code:

---

begin

Define the objective function $f(x)$: $x = \left(kp_n, ki_n, kd_n\right)^T$

$n=1,2$

  Generate initial population of fireflies y_i, ($i = 1,2.....15$)

  Light intensity **Ipidval**$_i$ at x_i is determined by $f(x_i)$ (Eq. 1.6).

  Define light absorption coefficient γ

  while (t < Max Generation)

      Tune the of attraction parameter β using Gaussian chaotic maps

        for $i = 1$ to s all s fireflies

         for $j = 1$ to i all s fireflies

          if Ipidval$_i$ < Ipidval$_j$

           Move firefly i towards j;

         end if

         Attractiveness varies with distance r_{ij} via $exp\left[-\gamma \times r_{ij}\right]$

        Evaluate new solutions and update light intensity

        end for j

        end for i

    Rank the fireflies and find current best.

  end while

    Post process on the best so far results and visualization

end

Proposed chaos-based FFA algorithm is considered and implemented for automatic generation control of two area interconnected thermal power system with energy storage unit and UPFC. The controller gain values are optimized by using GA, PSO, FFA and CFFA for three different cases that are shown in Tables 2, 3 and 4.

4 Results and Discussions

CFFA is applied to tune the controller parameters to keep better AGC behavior in the interconnected power system. Each control areas are equipped with a separate PID controller. The simulation and comparative performance analysis are made under

Table 2 PID controller gain values for case I

Controller gain values		Case I			
		GA-PID	PSO-PID	FFA-PID	CFFA-PID
K_p	K_{p1}	0.858864	0.9924	0.924788	0.9999
	K_{p2}	0.925162	0.9607	0.859708	0.9999
K_i	K_{i1}	0.976263	0.9797	0.999997	0.9999
	K_{i2}	0.811675	0.9312	0.811675	0.9999
K_d	K_{d1}	0.199156	0.4389	0.11024	0.035614
	K_{d2}	0.33417	0.6125	0.297011	0.9995
Fitness value	J	0.387433	0.3848	0.376957	0.35674

Table 3 PID controller gain values for case II

Controller gain values		Case II			
		GA-PID	PSO-PID	FFA-PID	CFFA-PID
K_p	K_{p1}	0.9999	0.9866	0.9998	0.99583
	K_{p2}	0.2951	0.7654	0.8102	0.9999
K_i	K_{i1}	0.3389	0.3581	0.3735	0.72966
	K_{i2}	0.7385	0.8536	0.7353	0.75775
K_d	K_{d1}	0.555	0.3372	0.5371	0.36609
	K_{d2}	0.8358	0.6438	0.1562	0.71399
Fitness value	J	1.8757	1.695	1.6569	1.25

Table 4 PID controller gain values for case III

Controller gain values		Case III			
		GA-PID	PSO-PID	FFA-PID	CFFA-PID
K_p	K_{p1}	0.5971	0.9984	0.9343	0.9988
	K_{p2}	0.3578	0.5333	0.5919	0.9988
K_i	K_{i1}	0.288	0.378	0.5922	0.81874
	K_{i2}	0.7246	0.785	0.5132	0.72936
K_d	K_{d1}	0.3083	0.1356	0.4067	0.53293
	K_{d2}	0.2365	0.7184	0.5313	0.98549
Fitness value	J	2.2952	1.6497	1.4797	1.2069

MATLAB Simulink environment. The upper and lower limits of controller parameters are 0 and 1, respectively, for all investigations for all optimization techniques (GA, PSO, FFA and CFFA). Because the higher gain values of controller gain affect the stability, so the system leads instability situation as per control theory view [2]. For all simulation analysis, delP₁ = 0.01 MW is applied. The study was carried out by the following three cases for verifying the ability of the CFFA optimization technique.

Case 1: Performance analysis of CFFA-PID with GA-PID, PSO-PID and FFA-PID controller
Case 2: Analysis of CFFA-PID in interconnected nonlinear power system
Case 3: Investigation of CFFA-PID controller in the interconnected system with UPFC and HAE energy storage unit

Case 1: Performance analysis of CFFA-PID with GA-PID, PSO-PID and FFA-PID controller

The effectiveness of the proposed CFFA tuned PID controller is examined by applying 1% SLP in area 1 of two area interconnected power system. Also, the effectiveness of CFFA tuned controller performance is comparatively compared with GA, PSO and FFA tuned PID controller under the same scenario. Figures 2 and 3 show deviations in area 1 and 2 frequency deviations; Fig. 4 represents deviations in tie-line power flow between area 1 and area 2.

The numerical values of settling time of different technique tuned PID controller equipped system performances are tabulated in Table 5.

By observing deviations in frequency, tie-line power flow in Figs. 2, 3 and 4 and numerical values in Table 5 of time domain specification parameter reveal that proposed CFFA-PID controller settled quicker over to GA, PSO and FFA-PID controller response.

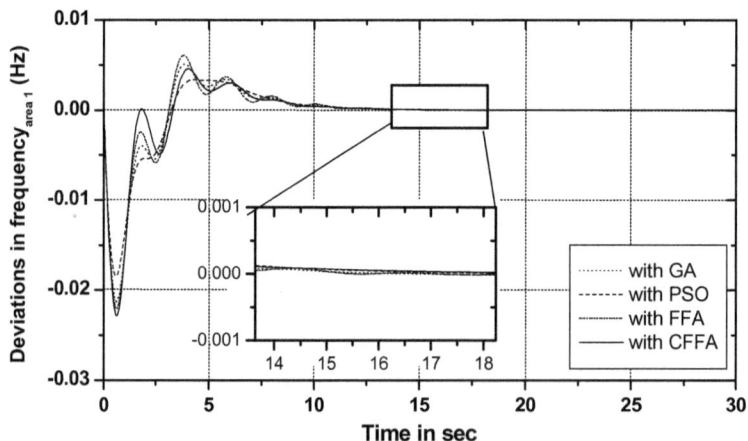

Fig. 2 Area 1 frequency deviations with different techniques tuned controller

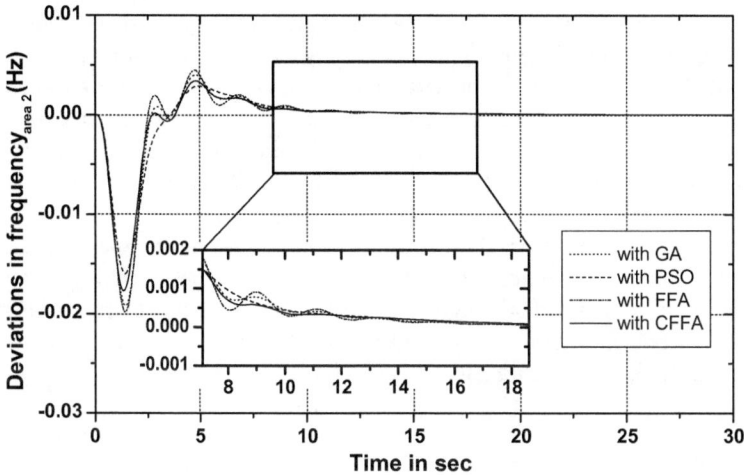

Fig. 3 Area 2 frequency deviations with different techniques tuned controller

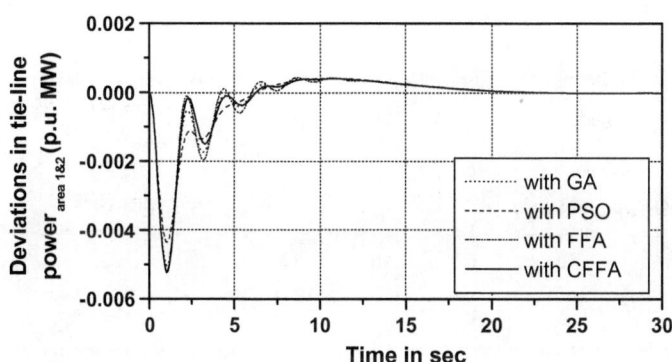

Fig. 4 Tie-line power flow deviation between area 1 and 2 with different techniques tuned controller

Table 5 Comparison of settling time in case I

Response	Settling time (s)			
	GA	PSO	FFA	CFFA
delF1	15.4	14.5	13.5	12.5
delF2	17	16	15	14.5
delPTie	22	21	20	18

Case 2: Analysis of CFFA-PID in interconnected nonlinear power system

The effectiveness of CFFA tuning is analyzed by adding nonlinear elements into the investigated power system. In this case, GRC and GDB nonlinear components are considered in area 1 and area 2. The performance (frequency and tie-line power flow) comparisons of the system by considering nonlinearity and various technique tuned controller are given in Figs. 5, 6 and 7. The settling time numerical value is depicted in Table 6.

From the system response, Figs. 5, 6 and 7 and settling time numerical values are depicted in Table 6. It is evident that the proposed optimization technique improves the response with minimal settling time when nonlinear elements are considered in the analyzed power system during sudden load disturbance compared to GA, PSO and FFA technique tuned PID controller response.

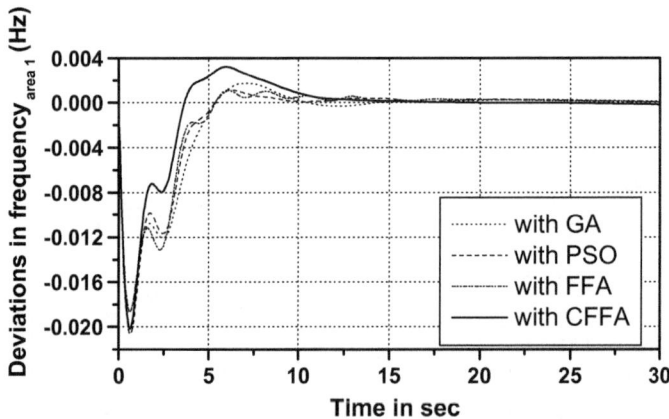

Fig. 5 Deviations in the frequency of area 1 response comparisons by considering GRC and GDB

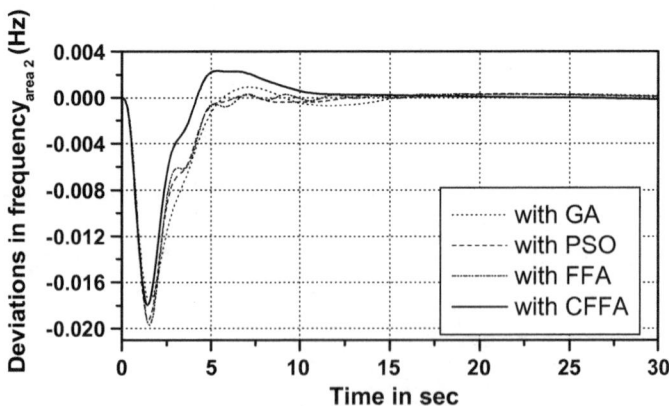

Fig. 6 Deviations in the frequency of area 2 response comparisons by considering GRC and GDB

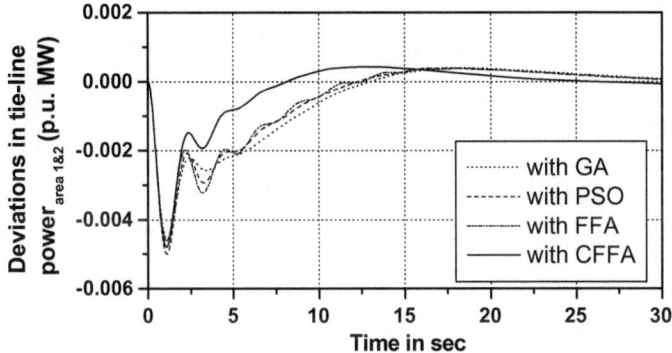

Fig. 7 Deviations in tie-line power flow deviations between area 1 and area 2 response comparisons by considering GRC & GDB

Table 6 Comparison of settling time in case II

Response	Settling time (s)			
	GA	PSO	FFA	CFFA
delF1	17.5	16.5	16	14
delF2	19	16	17	12
delPTie	30	30	30	24

Case 3: Investigation of CFFA-PID controller in the interconnected system with UPFC and HAE energy storage unit

In this case, the performance of PID controller is analyzed by tuning of CFFA optimization technique by a system equipped with UPFC and HAE unit for improving the response of system during sudden load demand condition. The system frequency deviations comparison of area 1 and 2, deviations in tie-line power flow are shown in Figs. 8, 9 and 10.

For case III, the numerical values of settling time in system responses are tabulated in Table 7.

By analyzing the above system response, numerical values reveal that CFFA technique effectively tuned the controller gain values-based controller and energy storage unit improves system response with minimal settling time compared to another optimization technique tuned controller performance (Figs. 11, 12 and 13).

From the response analysis of GA, PSO, FFA and CFFA tuned PID controller performance in interconnected two area thermal power system, numerical values of time domain specification parameters evident the proposed CFFA tuned PID controller give a more superior performance with minimal settling time during sudden load disturbance conditions.

The future work is further extended by increasing size of system and adding renewable energy sources. In addition, the effectiveness and robustness of proposed

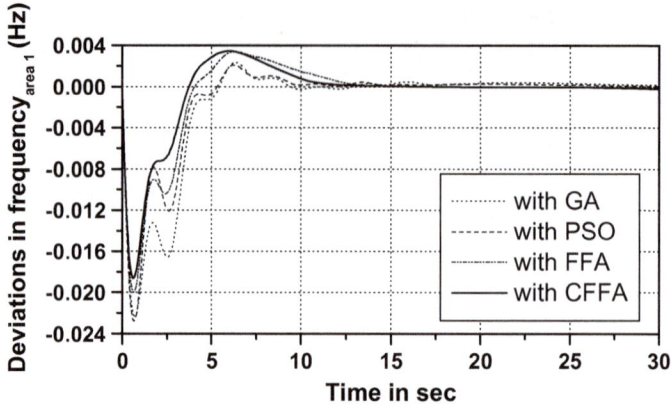

Fig. 8 Response comparison of a delF1 system equipped with UPFC and HAE unit

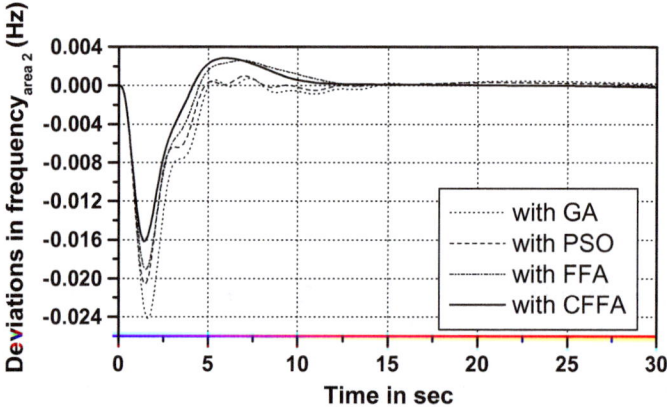

Fig. 9 Response comparison of a delF2 system equipped with UPFC and HAE unit

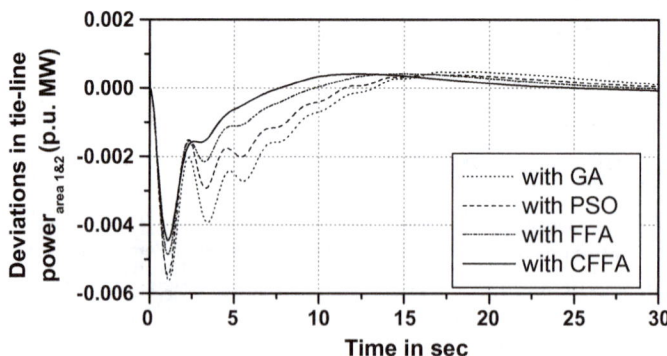

Fig. 10 Response comparison of a delPtie1and2 system equipped with UPFC and HAE unit

Table 7 Comparison of settling time in case III

Response	Settling time (s)			
	GA	PSO	FFA	CFFA
delF1	30	17	14.5	13
delF2	30	15	13.5	12
delPTie	30	29	28	24

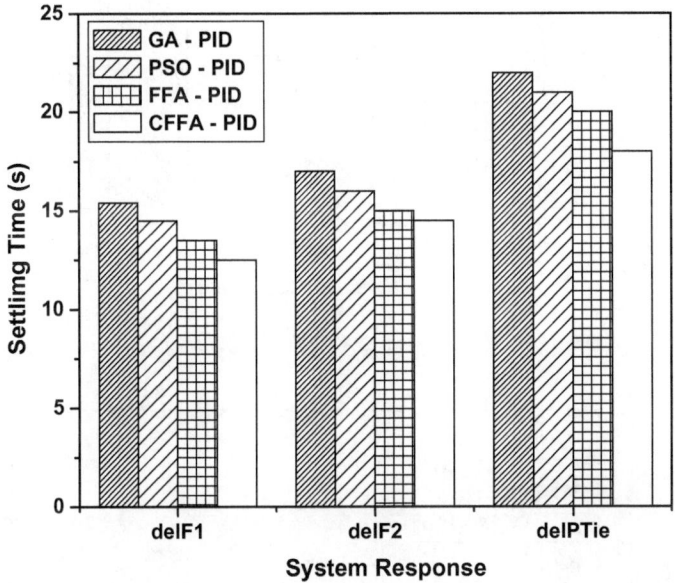

Fig. 11 Comparisons of settling time with GA, PSO, FFA and CFFA tuned PID controller in Case—I

research further verified by changing load disturbance patterns, variations in system parameters and considering nonlinearity components in the investigated power system.

5 Conclusion

In this work, an attempt is made to tune and design PID controller for automatic generation control in the interconnected thermal power system. The PID controller gain values are optimized by implementing chaos-based firefly algorithm. The supremacy of the proposed technique is examined by comparatively comparing the performance with GA-PID, PSO-PID and FFA-PID controller response in the analyzed power system. In addition, proposed approach is extended into the system equipped

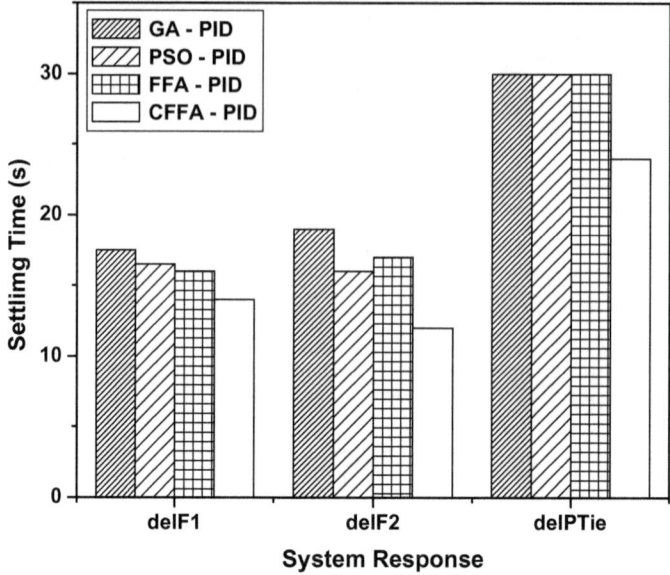

Fig. 12 Comparisons of settling time with GA, PSO, FFA and CFFA tuned PID controller in Case—II

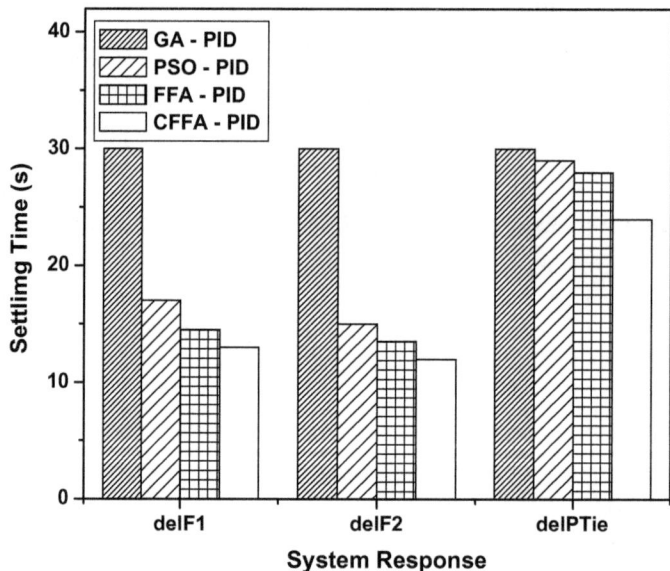

Fig. 13 Comparisons of settling time with GA, PSO, FFA and CFFA tuned PID controller in Case—III

with non-linearities for evaluating controller performance and energy storage unit for improving system response. From the simulation result analysis, it is obviously proved that projected CFFA-PID controller yields minimum settling time comparatively compared with other tuning methods (GA, PSO and FFA) by applying step load disturbance in area 1. In addition, controller gain values are effectively tuned when nonlinear components are included in the system. Further, the system response is effectively improved by considering HAE unit in area 2 of the interconnected thermal power system. Finally, it is observed that proposed optimization effectively tuned the controller gain value during normal operating condition, when nonlinear components are added into the system and energy storage unit comprised into power system effectively.

References

1. Kundur P, Balu NJ, Lauby MG (1994) Power system stability and control, vol 7. McGraw-Hill, New York
2. Yang TC, Cimen H, Zhu QM (1998) Decentralised load-frequency controller design based on structured singular values. IEE Gener Transm Distrib 145(1):7–14
3. Kundur P (1994) Power system stability and control. McGraw-Hill
4. Rerkpreedapong D, Hasanovic A, Feliachi A (2003) Robust load frequency control using genetic algorithms and linear matrix inequalities. IEEE Trans Power Syst 18(2):855–861
5. Zribi M, Al-Rashed M, Alrifai M (2005) Adaptive decentralized load frequency control of multi-area power systems. Int J Electr Power Energy Syst 27(8):575–583
6. Kocaarslan I, Çam E (2005) Fuzzy logic controller in interconnected electrical power systems for load-frequency control. Int J Electr Power Energy Syst 27(8):542–549
7. Pothiya S, Ngamroo I, Runggeratigul S, Tantaswadi P (2006) Design of optimal fuzzy logic based PI controller using multiple tabu search algorithm for load frequency control. Int J Control Autom Syst 4(2):155–164
8. Taher SA, Hematti R, Abdolalipour A, Tabei SH (2008) Optimal decentralized load frequency control using HPSO algorithms in deregulated power systems. Am J Appl Sci 5(9):1167–1174
9. Shayeghi H, Jalili A, Shayanfar HA (2008) Multi-stage fuzzy load frequency control using PSO. Energy Convers Manag 49(10):2570–2580
10. Sabahi K, Sharifi A, Sh MA, Teshnehlab M, Alisghary M (2008) Load frequency control in interconnected power system using multi-objective PID controller. J Appl Sci 8(20):3676–3682
11. Panda G, Panda S, Ardil C (2009) Automatic generation control of interconnected power system with generation rate constraints by hybrid neuro fuzzy approach. Int J Electr Power Energy Syst Eng 2(1):13–18
12. Ramesh S, Krishnan A (2009) Modified genetic algorithm based load frequency controller for interconnected power system. Int J Electr Power Eng 3(1):26–30
13. Aravindan P, Sanavullah MY (2009) Fuzzy logic based automatic load frequency control of two area power system with GRC. Int J Comput Intell Res 5(1):37–45
14. Ford JJ, Bevrani H, Ledwich G (2009) Adaptive load shedding and regional protection. Int J Electr Power Energy Syst 31(10):611–618
15. Roy R, Bhatt P, Ghoshal SP (2010) Evolutionary computation based three-area automatic generation control. Expert Syst Appl 37(8):5913–5924
16. Shankar R (2018) Sine-Cosine algorithm based PIDD 2 controller design for AGC of a multisource power system incorporating GTPP in DG unit. In: 2018 international conference on computational and characterization techniques in Engineering & Sciences (CCTES). IEEE, pp 197–201

17. Tasnin W, Saikia LC, Saha A, Saha D, Rajbongshi R (2018) Effect of different renewables and FACT device on an interconnected thermal system using SCA optimized fractional order cascade controllers. In: 2018 2nd international conference on power, energy and environment: towards smart technology (ICEPE). IEEE, pp 1–6

18. Khadanga RK, Kumar A (2019) Analysis of PID controller for the load frequency control of static synchronous series compensator and capacitive energy storage source-based multi-area multi-source interconnected power system with HVDC link. Int J Bio-Inspired Comput 13(2):131–139

19. Sahu RK, Sekhar GC, Priyadarshani S (2019) Differential evolution algorithm tuned tilt integral derivative controller with filter controller for automatic generation control. Evol Intell, 1–16

20. Åström KJ, Hägglund T (1995) PID controllers: theory, design, and tuning, vol 2. Instrument Society of America, Research Triangle Park, NC

21. Padhan S, Sahu RK, Panda S (2014) Application of firefly algorithm for load frequency control of multi-area interconnected power system. Electric Power Compon Syst 42(13):1419–1430

22. Pradhan PC, Sahu RK, Panda S (2016) Firefly algorithm optimized fuzzy PID controller for AGC of multi-area multi-source power systems with UPFC and SMES. Eng Sci Technol Int J 19(1):338–354

23. Sharma V, Naresh R, Pulluri H (2016) Automatic generation control of multi-source interconnected power system including DFIG wind turbine. In: 2016 IEEE 1st international conference on power electronics, intelligent control and energy systems (ICPEICES). IEEE, pp 1–6

24. Jagatheesan K, Anand B, Dey N, Ashour AS, Satapathy SC (2017) Performance evaluation of objective functions in automatic generation control of thermal power system using ant colony optimization technique-designed proportional–integral–derivative controller. Electr Eng, 1–17

25. Jagatheesan K, Anand B, Samanta S, Dey N, Santhi V, Ashour AS, Balas VE (2017) Application of flower pollination algorithm in load frequency control of multi-area interconnected power system with nonlinearity. Neural Comput Appl 28(1):475–488 (SCIE, SCOPUS, Impact Factor: 4.213)

26. Jagatheesan K, Anand B, Samanta S, Dey N, Ashour AS, Balas VE (2017) Design of a proportional-integral-derivative controller for an automatic generation control of multi-area power thermal systems using firefly algorithm. IEEE/CAA J Automat Sin

27. Jagatheesan K, Anand B, Ebrahim MA (2014) Stochastic particle swarm optimization for tuning of PID controller in load frequency control of single area reheat thermal power system. Int J Electr Power Eng 8(2):33–40

28. Jagatheesan K, Anand B, Nilanjan D, Ashour AS, Rajesh K (2019) Bat algorithm optimized controller for automatic generation control of interconnected thermal power system. Frontiers in artificial intelligence and applications, vol 314: information technology and intelligent transportation systems. Springer, Cham, pp 276–286

29. Sahu RK, Gorripotu TS, Panda S (2015) A hybrid DE-PS algorithm for load frequency control under deregulated power system with UPFC and RFB. Ain Shams Eng J 6:893–911

30. Francis R, Chidambaram IA (2015) Optimized PI + load frequency controller using BWNN approach for an interconnected reheat power system with RFB and hydrogen electrolyser units. Int J Electr Power Energy Syst 67:381–392

31. Yang XS (2008) Nature-inspired metaheuristic algorithms. Luniver Press

32. Yang XS (2009) Firefly algorithms for multimodal optimization. In: Stochastic algorithms: foundations and applications (SAGA 2009). Lecture notes in computer sciences, vol 5792, pp 169–178

33. Lukasik S, Zak S (2009) Firefly algorithm for continuous constrained optimization task. In: International conference on computational collective intelligence (ICCCI 2009). Lecture notes in artificial intelligence, vol 5796, pp 97–100

34. Yang XS (2010) Firefly algorithm, stochastic test functions and design optimization. Int J Bio-Inspired Comput 2(2):78–84

35. Dey N (ed) (2017) Advancements in applied metaheuristic computing. IGI Global

36. Dey N, Samanta S, Chakraborty S, Das A, Chaudhuri SS, Suri JS (2014) Firefly algorithm for optimization of scaling factors during embedding of manifold medical information: an application in ophthalmology imaging. J Med Imaging Health Inform 4(3):384–394

37. Kumar R, Rajan A, Talukdar FA, Dey N, Santhi V, Balas VE (2017) Optimization of 5.5-GHz CMOS LNA parameters using firefly algorithm. Neural Comput Appl 28(12):3765–3779
38. Kumar R, Talukdar FA, Dey N, Balas VE (2016) Quality factor optimization of spiral inductor using firefly algorithm and its application in amplifier. Int J Adv Intell Paradigms
39. Chakraborty S, Dey N, Samanta S, Ashour AS, Balas VE (2016) Firefly algorithm for optimized nonrigid demons registration. In: Bio-inspired computation and applications in image processing. Academic Press, pp 221–237
40. Samanta S, Mukherjee A, Ashour AS, Dey N, Tavares JMR, Abdessalem Karâa WB, Hassanien AE (2018) Log transform based optimal image enhancement using firefly algorithm for autonomous mini unmanned aerial vehicle: an application of aerial photography. Int J Image Graph 18(04):1850019
41. Gandomi AH, Yang X-S, Talatahari S, Alavi AH (2013) Firefly algorithm with chaos. Commun Nonlinear Sci Numer Simulat 18(2013):89–98
42. Yang XS, Deb S (2010) Eagle strategy using Levy walk and firefly algorithms for stochastic optimization. In: Nature inspired cooperative strategies for optimization (NISCO 2010). Studies in computational intelligence, vol 284. Springer, Berlin, pp 101–111
43. dos Santos Coelho L, de Andrade Bernert DL, Mariani VC (2011) A chaotic firefly algorithm applied to reliability-redundancy optimization. In: 2011 IEEE congress of evolutionary computation (CEC), New Orleans, LA, pp 517–521
44. Fister Jr I, Perc M, Kamal SM, Fister I (2015) A review of chaos-based firefly algorithms: perspectives and research challenges. Appl Mathe Comput 252(2015):155–165
45. Coelho L, Mariani VC (2008) Use of chaotic sequences in a biologically inspired algorithm for engineering design optimization. Expert Syst Appl 34:1905–1913

Chapter 9
Plant Biology-Inspired Genetic Algorithm: Superior Efficiency to Firefly Optimizer

Neeraj Gupta, Mahdi Khosravy, Om Prakash Mahela and Nilesh Patel

1 Introduction

The pioneering work in 1950s [1] was the first time that an evolutionary theory solved a problem. After that evolutionary computation showed its increasing power as it was demonstrated in the work of Nils et al. in 1960 as well the Rechenberg et al. [2]. Academia was highly drawn to these new approaches inspired by nature for solving complex practical problems [3]. The John Holland (1962) work is indeed the very first population-based algorithm [4]. At this stage, the genetic algorithm (GA) was widely used over different application results [5–7]. Later on a variety of population-based optimizers by inspiration from nature came as listed below: artificial immune systems (1994) [8], ant colony optimization (1992) [9], bat algorithm (2010) [10], cultural algorithms (1994) [11], DNA computing similar to parallel computing (1994) [12], hybrid algorithms (1996) [13], particle swarm optimization (PSO, 1995) [14, 15], memetic algorithms (2004) [16].

The above-mentioned techniques present a range of different approaches as optimizers. Recent year, the focus of invention of nature-inspired optimizers is on improving the accuracy, robustness of the techniques, for example, solving the com-

N. Gupta · N. Patel
Department of Computer Science and Engineering,
Oakland University, Rochester, Oakland, MI, USA
e-mail: neerajgupta@oakland.edu

M. Khosravy (✉)
Electrical Engineering Department, Federal University of Juiz de Fora, Juiz de Fora, Brazil
e-mail: mahdi.khosravy@ufjf.edu.br; dr.mahdi.khosravy@ieee.org

Electrical Engineering Department, University of the Ryukyus, Nishihara, Japan

O. P. Mahela
Rajasthan Rajya Vdhyut Prasaran Nigam Ltd., Jodhpur, Rajasthan 342001, India
e-mail: opmahela@gmail.com

© Springer Nature Singapore Pte Ltd. 2020
N. Dey (ed.), *Applications of Firefly Algorithm and its Variants*,
Springer Tracts in Nature-Inspired Computing,
https://doi.org/10.1007/978-981-15-0306-1_9

193

plex problems of finding the optimum location of transmission lines and generator plants in power systems [17, 18]. By growth of the computation power, more complex methods with higher accuracy were born; probabilistic genetic algorithm and robust stochastic GA [19], shuffled frog leap algorithm (SFLA) [20], tabu search [21], and hunting search algorithms such as prey–predator, wolf–grey [22], biogeography-based optimization (BBO) [23], novel bat algorithm (NBA) [24], fish swarm algorithms (FSA) [25], bacterial foraging optimization [26], comprehensive learning PSO (CLPSO) [27], cooperative PSO (CPSO) [28], and social learning PSO (SLPSO) [29].

As the power and popularity of meta-heuristic optimization techniques grow [30], industries and researchers get more confidence for taking the highest possible efficiency by their application to a wide range of already available techniques as data mining [31] big data [32, 33], signal processing [34, 35], image processing [36], image adaptation [37], image enhancement [38], telecommunications [39–42], quality assessment [43], noise cancellation [44], morphological kernel design [45], morphological filtering [46], blind source separation [47–51], blind component processing [52], Acoustic data embedding [53], etc. A strong swarm intelligent optimizer is firefly algorithm (FA) [54–60]. FA animates the behaviors of fireflies population in being attracted by brighter partners. They have communication and warning for the risk of a predator amongst themselves. FA deals better with multimodal functions than PSO and GA algorithms [61]. Recently, the authors were introduced a novel meta-heuristic optimizer by inspiration from plant biology [62]. This chapter presents a critical comparison of plant biology-inspired GA (PBGA) and FA.

2 Plant Biology-Inspired Genetic Algorithm

In a very recent work, by inspiration from plant biology a variation of GA has been invented by the authors which overcome lots of fundamental problems [62], as we call it here "plant biology-inspired genetic algorithm" (PBGA). PBGA differentiates the local and global optima via its intrinsic routines and overcomes the very basic classic issue of searching for the global optimum solution. PBGA inspiration from plant biology is simply discussed here, and for further information, we refer the reader to [62].

2.1 Mendelian Evolutionary Theorem

Mendel is well-known as the father of inheritance laws by his famous experiments on pea plants. He gave a theory that the genes are transferred from a parent to offspring as dominant or recessive traits. Mendel laws are briefly as follows: (i) Segregation: Each pair of genes involves an inherited trait as a particular physical characteristic of the parents. (ii) Independent assortment: Inheritance of one trait is not dependent on the inheritance of another when more than one characteristics are

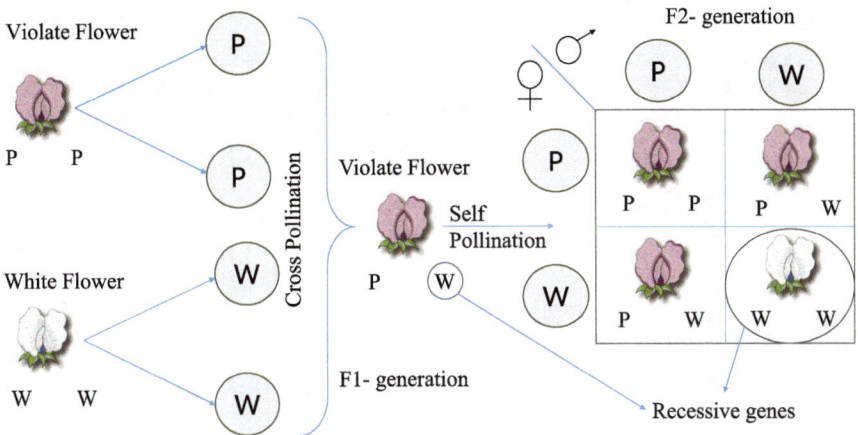

Fig. 1 Mendel's law of inheritance

inherited. (iii) Dominance: One of the factors for a pair of inherited traits is subjected to be dominant and the other recessive unless both factors are recessive. PBGA uses these rules of natural selection of the genes from parents and develops an efficient meta-heuristic optimizer.

Figure 1 shows the Mendel law of inheritance.

2.2 The Biological Structure of DNA

In Mendel theorem, the chromosomes related to each plant contain the unique codes of genetics and are building blocks of the DNA. DNA is a double-helix structure of strands/chromosomes which are running in opposite direction and appeared like a twisted ladder as in Fig. 2.

In a plant, the reproduction is by "pollination" process where pollens are transferred from male to female. There are two types of pollination, "self-pollination" and "cross-pollination," wherein the case of first one—the pollens are transferred to the flower's stigma of the same plant, and in latter case—the transfer is to another plant flower. 80% of the plant pollination happens through living creature and is called biotic pollination. Mendel fertilized two breeds of pea plants in the laboratory by biotic pollination, where he dropped pollen from male reproductive part to the female reproductive part of different, as cross-pollination, and same flowers, as self-pollination. This process inspired to model pollination operation, where plants are selected from different species to fertilize.

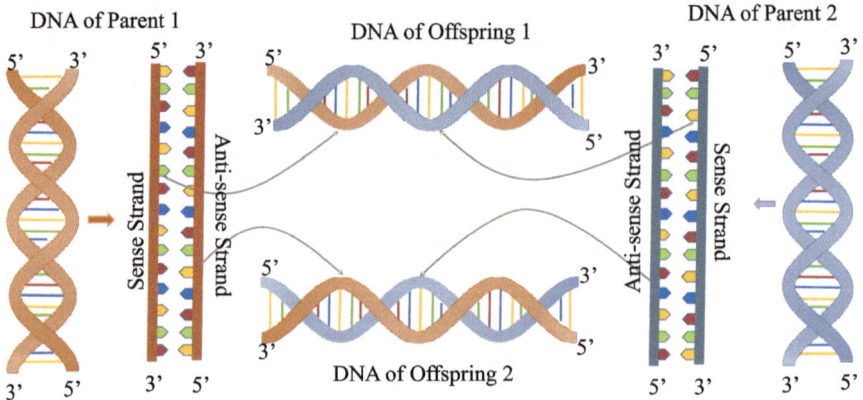

Fig. 2 Anti-parallel DNA structure and formation of offspring DNA

2.3 Epimutation and Rehabilitation as Self-organizing Behaviors

The processes in nature are self-organizing, which means when the output of a process goes in the wrong way, the process takes a rehabilitation step to improve the output. A cycle of rehabilitations after a mutation makes an evolution where all this process is called "epimutation." Because of that genes are subjected to change as they exposed by environmental effects like fertilizers, UV light, chemicals, and X-ray. Under an evolution, the DNA is synthesized, and if the change results in an appropriate state, it tries self-recovery. Figure 3 shows the concept of epimutation. The main effect of epimutation for an organism is self-improvement, maintenance, regaining physical power, and strength. In Fig. 3, five phases of an organism are depicted; p1, that is the current state; p2, p3, and p4 are the phases which are not better than the current phase and organism should return to p1 by a rehabilitation process, and finally, p5 has a better phase where the evolution is accepted.

In the optimization process, the epimutation results in a better neighbor of the pseudo-best solution (PBS), as called here "fine-tuning" of the global best point.

3 Developing Building Block

Although PBGA is inspired by GA, it has totally different structure. Despite GA, PBGA deploys the exchange of information between chromosomes. The optimizer as a computer program uses an inspiration of the above-mentioned process by five inspired operators as flipper, pollination, breeding, discrimination, and epimutation. Flipper is necessary for completing the double-strand DNA. The DNA strands are

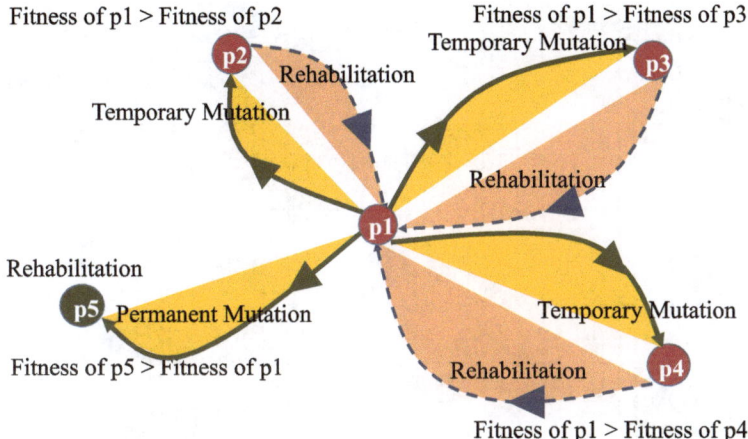

Fig. 3 Epimutation process

Fig. 4 Genotype and phenotype representation of a point

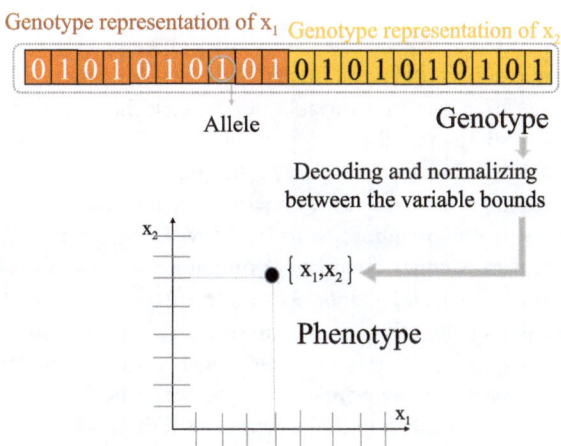

complementary to each other. The details of the operators are given in [62]. Critical features of PBGA are as follows:

Due to Mendel's law binary nature, it becomes feasible to mimic it as a binary coded algorithm. Each gene in the chromosome of double-strand DNA corresponds to visible characteristics of the plant, like size and color of the flower; that is, in genotype representation, genotype G is a binary sequence of bits $g[l] \in \{0, 1\}$, and here, l is the position of the corresponding gene $g[l]$ in the chromosome. Decoded value of G refers to the genetic contribution to the phenotype, $H = f(G)$. In [62], PBGA algorithm is based on binary codes. Figure 4 illustrates the relation between genotype and phenotype spaces. In this figure, the chromosome of 20 bits is used to represent two variables x_1 and x_2, where every ten bits (genes) codes one variable. PBGA is a process in genome space with binary numbers. It needs a coding–decoding

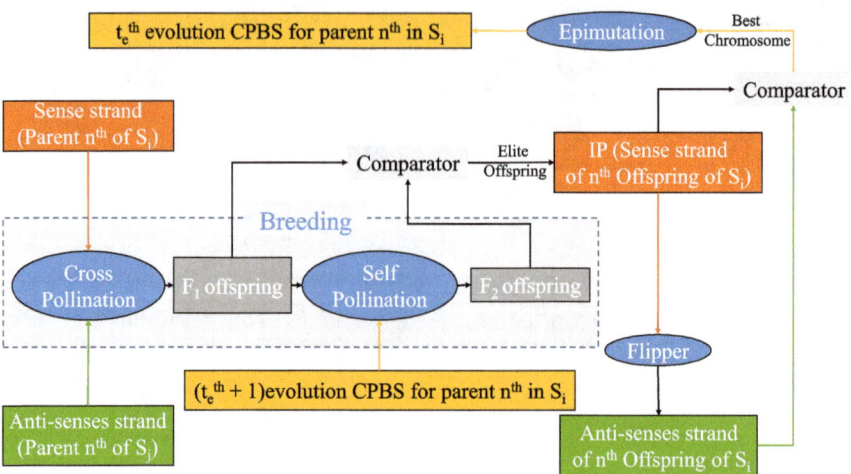

Fig. 5 Block diagram of METO procedure in one evolution

approach to transfer a binary code into real space; see Fig. 4. The number of genes corresponding to a variable depends on the required accuracy of the solution. The block diagram in Fig. 5 describes the PBGA process of evolution. PBGA inspired by Mendel's law performs over multiple species. First, two parents DNAs are acquired as shown in Fig. 5. DNA uncoils and separates the strands, and two strands from two parents integrate to form a new DNA for the offspring. Based on the experiment of Mendel, first cross-pollination is carried out which cross-fertilize the two different species plants. Resulting offspring are F_1-offspring and have hereditary characteristics from their parents, only. After getting F_1-generation offspring, we follow self-pollination scheme where resulting F_1-plants are self-fertilized. According to the Mendel principles, in the resulting F_2-generation, heredity characteristics are transferred from the ancestors. These genetic traits called as recessive elements which appear in F_2-generation with some probability. After the above process, from the two-generation offspring, we select elite plants to go in the next evolution phase. At this stage, we acquire sense strands of offspring, whereby using flipper operator, we produce associated anti-sense strands. The best offspring at this stage is compared with the in-memory best solution from the past evolution, also known as a recessive chromosome. If the new solution is found better, the previous one is replaced with the new best. Concludes the solution once one of the following termination criteria met:

1. Reaching a maximum number of iterations.
2. If the average of the species does not change in a certain number of iterations.
3. The same answer comes after each successive generation for m times.
4. The error goes below the desired value.

Condition 2 is for the sake of diversity. Pseudo-code for producing F1-generation offspring is given by Algorithm 1 in Ref. [62]. Fertilization as a result of cross-

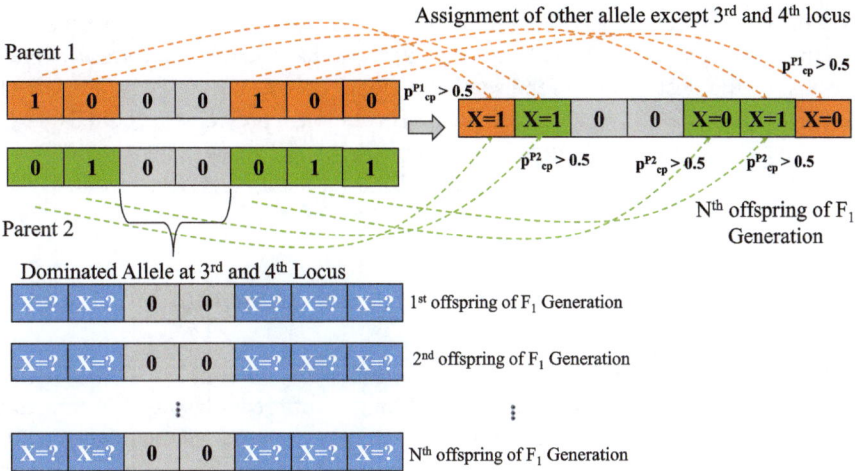

Fig. 6 Production of F_1 generation offspring

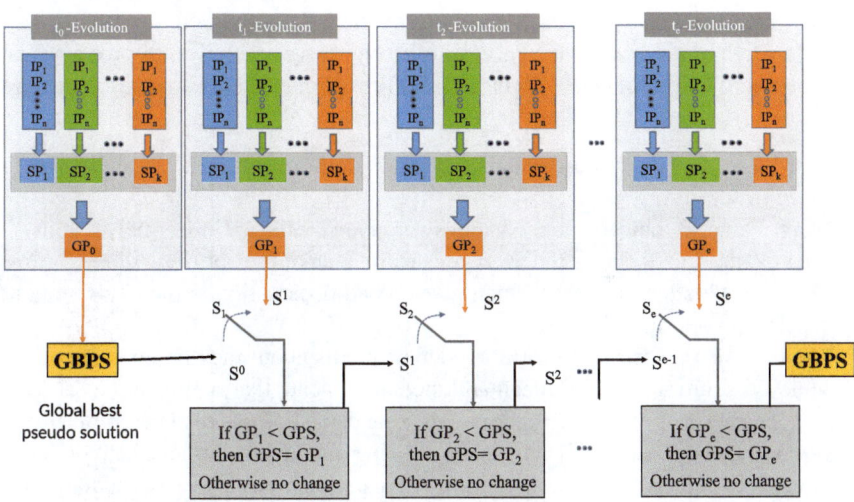

Fig. 7 Flow diagram for finding global best pseudo-solution (GBPS)

pollination can give multiple offspring as shown in Fig. 6. In this figure, the genes represented by "X" can have characteristics of any parents based on the transfer probability of heredity.

As we have described that multiple species are required to apply the Mendelian theory of plant genetics, Fig. 7 shows the procedure to extract the global best pseudo-solution (GBPS) after each evolution. Figure 7 shows evolution blocks from t_0 to t_e, where each block contains the parallel populations entitled to respective multiple species. Each population holds the n number of plants, whose fitness is represented by independent pseudo-best IP_n. From all IPs in each k species, we extract the best as kth species pseudo-best (SP) solution. Now, we have k solution form available species, so the best of them represents the global pseudo-best (GP) as a result of current evolution GBPS. Under the process of evolution, if previously acquired GP_{e-1} is inferior to the current GP_e, immediately replaced by it. This procedure is continued until the last evolution, and the resulting GBPS is the solution as described in Fig. 7.

4 PBGA Efficiency Compared to Firefly Algorithm

As the main target of the chapter is the comparison of efficiency of PBGA with firefly algorithm (FA), we have used a set of well-known different classes of benchmark test functions from simple to complicated one. The formulas of each benchmark function are given together with comparison results. Both optimizers are evaluated over a medium number of 30 variable initially over each benchmark function where the statistical results of their efficiency over the corresponding benchmark are given through individual tables for each benchmark function. In addition, for each benchmark function, the convergence curves of the optimizers are given to give a visual comparison of their efficiency. Eight statistical performance indicators are used here as follows:

1. Best Value: It indicates the obtained solution with maximum accuracy in Monte Carlo (MC) runs.
2. Worst Value: It indicates the worst answer given by the optimizer in MC runs.
3. Mean Value (μ): It shows the average performance of the optimizers in MC runs. This is a statistical property which gives broad observation to the readers about to select an optimizer.
4. Median: As a measure of central tendency, it tells about the middle value of the achieved solution. Similar mean and median indicate that optimizer has at least 50% chance to give this as output. Mean and median have some similar role, while an average has the disadvantage of being affected by outliers, i.e., too high- or too low-valued solutions than others. This is the reason we use median for a better measure of a midpoint.
5. Mode: It identifies that how much time a particular solution comes in multiple runs of the optimizer. This performance indicator shows the reliability of the optimizer. This central tendency is the two-tuple number where one represents the most reliable solution, and the other is a number of its counts in multiple runs and represented by MC in result tables.
6. Standard deviation (Std): It is a measure to quantify the amount of variation or dispersion in the achieved solution from the optimizer in numerous runs. The lower the standard deviation, the better the performance and reliability of the optimizer.

7. Consistency ($\mathscr{C}$): This indicator presents how many times an optimizer qualifies a threshold value. For this, we decide to mean of METO as the threshold value to show comparative analysis with other algorithms.

Parameters associated with the algorithms are described below, inspired by their respective publications.

- Plant biology-inspired GA (PBGA):

    1. Number of bits to represent one variable is 35,
    2. Lower limit Mendelian probability is 0,
    3. Upper limit of Mendelian probability is 1,
    4. Min limit of mutation for global best is 0.005,
    5. Max limit of mutation for global best is 1,
    6. Rehabilitation against epimutation is 5,
    7. Number o f species is 2.

- Firefly Algorithm (FA) [54]:

    1. Light absorption coefficient = 1,
    2. Attraction coefficient base value = 2,
    3. Mutation coefficient = 0.2,
    4. Mutation coefficient damping ratio = 0.98,
    5. Uniform mutation range = 0.05 * (VarMax − VarMin),
    6. m = 2.

4.1 Deceptive Function (DCF)

This function belongs to the global optimization problem having the property of multi-modularity for minimization. This function is defined as

$$f_{\text{deceptive}}(x) = -\left[\frac{1}{n}\sum_{i=1}^{n} g_i(x_i)\right]^{\beta} \tag{1}$$

Here, nonlinearity factor β is fixed to 2 and decision function $g_i(x_i)$ is defined for $i = 1, \ldots, n$ as

$$g_i(x_i) = \begin{cases} -\frac{x}{\alpha} + \frac{4}{5} & \text{if } 0 \le x_i \le \frac{4}{5}\alpha_i \\ \frac{5x}{\alpha} - 4 & \text{if } \frac{4}{5}\alpha_i \le x_i \le \alpha_i \\ \frac{5(x-\alpha_i)}{\alpha_i-1} & \text{if } \alpha_i \le x_i \le \frac{1+4\alpha_i}{5} \\ \frac{x-1}{1-\alpha_i} & \text{if } \frac{1+4\alpha_i}{5} \le x_i \le 1 \end{cases} \tag{2}$$

In the above equation, n represents the number of dimensions and the range of each variable is defined as $x_i \in [0, 1]$. This function is a complicated one to solve optimally to the number of deceptive points inherent, thus complication in finding the global solution. The comparative results are shown in Table 1 and Figs. 8 and 9.

Table 1 Comparison of the optimizers in statistical properties for deceptive function

EOA	Best	Mean	Median	Mode	Std	Consist (%)	Worst
PBGA	−9.96E−01	−9.60E−01	−9.59E−01	−9.96E−01	1.94E−02	49	−8.87E−01
FA	−7.86E−01	−7.31E−01	−7.28E−01	−7.86E−01	2.28E−02	0	−6.83E−01

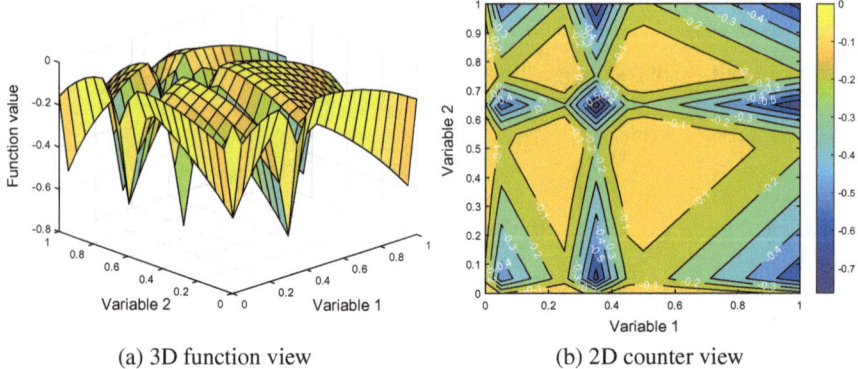

(a) 3D function view   (b) 2D counter view

Fig. 8 Deceptive function

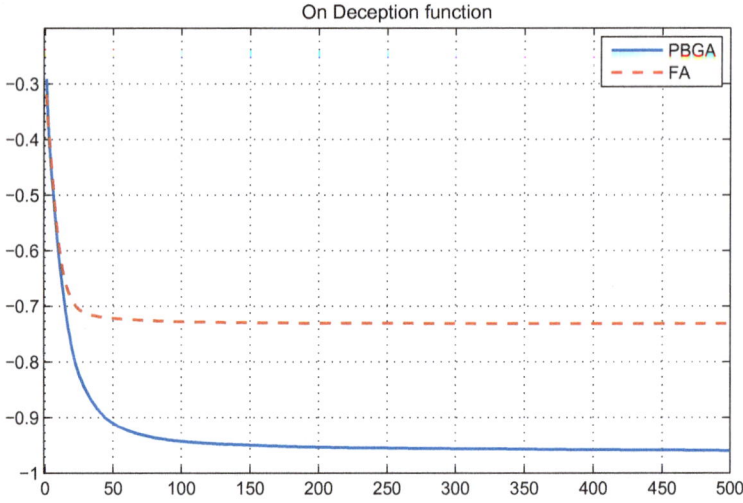

Fig. 9 Average convergence curve of optimizers in 500 evolutions for deceptive function

4.2   *Schwefel Function No 26 (SF26):*

This benchmark function is a complex global optimization problem with many local minima. This multimodal problem is evaluated on the hypercube of n dimensions where each dimension has the range as $x_i \in [-500, 500]$. The Schwefel 26 function is defined as

$$f_{\text{Schwefel26}}(\mathbf{x}) = 418.9829n - \sum_{i=1}^{n} x_i \sin(\sqrt{|x_i|}) \tag{3}$$

The pictorial representation in 3-D is given in Fig. 10, right sub-figure, where contour diagram is shown by right sub-figure. In this figure, we can observe the multi-modularity of the function with multiple global minima. Although the functions with multiple global minima are not very hard to solve, here we compare the efficiency and consistency of optimizers as shown in Table 2. To see more clear results, please follow Fig. 11. In this figure, we can observe that PBGA converges faster with improved results compared to FA. Moreover, the consistency of the PBGA is 92% while for FB is zero which can be seen in Table 2, with very low standard deviation. The best achieved by PBGA is $-1.26E+04$, where the average performance is represented by its mean equal to $-1.26E+04$. The median $-1.26E+04$ is similar to the mean which represents that very low level of uncertainty in searching the best solution. This can be validated by the achieved similar values for best, mean, and median. We can also observe that for this function FA is much far for the worst value acquired by PBGA.

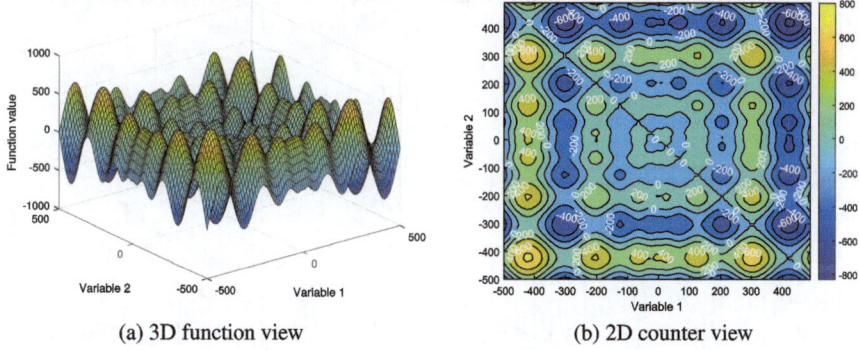

(a) 3D function view   (b) 2D counter view

Fig. 10   Schwefel function No 226

Table 2   Statistical comparison of optimizers on Schwefel function No. 26

EOA	Best	Mean	Median	Mode	Std	Consist (%)	Worst
PBGA	**−1.26E+04**	−1.26E+04	−1.26E+04	−1.26E+04	9.34E+00	92	−1.25E+04
FA	**−1.10E+04**	−9.88E+03	−9.94E+03	−1.10E+04	5.26E+02	0	−8.54E+03

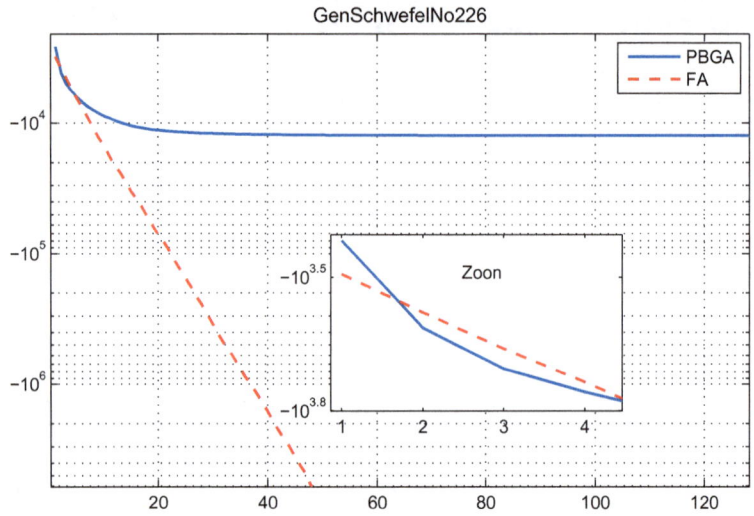

Fig. 11 Average convergence in 500 evolutions for Schwefel function No. 26

4.3 Inverted Cosine Wave Function (ICWF):

This function is taken for experimentation because of its unique property as a global optimization problem. This n-dimensional function ranges as $-5 \leq x_i \leq 5$, with one global optimum. Its complex shape restricts most of the optimizers to find its optima. Mathematical equation of inverted cosine wave function is given below

$$f(X) = -\sum_{i=1}^{n-1} \left\{ e^{\left[\frac{-\left(x_i^2 + x_{i+1}^2 + 0.5x_i x_{i+1}\right)}{8} \right]} \cos\left(4 \times \sqrt{x_i^2 + x_{i+1}^2 + 0.5x_i x_{i+1}}\right) \right\} \tag{4}$$

3-D representation of this function for two variables is shown in Fig. 12 left-side sub-figure, where its contour plot can be seen on right-side sub-figure. Figure 12 shows its complication to be solved by the optimizers. Multiple rings around the global optimization solution deviated the optimizers to converge to its optimal solution.

Average results and other statistical characteristics achieved can be observed in Table 3. The average convergence curve achieved in 500 generations is shown in Fig. 13. Here we can observe that the PBGA outperform FA with the big gap. The higher average performance μ is achieved by PBGA as usual, where similar median shows a lower shift from the mean. Thus, the consistency is about 48% and shown in Table 3.

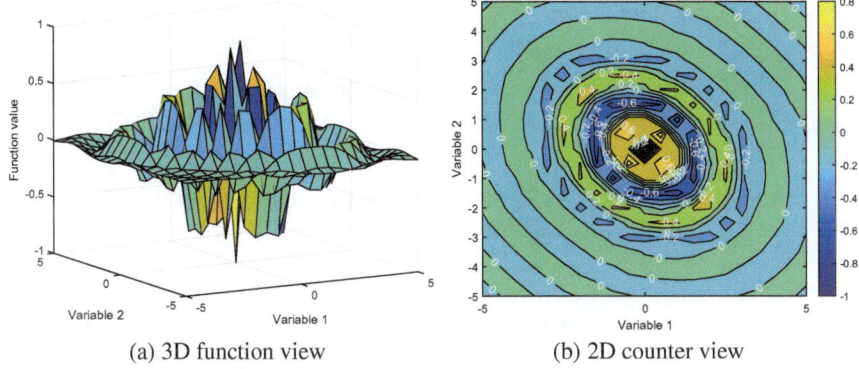

(a) 3D function view   (b) 2D counter view

Fig. 12 Inverted cosine wave function

Table 3 Comparision of optimizers on statistical properties for an inverted cosine wave function

EOA	Best	Mean	Median	Mode	Std	Consist (%)	Worst
BPGA	**−2.77E+01**	−2.41E+01	−2.40E+01	−2.77E+01	1.62E+00	48	−2.14E+01
FA	**−2.22E+01**	−1.88E+01	−1.90E+01	−2.22E+01	1.78E+00	0	−1.38E+01

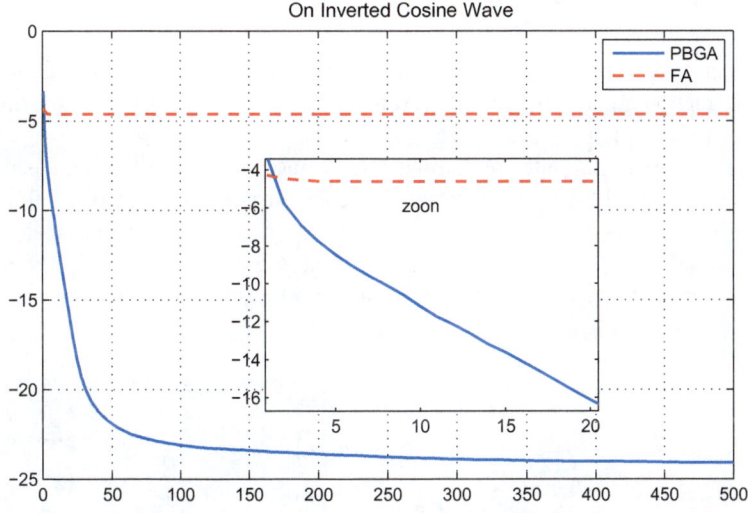

Fig. 13 Average convergence curve of optimizers in 500 evolutions for inverted cosine wave function

4.4 Rastrigin Function (RSTF):

The Rastrigin benchmark function is a global optimization problem defined on n dimensions as follows

$$f_{\text{Rastrigin}}(\mathbf{x}) = 10n \sum_{i=1}^{n} \left[x_i^2 - 10 \cos(2\pi x_i) \right] \tag{5}$$

This is a multimodal minimization problem where each dimension ranges between -5.12 and 5.12. This problem is encountered as one of the hard problems due to its modularity with curvature. Multiple local minima restrict the optimizers to converge at global minima. Pictorial representation of this function in 3-D is shown in Fig. 14. Here, left-side sub-figure shows its complexity, which can be verified by its contour diagram on right-side sub-figure. Solution achieved by optimizers can be observed in Table 4. Here we can see the consistency of the PBGA is higher as achieves $3.59E^{-11}$(Fig. 15).

4.5 Ackley Function (ACKF)

This function is defined in n-dimensions and widely used for testing evolutionary optimization algorithms. Figure 16 represents its two variables plot, showing its global solution region in the 2-D contour plot.

It has the nearly flat outer region with less valued many local minima as disturbance, and a large hole at the center which is characterized by a global solution. The function poses a risk, to be trapped an optimization algorithm at one of its many local minima. This can be seen in Table 5 for the results acquired by algorithms. In the resulting table, we can observe that PBGA best answer is 10,000 times smaller

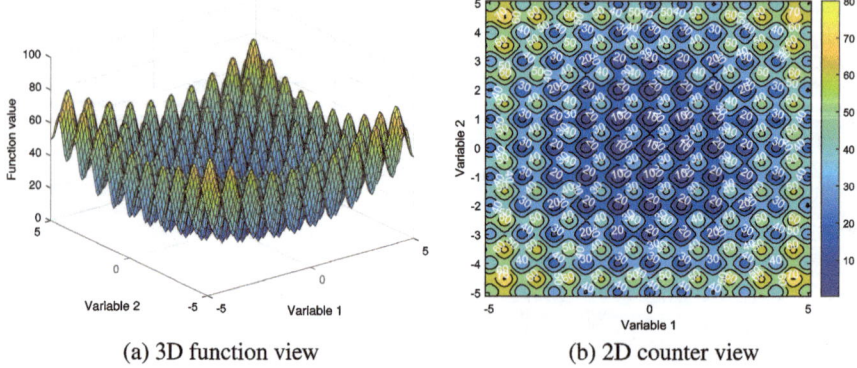

(a) 3D function view   (b) 2D counter view

Fig. 14 Rastrigin function

Table 4 Comparision of optimizers on statistical properties for Rastrigin function

EOA	Best	Mean	Median	Mode	Std	Consist (%)	Worst
PBGA	**3.59E−11**	3.27E+00	3.05E+00	3.59E−11	1.97E+00	64	1.12E+01
FA	**1.69E+01**	3.83E+01	3.98E+01	1.69E+01	1.13E+01	0	6.77E+01

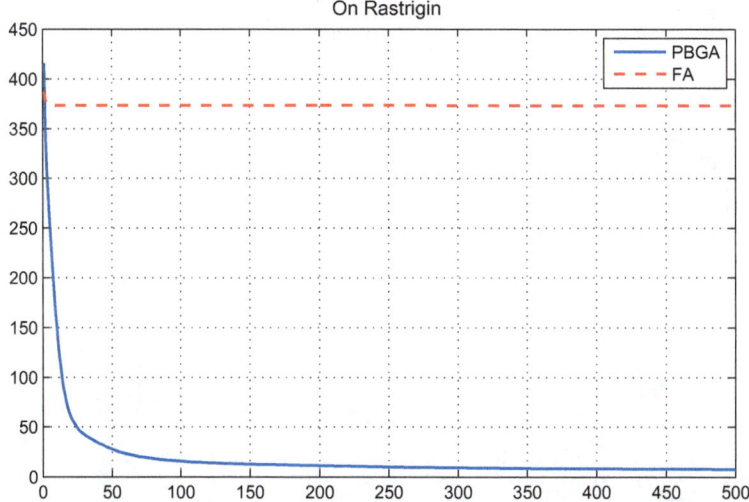

Fig. 15 Average convergence curve of optimizers in 500 evolutions for Rastrigin function

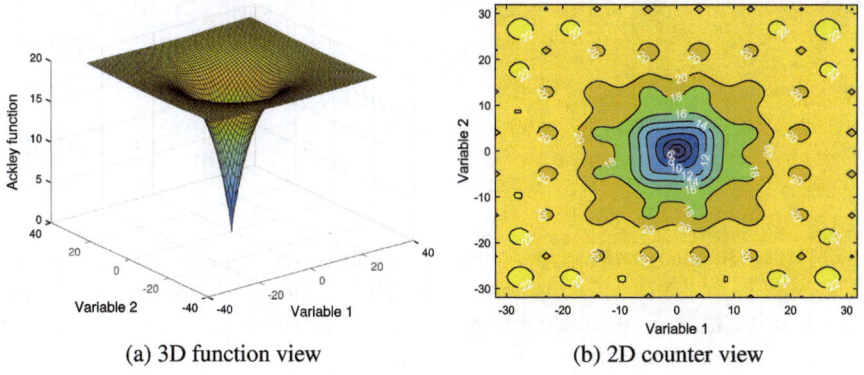

(a) 3D function view   (b) 2D counter view

Fig. 16 Ackley function

Table 5 Comparision of optimizers on statistical properties for Ackley function

EOA	Best	Mean	Median	Mode	Std	Consist (%)	Worst
PBGA	**8.35E−09**	4.36E−05	1.04E−06	8.35E−09	1.05E−04	76	7.37E−04
FA	**5.28E−05**	6.68E−05	6.71E−05	5.28E−05	3.92E−06	0	7.43E−05

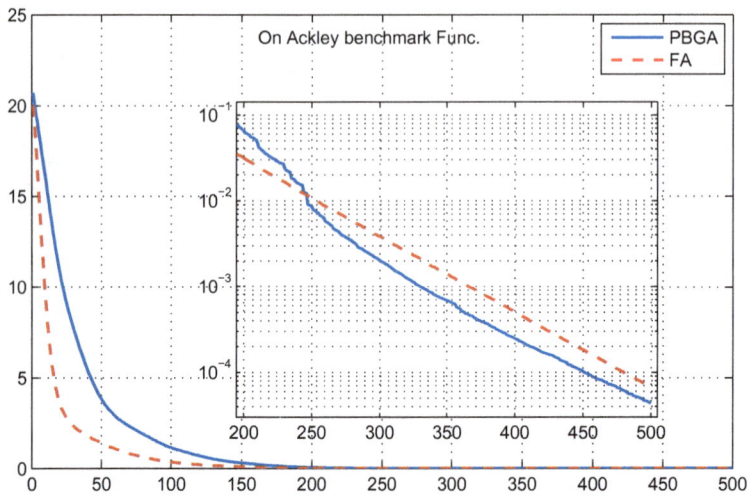

Fig. 17 Average convergence curve of optimizers in 500 evolutions on Ackley function

Table 6 Comparison of optimizers on statistical properties for Rosenbrock Leone function

EOA	Best	Mean	Median	Mode	Std	Consist (%)	Worst
PBGA	**2.33E+00**	2.00E+01	1.72E+01	2.33E+00	1.69E+01	65	7.26E+01
FA	**8.20E+00**	1.14E+01	1.14E+01	8.20E+00	1.34E+01	100	1.45E+01

than FA which means big distance in being better. Averaged convergence curve can be seen in Fig. 17. Interestingly until around 250 iterations, FA performs better, but after that PBGA surpass the FA and performs better than its mean value reaches to 4.26×10^{-5} while the mean value for FA is 6.68×10^{-5} (Table 6).

4.6 Rosenbrock Function (RBF):

Mathematical representation of Rosenbrock function is as follows:

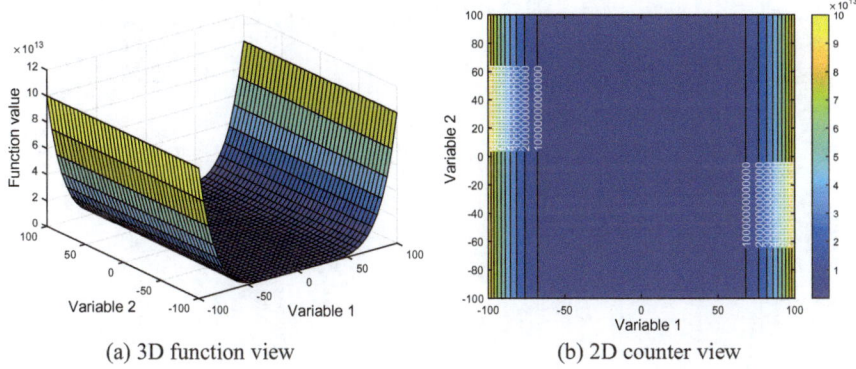

(a) 3D function view   (b) 2D counter view

Fig. 18 Rosenbrock Leone function

$$f(x, y) = \sum_{i=1}^{n}[b(x_{i+1} - x_i^2)^2 + (a - x_i)^2] \tag{6}$$

The parameters a and b are generally have the value 1 and 100, respectively. The function is continuous, convex and defined on n-dimensional space, where non-separable and multimodal characteristics make it complicated to solve. However, this function is differentiable.

3-D view and associated 2-D contour are given in Fig. 18, where we can see its valley shape which offers multimodal characteristic to the optimizers. It is evident from Table 6 that FA has better performance over these types of benchmark functions. It converges faster, and it reaches a better mean value compared to PBGA. It is due to the smooth nature of Rosenbrock function that exceptionally FA outperforms PBGA (Fig. 19).

4.7   Weierstrass Function (WSF):

Results of this function can be seen in Fig. 21 and Table 7. The mathematical definition of this function can be seen as:

$$f(X) = \sum_{i=1}^{n} \left(\sum_{k=0}^{kmax} \left[a^k \cos(2\pi b^k (x_i + 0.5)) \right] \right) - n \sum_{k=0}^{kmax} \left[a^k \cos(2\pi b^k * 0.5) \right] \tag{7}$$

Here, the parameters are as: $a = 0.5$, $b = 3$, and kmax = 20. 3-D view and associated 2-D contour of WSF are given in Fig. 20 between the range from -0.5 to 0.5 for each variable. From Table 7, one can observe that PBGA is better for this function too.

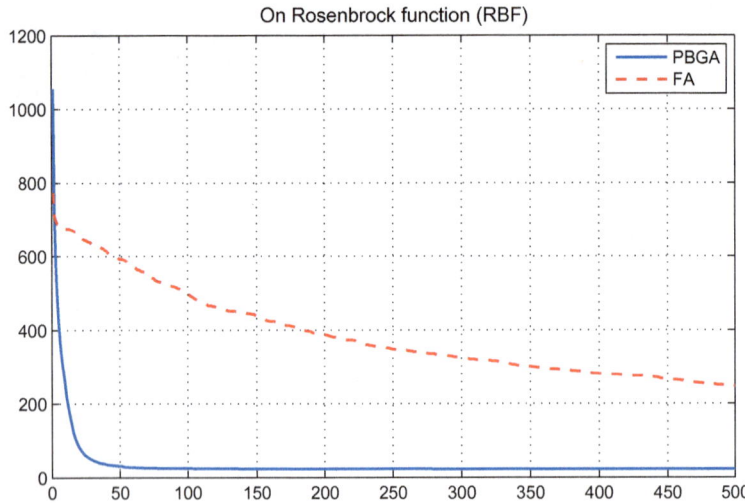

Fig. 19 Average convergence curve of optimizers in 500 evolutions for Rosenbrock Leone function

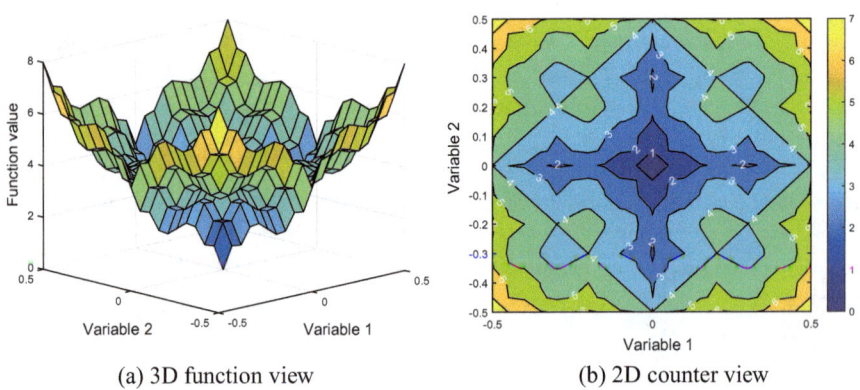

(a) 3D function view   (b) 2D counter view

Fig. 20 Weierstrass function (WSF)

Table 7 Comparison of optimizers on statistical properties for Weierstrass function

EOA	Best	Mean	Median	Mode	MC	Std	Consist (%)	Worst
PBGA	**2.33E+00**	2.09E+01	1.72E+01	2.33E+00	1	1.69E+01	65	7.26E+01
FA	**1.25E+02**	2.48E+02	2.35E+02	1.25E+02	1	8.78E+01	0	5.41E+02

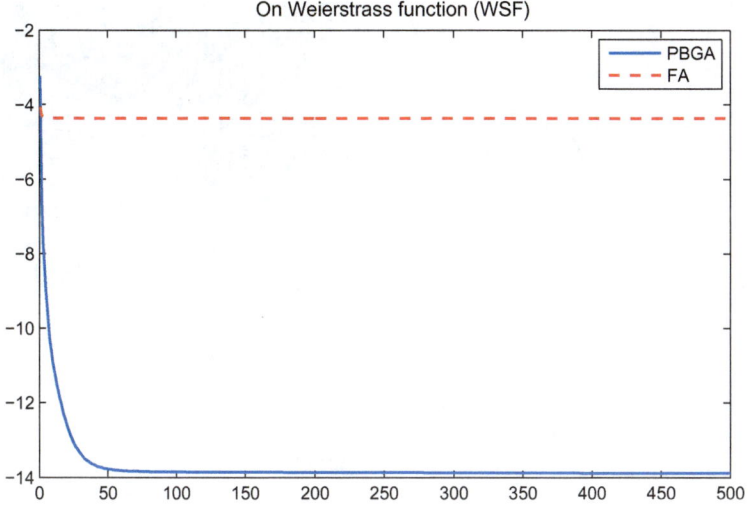

Fig. 21 Average convergence curve of optimizers in 500 evolutions for Weierstrass function

4.8 Sphere Function (SF)

Mathematically, this function is as follows:

$$f(\mathbf{x}) = \sum_{i=1}^{n} x_i^2 \tag{8}$$

The function is continuous, convex, separable and is defined on n-dimensional space. The differentiability of this unimodal function makes it easier to solve by conventional optimization methods as well. 3-D representation for two variables is shown in Fig. 22 with its 2-D contour plot. Here, we can see its smoother surface with the blue region at the center indicating the global solution. The function can be defined on any input domain, but it is usually evaluated on the hypercube $x_i \in [-100, 100]$.

The results obtained by 100 individual runs of this function are shown in Fig. 23 and Table 8. In this case also, similar to the Rosenberg function, due to smoothness of search area the FA has better performance than PBGA on average, but still PBFA performance is comparable with FA. Both show very high consistency over this function.

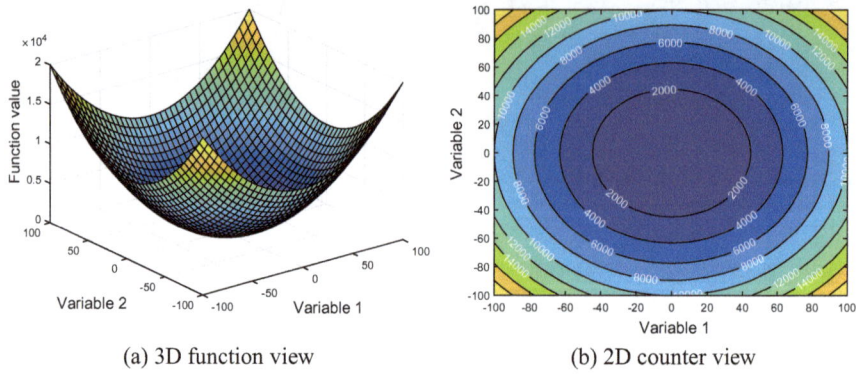

(a) 3D function view   (b) 2D counter view

Fig. 22 Sphere function

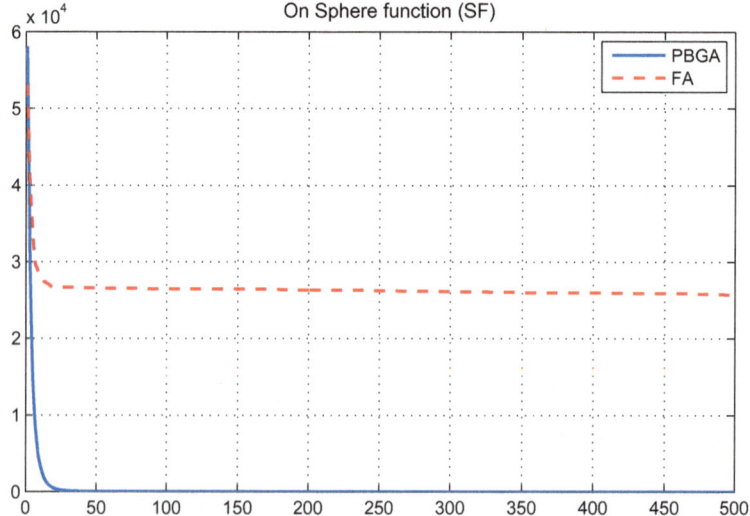

Fig. 23 Average convergence curve of optimizers in 500 evolutions for sphere function

Table 8 Comparison of optimizers on statistical properties for sphere function

EOA	Best	Mean	Median	Mode	Std	Consist (%)	Worst
PBGA	**0.00E+00**	2.46E−07	0.00E+00	0.00E+00	1.23E−06	96	7.58E−06
FA	**7.53E−10**	2.21E−09	2.25E−09	7.53E−10	5.59E−10	100	3.39E−09

4.9   Keane's Bump Function (KBF)

It is a complicated function. The Keane global optimization problem is a multimodal minimization problem defined as follows:

$$f(x) = -\left| \frac{\sum_{i=1}^{n} \cos^4(x_i) - 2 \prod_{i=1}^{n} \cos^2(x_i)}{\left(\sum_{i=1}^{n} i x_i^2 \right)^{0.5}} \right|$$

subject to:

$$g_1(x) = 0.75 - \prod_{i=1}^{n} x_i < 0 \tag{9}$$

$$g_2(x) = \sum_{i=1}^{n} x_i - 7.5n < 0$$

Its global value exists at the nonlinear boundary and thus shows complexity in finding the optima. The 3-D views for two variables are shown in Fig. 24. With its 2-D contour plot, here, we can see that it has one global solution surrounded by many local minima. This local solution traps most of the evolutionary optimizers. This can be seen in Table 9 and convergence curves are shown in Fig. 25. For this function, we can see the outperformance of PBGA over FA.

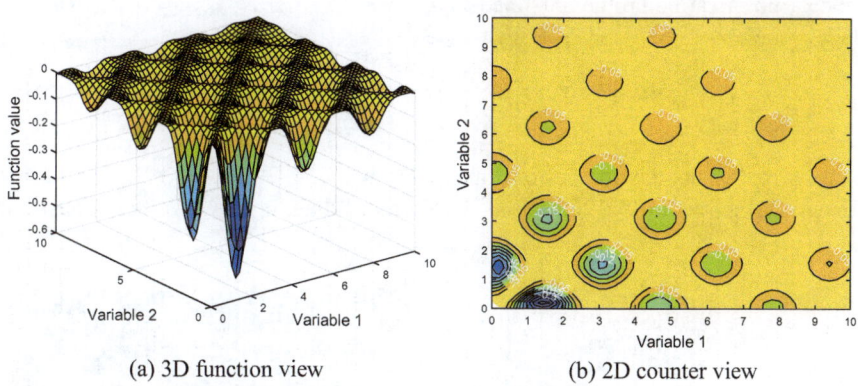

(a) 3D function view   (b) 2D counter view

Fig. 24   Keane's bump function

Table 9   Comparision of optimizers on statistical properties for Keane's bump function

EOA	Best	Mean	Median	Mode	Std	Consist (%)	Worst
PBGA	**−8.20E−01**	−8.09E−01	−8.11E−01	−8.20E−01	7.87E−03	63	−7.77E−01
FA	−8.22E−01	−4.63E−01	−4.44E−01	−4.44E−01	7.88E−02	1	−4.44E−01

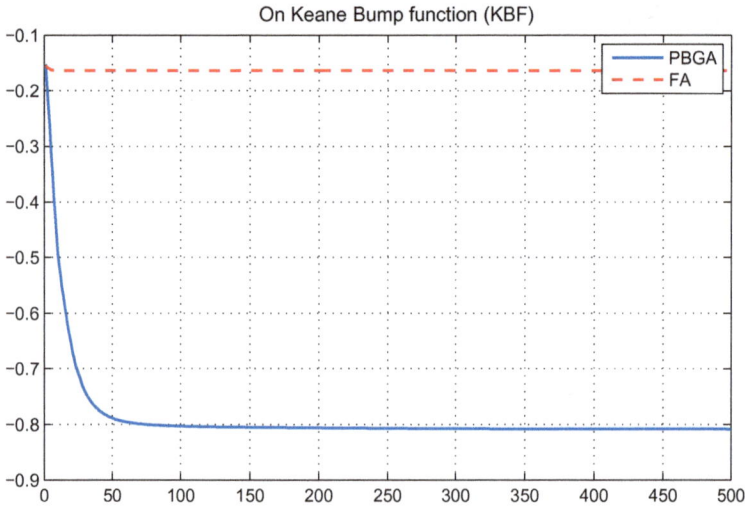

Fig. 25 Average convergence curve of optimizers in 500 evolutions for Keane's bump function

4.10 Dixon and Price Function (DPF)

This class defines the Dixon and Price global optimization problem. This is a multi-modal minimization problem defined as follows:

$$f_{\text{DixonPrice}}(\mathbf{x}) = (x_i - 1)^2 + \sum_{i=2}^{n} i(2x_i^2 - x_{i-1})^2 \tag{10}$$

Here, n represents the number of dimensions and $x_i \in [-10, 10]$ 3-D view of this function is shown in Fig. 26 with its 2-D contour diagram. Due to its flat basin, it

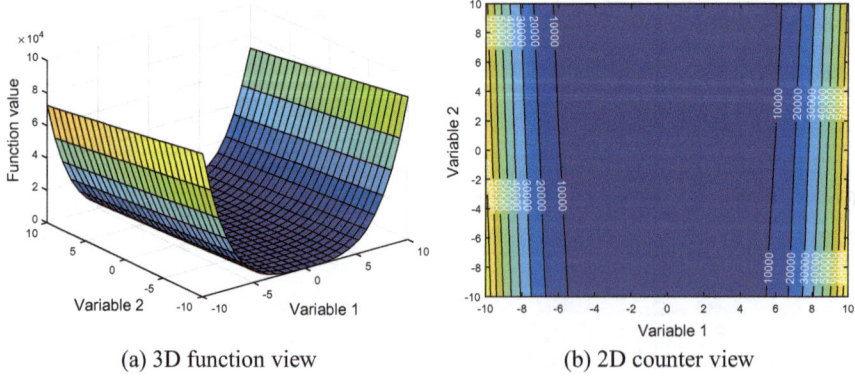

(a) 3D function view   (b) 2D counter view

Fig. 26 Dixon and Price function (DPF)

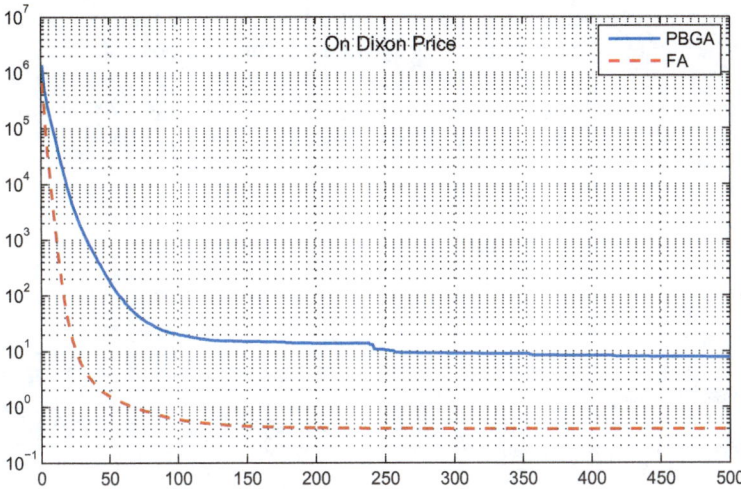

Fig. 27   Average convergence curve of optimizers in 500 evolutions for Dixon and Price function

Table 10   Comparison of optimizers on statistical properties for Dixon and Price function

EOA	Best	Mean	Median	Mode	Std	Consist (%)	Worst
PBGA	**2.35E−01**	7.79E+00	7.76E+00	2.35E−01	6.46E+00	51	2.89E+01
FA	**2.24E−01**	3.99E−01	2.25E−01	2.24E−01	7.48E−01	100	4.01E+00

is not easy to solve. The results of all optimizers on 100 multiple runs are shown in Fig. 27 and Table 10. In the case of Dixon and Price, we can see again that due to the smooth nature of the function, FA has better performance; however, still their performance is comparable and close.

5   Concluding Remarks

The chapter gives a brief introduction to plant biology-inspired genetic algorithm (PBGA) optimizer, and then it compares its efficiency to the well-known method of firefly algorithm. The comparative study is over a variety of standard and complex benchmark functions. The statistical study of the results by several indices approves the higher efficiency of PBGA compared to FA. Besides, the chapter provides convergence curves to the solution by both optimizers for each benchmark function. The convergence curves visually demonstrate the advantages of PBGA in fast reaching the solution, stability, and higher accuracy. In some exceptional case that the benchmark function has a smooth nature, we observed that FA performed better than PBGA, like Rosenbrock, Sphere, and Dixon and Price functions, wherein PBGA converges faster and reaches a better mean value compared to FA. The authors suggest PBGA as a new accurate and stable optimizer for future complex problems.

Acknowledgements Our very special acknowledgment goes to Professor Ishwar Sethi in the Department of Computer Science and Engineering, Oakland University, Rochester, Michigan, the USA, for his very worthwhile advices during this work.

References

1. Dasgupta D, Michalewicz Z (2013) Evolutionary algorithms in engineering applications. Springer Science & Business Media
2. Fogel DB (2006) Foundations of evolutionary computation. In: Modeling and simulation for military applications. International Society for Optics and Photonics, vol 6228, p 622–801
3. Binitha S, Sathya SS et al (2012) A survey of bio inspired optimization algorithms. Int J Soft Comput Eng 2(2):137–151
4. Crainic TG, Toulouse M (2003) Parallel strategies for meta-heuristics. In: Handbook of meta-heuristics. Springer, pp 475–513
5. Tomassini M (1995) A survey of genetic algorithms. In: Annual reviews of computational physics III. World Scientific, pp 87–118
6. Gupta N, Patel N, Tiwari BN, Khosravy M (2018) Genetic algorithm based on enhanced selection and log-scaled mutation technique. In: Proceedings of the future technologies conference. Springer, pp 730–748
7. Singh G, Gupta N, Khosravy M (2015) New crossover operators for real coded genetic algorithm (RCGA). In: 2015 international conference on intelligent informatics and biomedical sciences (ICIIBMS). IEEE, pp 135–140
8. Moscato P, Cotta C, Mendes A (2004) Memetic algorithms. In: New optimization techniques in engineering. Springer, pp 53–85
9. Dorigo M, Birattari M (2010) Ant colony optimization. Springer
10. Moraes CA, De Oliveira EJ, Khosravy M, Oliveira LW, Honório LM, Pinto MF (2020) A hybrid bat-inspired algorithm for power transmission expansion planning on a practical brazilian network. In: Applied nature-inspired computing: algorithms and case studies (pp 71–95). Springer, Singapore
11. Jin X, Reynolds RG (1999) Using knowledge-based evolutionary computation to solve nonlinear constraint optimization problems: a cultural algorithm approach. In: Proceedings of the 1999 congress on evolutionary computation-CEC99 (Cat. No. 99TH8406), vol 3. IEEE, pp 1672–1678
12. Pelikan M, Goldberg DE, Lobo FG (2002) A survey of optimization by building and using probabilistic models. Comput Optim Appl 21(1):5–20
13. Michalewicz Z, Schoenauer M (1996) Evolutionary algorithms for constrained parameter optimization problems. Evol Comput 4(1):1–32
14. Shi Y, Eberhart RC (1999) Empirical study of particle swarm optimization. In: Proceedings of the 1999 congress on evolutionary computation-CEC99 (Cat. No. 99TH8406), vol 3. IEEE, pp 1945–1950
15. Khosravy M, Gupta N, Patel N, Senjyu T, Duque CA (2020) Particle swarm optimization of morphological filters for electrocardiogram baseline drift estimation. In: Applied nature-inspired computing: algorithms and case studies, Springer, In Press
16. Buriol L, França PM, Moscato P (2004) A new memetic algorithm for the asymmetric traveling salesman problem. J Heuristics 10(5):483–506

17. Gupta N, Shekhar R, Kalra PK (2014) Computationally efficient composite transmission expansion planning: a pareto optimal approach for technoeconomic solution. Int J Electr Power Energy Syst 63:917–926
18. Authors (2012) Tepaccess. J 2(2):137–151
19. Tu Z, Lu Y (2004) A robust stochastic genetic algorithm (stga) for global numerical optimization. IEEE Trans Evol Comput 8(5):456–470
20. Eusuff M, Lansey K, Pasha F (2006) Shuffled frog-leaping algorithm: a memetic meta-heuristic for discrete optimization. Eng Optim 38(2):129–154
21. Montané FAT, Galvao RD (2006) A tabu search algorithm for the vehicle routing problem with simultaneous pick-up and delivery service. Comput Oper Res 33(3):595–619
22. Oftadeh R, Mahjoob M, Shariatpanahi M (2010) A novel meta-heuristic optimization algorithm inspired by group hunting of animals: hunting search. Comput Math Appl 60(7):2087–2098
23. Simon D (2008) Biogeography-based optimization. IEEE Trans Evol Comput 12(6):702–713
24. Meng X-B, Gao XZ, Liu Y, Zhang H (2015) A novel bat algorithm with habitat selection and doppler effect in echoes for optimization. Expert Syst Appl 42(17–18):6350–6364
25. Shen W, Guo X, Wu C, Wu D (2011) Forecasting stock indices using radial basis function neural networks optimized by artificial fish swarm algorithm. Knowl-Based Syst 24(3):378–385
26. Passino KM (2010) Bacterial foraging optimization. Int J Swarm Intell Res (IJSIR) 1(1):1–16
27. Liang JJ, Qin AK, Suganthan PN, Baskar S (2006) Comprehensive learning particle swarm optimizer for global optimization of multimodal functions. IEEE Trans Evol Comput 10(3):281–295
28. Van den Bergh F, Engelbrecht AP (2004) A cooperative approach to particle swarm optimization. IEEE Trans Evol Comput 8(3):225–239
29. Cheng R, Jin Y (2015) A social learning particle swarm optimization algorithm for scalable optimization. Inf Sci 291:43–60
30. Dey N (2017) Advancements in applied metaheuristic computing. IGI Global
31. Gutierrez CE, Alsharif MR, Cuiwei H, Khosravy M, Villa R, Yamashita K, Miyagi H (2013) Uncover news dynamic by principal component analysis. Shanghai, China, ICIC Express Lett 7(4):1245–1250
32. Gutierrez CE, Alsharif PMR, Khosravy M, Yamashita PK, Miyagi PH, Villa R (2014) Main large data set features detection by a linear predictor model. In: AIP conference proceedings, vol 1618. AIP, pp 733–737
33. Gutierrez CE, Alsharif MR, Yamashita K, Khosravy M (2014) A tweets mining approach to detection of critical events characteristics using random forest. Int J Next-Gener Comput 5(2):167–176
34. Sedaaghi MH, Khosravi M (2003) Morphological ECG signal preprocessing with more efficient baseline drift removal. In: 7th. IASTED international conference, ASC, pp 205–209
35. Khosravi M, Sedaaghi MH (2004) Impulsive noise suppression of electrocardiogram signals with mediated morphological filters. In: 11th Iranian conference on biomedical engineering, ICBME, pp 207–212
36. Khosravy M, Asharif MR, Sedaaghi MH (2008) Medical image noise suppression using mediated morphology. IEICE Tech IEICE Rep 265–270
37. Khosravy M, Gupta N, Marina N, Sethi I, Asharifa M (2017) Perceptual adaptation of image based on chevreulmach bands visual phenomenonn. IEEE Signal Process Lett 24(5):594–598
38. Khosravy M, Gupta N, Marina N, Sethi I, Asharif M (2017) Brain action inspired morphological image enhancement. In: Nature-inspired computing and optimization. Springer, Cham, pp 381–407
39. Khosravy M, Alsharif MR, Guo B, Lin H, Yamashita K (2009) A robust and precise solution to permutation indeterminacy and complex scaling ambiguity in BSS-based blind MIMO-OFDM receiver. In: International conference on independent component analysis and signal separation, Springer, pp 670–677
40. Asharif F, Tamaki S, Alsharif MR, Khosravy M, Ryu H (2013) Performance improvement of constant modulus algorithm blind equalizer for 16 QAM modulation. Int J Innov Comput, Inf Control 7(4):1377–1384

41. Khosravy M, Alsharif MR, Yamashita K (2009) An efficient ICA based approach to multiuser detection in MIMO OFDM systems. In: Multi-carrier systems and solutions. Springer, pp 47–56
42. Khosravy M, Alsharif MR, Khosravi M, Yamashita K (2010) An optimum pre-filter for ICA based mulit-input multi-output OFDM system. In: 2010 2nd international conference on education technology and computer, vol 5. IEEE, pp V5–129
43. Khosravy M, Patel N, Gupta N, Sethi I (2019) Image quality assessment: a review to full reference indexes. In: Recent trends in communication, computing, and electronics. Springer, pp 279–288
44. Khosravy M, Asharif MR, Sedaaghi MH (2008) Morphological adult and fetal ECG preprocessing: employing mediated morphology. IEICE Tech Rep IEICE 107:363–369
45. Sedaaghi MH, Daj R, Khosravi M (2001) Mediated morphological filters. In: Proceedings 2001 international conference on image processing, vol 3. IEEE, pp 692–695
46. Khosravy M, Gupta N, Marina N, Sethi IK, Asharif MR (2017) Morphological filters: an inspiration from natural geometrical erosion and dilation. In: Nature-inspired computing and optimization. Springer, Cham, pp 349–379
47. Khosravy M, Asharif MR, Yamashita K (2009) A pdf-matched short-term linear predictability approach to blind source separation. Int J Innov Comput Inf Control (IJICIC) 5(11):3677–3690
48. Khosravy M, Alsharif MR, Yamashita K (2009) A PDF-matched modification to stones measure of predictability for blind source separation. In: International Symposium on Neural Networks. Springer, Berlin, pp 219–222
49. Khosravy M, Asharif MR, Yamashita K (2011) A theoretical discussion on the foundation of stones blind source separation. Signal, Image Video Process 5(3):379–388
50. Khosravy M, Asharif M, Yamashita K (2008) A probabilistic short-length linear predictability approach to blind source separation. In: 23rd international technical conference on circuits/systems, computers and communications (ITC-CSCC 2008). Yamaguchi, Japan, pp 381–384
51. Khosravy M, Alsharif MR, Yamashita K (2009) A pdf-matched modification to stones measure of predictability for blind source separation. In: International symposium on neural networks, Springer, pp 219–228
52. Khosravy M, Gupta M, Marina M, Asharif MR, Asharif F, Sethi I (2015) Blind components processing a novel approach to array signal processing: a research orientation. In: 2015 international conference on intelligent informatics and biomedical sciences, ICIIBMS, pp 20–26
53. Khosravy M, Punkoska N, Asharif F, Asharif MR (2014) Acoustic OFDM data embedding by reversible walsh-hadamard transform. In: AIP conference proceedings. AIP vol 1618, pp. 720–723
54. Yang X-S (2010) Firefly algorithm, stochastic test functions and design optimisation, arXiv preprint arXiv:1003.1409
55. Dey N, Samanta S, Chakraborty S, Das A, Chaudhuri SS, Suri JS (2014) Firefly algorithm for optimization of scaling factors during embedding of manifold medical information: an application in ophthalmology imaging. J Med Imaging Health Inform 4(3):384–394
56. Kumar R, Rajan A, Talukdar FA, Dey N, Santhi V, Balas VE (2017) Optimization of 5.5-ghz cmos lna parameters using firefly algorithm. Neural Comput Appl 28(12):3765–3779
57. Jagatheesan K, Anand B, Samanta S, Dey N, Ashour AS, Balas VE (2017) Design of a proportional-integral-derivative controller for an automatic generation control of multi-area power thermal systems using firefly algorithm. IEEE/CAA J Autom Sin
58. Kumar R, Talukdar FA, Dey N, Balas VE (2016) Quality factor optimization of spiral inductor using firefly algorithm and its application in amplifier. Int J Adv Intell Parad
59. Chakraborty S, Dey N, Samanta S, Ashour AS, Balas VE (2016) Firefly algorithm for optimized nonrigid demons registration. In: Bio-inspired computation and applications in image processing, Elsevier, pp 221–237
60. Samanta S, Mukherjee A, Ashour AS, Dey N, Tavares JMR, Abdessalem Karâa WB, Taiar R, Azar AT, Hassanien AE (2018) Log transform based optimal image enhancement using firefly algorithm for autonomous mini unmanned aerial vehicle: an application of aerial photography. Int J Image Graph 18(04):1850019

61. Fister I, Fister J, Yang XS, Brest J (2013) A comprehensive review of firefly algorithms. Swarm Evol Comput 13:34–46
62. Gupta N, Khosravy M, Patel N, Sethi I (2018) Evolutionary optimization based on biological evolution in plants. Procedia Comput Sci 126:146–155

Chapter 10
Firefly Algorithm-Based Kapur's Thresholding and Hough Transform to Extract Leukocyte Section from Hematological Images

Venkatesan Rajinikanth, Nilanjan Dey, Ergina Kavallieratou and Hong Lin

1 Introduction

In recent years, a significant number of traditional and soft computing-assisted techniques are extensively considered to preprocess and post-process and the digital pictures are existing in various domains, such as image enhancement [1], biometrics [2], document processing [3–7], and clinical level disease evaluation [7–11]. Assessment of a chosen image using a selected traditional technique may require more computations, and sometimes the traditional procedure may not be suitable to build an automated image evaluation tool (AIET).

AIET is always necessary to implement an automated evaluation of the digital images, since it needs very minimal operator involvement. Most of the AIET techniques are employed with the recent soft computing methods, which will work independently with lesser operator participation.

V. Rajinikanth (✉)
St. Joseph's AI Group, St. Joseph's College of Engineering, Chennai, Tamilnadu 600119, India
e-mail: v.rajinikanth@ieee.org

N. Dey
Department of Information Technology, Techno India College of Technology, Kolkata,
West Bengal 700156, India
e-mail: neelanjan.dey@gmail.com

E. Kavallieratou
Department of Information and Communication Systems Engineering, University of the Aegean,
Samos, Greece
e-mail: kavallieratou@aegean.gr

H. Lin
Department of Computer Science & Engineering Technology, University of Houston-Downtown,
Houston, USA
e-mail: linh@uhd.edu

© Springer Nature Singapore Pte Ltd. 2020
N. Dey (ed.), *Applications of Firefly Algorithm and its Variants*,
Springer Tracts in Nature-Inspired Computing,
https://doi.org/10.1007/978-981-15-0306-1_10

The recent works in the literature authenticate the availability of a considerable number of traditional and modern image evaluation procedures for a class of clinical level disease evaluation process. In medical clinics, the usage of the automated disease evaluation tool will reduce the examination time and it also serves as an aiding tool for the doctors while taking the decision during the disease identification and treatment planning. Further, the automated disease evaluation procedure will ease the medical image examination task. Development of a disease evaluation tool for grayscaled medical imagery, such as computed tomography (CT), magnetic resonance imaging (MRI), digital mammogram (DM), and ultrasonic image (UI), is quite easy compared to the RGB scale pictures [12–14]. The RGB scale pictures, such as the fundus retinal image (FRI), thermal images, dermoscopy pictures, and hematological images (HI) [15–18], are quite difficult due to its complex nature.

The chief aim of this research work is to develop an automated hematological image evaluation procedure based on a soft computing-assisted technique [19–21]. The proposed approach initially implements an image enhancement with Kapur's multi-thresholding and later employs a Hough transform and morphology-based circle detection practice to extract the leukocyte region with greater accuracy. Implementation of a traditional Kapur's multi-thresholding is really a challenging task, and it requires more computation time compared to soft computing-based techniques. Since, in traditional multi-thresholding, the threshold values are randomly varied and the corresponding Kapur's entropy is to be computed manually. To reduce the complexity, in recent years, soft computing techniques, such as swarm-based and evolutionary-based methods, are commonly implemented by the researchers [19–22].

In the literature, a substantial amount of nature motivated algorithms are discussed and implemented to find the answer for a class of optimization problems. Every algorithm has its own merit and demerits and choosing a suitable method to solve the optimization task depends on (i) success rate, (ii) ease in realization, and (iii) reputation. Firefly algorithm (FA) is one of the nature-inspired, swarm-related metaheuristic techniques, proposed its optimal success for a class of optimization task over a decade [23]. The added advantage of the FA is, its mathematical expression is simple and the possible modification/enhancement is quite simple compared to other metaheuristic techniques. Further, the success rate, real-time implementation, and reputation of the FA are better compared to few metaheuristic approaches available in the literature [24–29].

Due to these reasons, the proposed research adopts the modified version of FA called the Brownian walk FA (BFA). This algorithm is quite sluggish compared to the traditional Lévy flight FA (LFA); but it tenders the best success rate during the optimization search [30]. The proposed AIET employs the BFA-based Kapur's thresholding as the first stage to preprocess the hematological images and Hough transform-based circle detection with morphological operation as the second stage to extract the leukocyte region from the thresholded hematological images.

A comparative evaluation among the mined leukocyte region and the Ground truth (GT) is then performed and the necessary image performance parameters (IPP), like accuracy, precision, sensitivity, specificity, and F1 score, are computed to confirm the supremacy of proposed technique. Further, this approach is validated against the recent techniques, such as Shannon's entropy-based Chan-Vese (SECV) [17] and Shannon's entropy-based level-set (SELS) methods [18]. The experimental outcome of the proposed technique confirms that proposed AIET offered better accuracy contrast with the alternative approaches.

The remaining part of this chapter is organized as follows: Sect. 2 presents the details of the FA and leukocyte segmentation techniques; Sect. 3 discusses the methods applied in this chapter; solutions and recommendations are discussed in Sect. 4; future research path and conclusion of this study are discussed in Sects. 5 and 6, correspondingly.

2   Related Works

Firefly algorithm (FA) is one of the most successful nature-inspired metaheuristic technique extensively considered by the researchers to resolve a selection of optimization problems ranging from a numerical optimization to real-world problems [25, 26]. Further, a variety of modifications have been incorporated in the traditional FA to enhance its outcome in search time, optimization accuracy, utilizing the benefit of other algorithms and mathematical concepts, and developing new techniques by keeping the traditional FA as the base [27, 28].

The recent review articles also confirm the performance of FA in solving the variety of engineering optimization task [25–27] and the recent article by Yang and He [28] confirms how and why the FA works well in a class of optimization tasks. Due to these reasons, the FA-assisted procedure is implemented in the proposed work to preprocess the RGB-scaled test image. The work adopted in this research implements the BFA discussed in [29, 30], in which every firefly is driven along with a Brownian operator during the search space exploration. The previous work of Raja et al. [31] confirmed that BFA offered superior result on Otsu-based grayscale picture multi-thresholding process. Recently, a considerable number of circle detection approaches are proposed to detect the circular/elliptical objects existing in the test image [32–35]. The major aim of the circle detection technique is to extract the circular shapes existing in the test images. Venkatalakshmi and Thilagavathi [36] implemented HT method to extract and evaluate the red blood cells (RBC). The work of Bagui and Zoueu [37] also implemented HT to extract the RBC from the clinical grade images. Cuevas et al. [38] implemented an approach to extract and evaluate the white blood cells from the blood smear images. A detailed review on the HT-based edge detection can be found in Illingworth and Kittler [34] and Mukhopadhyay and Chaudhuri [35]. All the above works confirm that the HT technique is efficient in segmenting the circular/elliptical section from the test image. The work of Prinyakupt and Pluempitiwiriyawej [39] also verifies that the detection and evaluation of the leukocyte using a suitable image processing practices are essential, and this technique helps in identifying the disease based on the analysis performed by the leukocyte section.

The proposed work aims to develop an automated leukocyte extraction tool using the BFA and KE-based thresholding and HT-based segmentation.

3 Main Focus of the Chapter

The proposed work aims to implement an automated tool to extract the leukocyte section from an RGB-scaled hematological image with an improved accuracy. Figure 1 depicts the various stages existing in the proposed AIET.

Fig. 1 Stages in the proposed AIET

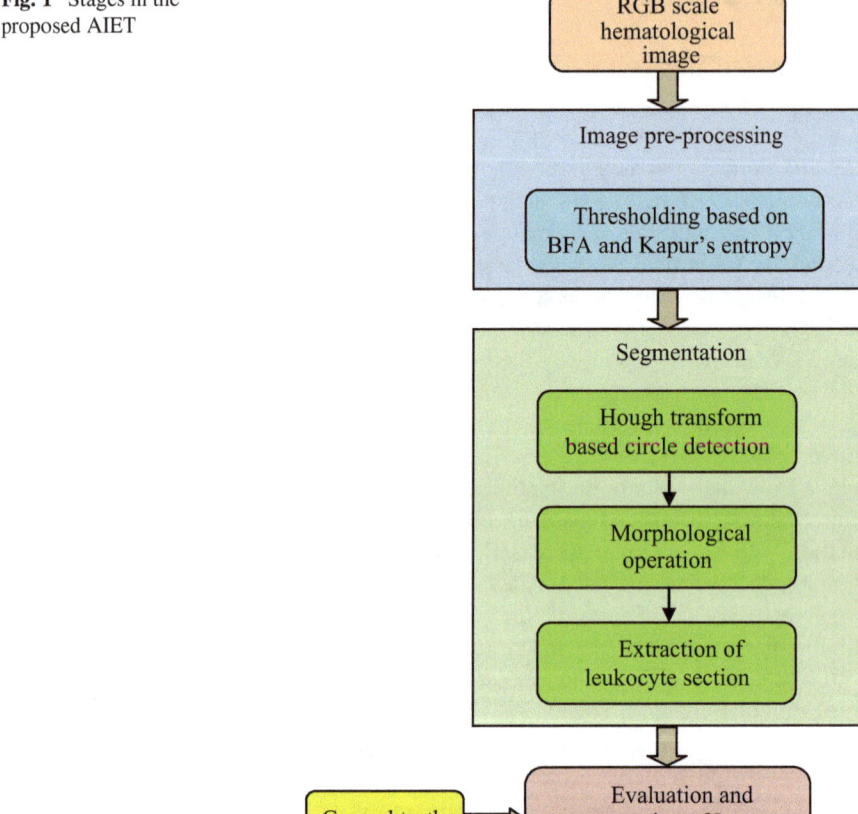

The benchmark test image collected from the database is initially preprocessed with the adopted thresholding technique. This thresholding will enhance the leukocyte section by eliminating the other blood cells existing in the background. Normally, the leukocyte is stained with a chosen coloring agent, and identification and enhancement of the leukocyte pixels of the chosen digital hematological image are achieved with the BFA+KE thresholding. After the enhancement, automated extraction of the leukocyte is achieved in post-processing technique, which involves in Hough transform-based circle detection, morphological operation to dilate/eliminate the circles with lesser radius, mining of the leukocyte section. The mined leukocyte region is in binary form (black background and white leukocyte section), which helps to perform a relative evaluation with the related ground truth (GT) image. After this study, essential image performance parameters (IPP) are computed, and based on this value, the performance of the proposed AIET is verified. Further, the proposed AIET is also validated with the other existing techniques in the literature.

3.1 LISE Database

The hematological images available in 'leukocyte images for segmentation and classification' (LISC) database are used for the examination. This dataset consists of the leukocytes, such as basophil (53 images), eosinophil (39), lymphocyte (52 images), monocyte (48 images), neutrophil (50 images), and mixed cases (8 images). All these images are in RGB scale with a dimension of 720×576 pixels. This is one of the clinical grade thin blood smear image widely adopted to test the computerized leukocyte extraction and identification task. This image dataset was prepared and contributed by Rezatofighi and Soltanian-Zadeh [40]; who executed a computerized technique to extract and classify the different leukocyte sections, and this dataset and its explanation can be found in [41].

3.2 Firefly Algorithm

The initial version of the FA was invented by Yang in 2008 [42] to find optimal solutions for a class of numerical problems. Later, due to its optimal accuracy and real-time realization, a number of researchers are considered the traditional FA (TFA) to find solutions for their chosen problems [26–29] and modified the TFA with various methods to further enhance its optimal accuracy [30].

The BFA is one of the modified forms of the TFA initially proposed by Raja et al. [30], in which the fireflies are driven by a physical law known as Brownian walk. This BFA is then adopted in image processing [27] and process control [38, 43] applications and the outcome of these studies confirmed that BFA offers the better result for the chosen problem compared to the TFA and approaches, like genetic algorithm, particle swarm, and bacterial foraging techniques.

In the BFA, the Lévy operator is modified by the Brownian operator and the relation between the Lévy and Brownian is clearly discussed in the work of Raja et al. [31].

To express BFA, consider two fireflies (agents) A and B. During the image thresholding task, the light emitted (flash) in most fitting agent is more contrast to other agent and the movement of the attracted agent 'A' toward the glittering agent 'B' can be mathematically calculated with the help of Eq. (1):

$$F_A^{t+1} = F_A^t + \beta_0 e^{-\gamma\, d_{AB}^2}(F_B^t - F_A^t) + \alpha_1.\text{sign}(\Re - 1/2) \oplus W(s) \qquad (1)$$

where F_A^{t+1} = restructured position of an agent, F_A^t = initial position of an agent, $\beta_0 e^{-\gamma\, d_{AB}^2}(F_B^t - F_A^t)$ = attractive energy among agents, $\Re$ = random number, $W(s) = V.|s|^{\alpha/2}$ = Brownian walk expression, V = subjective variable, β = spatial factor and α = temporal factor. In Brownian walk, β is assigned as two and α is fixed as one. Additional details on Brownian walk can be found in [44–46].

3.3 Kapur's Thresholding

Recently, entropy-assisted image examination schemes are widely adopted by the researchers, due to its superiority [47–50]. Kapur's is one of the entropy-assisted approaches considered to threshold digital pictures [51, 52].

This approach is mathematically expressed as follows:

Let, $Th = [th_1, th_2, ..., th_{k-1}]$ denotes threshold vector.

For a given image, the thresholds 'Th' are arbitrarily varied manually or with the help of a chosen soft computing technique until the entropy reaches the maximal limit '$J_{\max}$.' In this work, the BFA is employed to vary the thresholds, till '$J_{\max}$' is reached.

Kapur's function for a preferred threshold can be denoted as

$$J_{\max} = f_{\text{kapur}}(Th) = \sum_{j=1}^{k} H_j^C \qquad (2)$$

In RGB image, the entropy is separately computed for every color (R, G, and B for a chosen Th). In the proposed work, the 'Th' is selected as three; hence for every test image, the BFA is employed to three different thresholds for R, G, and B pixel group, which enhances the leukocyte section. In this work, the following initial values are assigned for the BFA number of agents = 30, search dimension = 3, number of iterations = 2000, and stopping function = $J_{\max}$.

3.4 Hough Transform Implementation

Hough transform (HT) is one of the commonly used procedures to detect the circles/circular-shaped objects from the chosen digital image [32–34].

The HT implemented in this study is discussed in [53].

Let, r = radios, X and Y = image axis, A and B = center of an arbitrary circle. Then, it can be expressed as

$$(X - A)^2 + (Y - B)^2 = r^2 \tag{3}$$

In this, the HT is responsible to identify the X and Y based on the chosen A, B, and r. The HT is allowed to locate and trace the binary pixels (1's) existing in the test picture after a possible border detection process. The traditional HT traces and extracts all the existing highly visible binary pixels (1's) in the digital image. In most of the cases, this HT fails to extract the leukocyte section from the hematological image. HT-based circle recognition process can be found in [35, 39]. In order to improve the success rate in HT-based circle detection, the morphological operation is included in this tool to eliminate the unwanted visible pixels in the preprocessed image.

The following steps present the modifications implemented in HT to improve its segmentation accuracy:

Step 1: Consider the preprocessed hematological image and implement the border detection
Step 2: Execute the HT with a chosen circle radius and identify the entire pixel groups with a magnitude unity
Step 3: Extract the Haralick features and identify the major axis from the identified pixel groups
Step 4: Implement the morphological operation to eliminate the pixel groups, which is < the identified major axis
Step 5: Enhance the leukocyte section with a pixel-level comparison
Step 6: Extract the existing pixel group and validate it with the traces made by the HT.

3.5 Appraisal

The performance of AIET configuration was authenticated by calculating the image performance parameters (IPP) such as accuracy, precision, sensitivity, specificity, and F1 score, and its mathematical expressions are depicted below [47–49, 54, 55];

$$\text{Accuracy} = (\text{TP} + \text{TN})/(\text{TP} + \text{TN} + \text{FP} + \text{FN}) \tag{4}$$

$$\text{Precision} = \text{TP}/(\text{TP} + \text{FP}) \tag{5}$$

$$\text{Sensitvity} = \text{Recall} = \text{TP}/(\text{TP} + \text{FN}) \tag{6}$$

$$\text{Specificity} = \text{TN}/(\text{TN} + \text{FP}) \tag{7}$$

$$\text{F1 Score} = 2\text{TP}/(2\text{TP} + \text{FP} + \text{FN}) \tag{8}$$

$$\text{True-Positive-Rate} = \text{TP}_{\text{rate}} = \text{TP}/P \tag{9}$$

$$\text{True-Negative-Rate} = \text{TN}_{\text{rate}} = \text{TN}/N \tag{10}$$

$$\text{False-Positive-Rate} = \text{FP}_{\text{rate}} = \text{FP}/N \tag{11}$$

$$\text{False-Negative-Rate} = \text{FN}_{\text{rate}} = \text{FN}/P \tag{12}$$

where TN, TP, FN, and FP represent true-negative, true-positive, false-negative, and false-positive, respectively.

4   Solutions and Recommendations

This division of the chapter presents the experimental outcomes and its discussions. This work is executed with the RGB-scaled hematological images of LISC database [41]. Figure 2 depicts the sample test images adopted in this work, in which Fig. 2a illustrates the sample examination pictures and Fig. 2b depicts the associated GT.

Initially, the BFA+KE-based thresholding is implemented for this image, which helps to enhance the stained leukocyte section of the test picture considerably. After the possible enhancement, the post-processing method is employed to mine the leukocyte image. This step consists of the procedures, such as (i) detection of all the possible regions using Hough transform, (ii) evaluation of the region with large major axis using the Haralick algorithm, (iii) implementing the morphological dilation and erosion of the pixel groups whose dimension is lesser than the major axis, (iv) detection and extraction of the binary form of leukocyte. The above-said technique is repeatedly implemented on all the images existing in the LISC database. After extracting the leukocyte, the IPPs are then computed by means of a comparison with the GT. Figure 3 presents the outcome attained with the AIET.

Figure 3a, b depicts the test image adopted for the demonstration and its related GT. Normally, the GT will be in the form of binary (background is assigned with 0's, and the leukocyte section is assigned with 1's). HT is then implemented on the threshold picture. Before implementing the HT, the edge detection is executed and

(a)   **(b)**   **(c)**   **(d)**

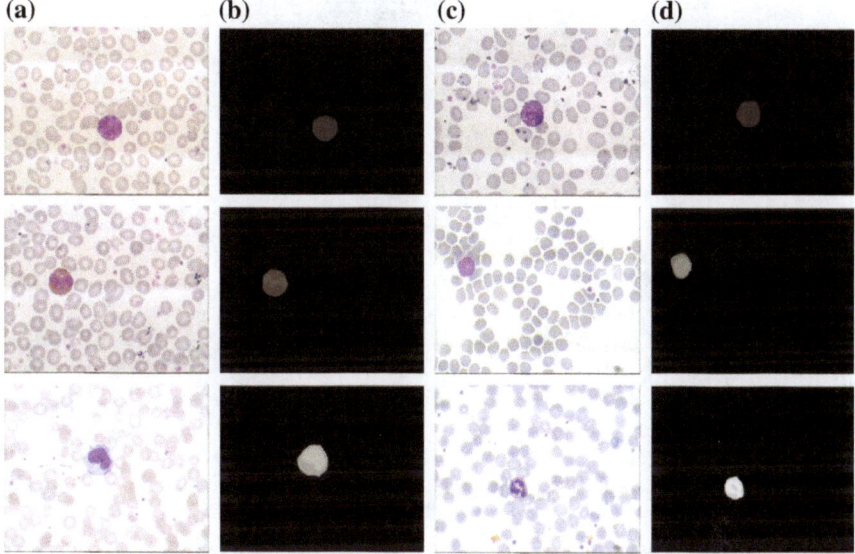

Fig. 2 Sample test images of LISC database

it will trace all the possible edges of Fig. 3c and offers a binary image structure. The HT with an assigned circle radius (radius = 5) then traces all the pixel groups of Fig. 3c and provides the image Fig. 3d. From Fig. 3d, it can be noted that the traditional HT traces all the pixel groups along with the existing leukocyte section. If this image is considered as it is for the relative study with the GT, the IPP values will be poor. Hence, it is necessary to implement the Haralick texture evaluation to compute the major axis of the detected circle-like structure. After finding the major axis, a morphological procedure is to be implemented to dilate/eliminate the pixel groups of Fig. 3d, which is lesser than the major axis.

Initially, the test picture is preprocessed with the BFA + KE-based three-level thresholding process and the enhanced picture is depicted in Fig. 3c. The outcome of this operation is shown in Fig. 3e. After the morphological elimination, the circle detection based on the HT is repeated and its outcome is presented in Fig. 3f. After the detection, a pixel-level enhancement (multiplication and color fill) is then implemented and the obtained result is shown in Fig. 3g. After the enhancement, the leukocyte section is extracted and its binary form is presented in Fig. 3h.

Later, a relative assessment among Fig. 3b, h is implemented and the obtained pixel-level values are as follows: TP = 6514, TN = 406,973, FP = 0, FN = 1233, accuracy = 99.70%, precision = 100%, sensitivity = 84.08%, specificity = 100%, and F1 score = 91.35%.

The performance of this comparison is then confirmed with a confusion matrix as shown in Fig. 4. This figure confirms that the proposed technique extracts the leukocyte region exactly from the adopted test picture. The proposed confusion

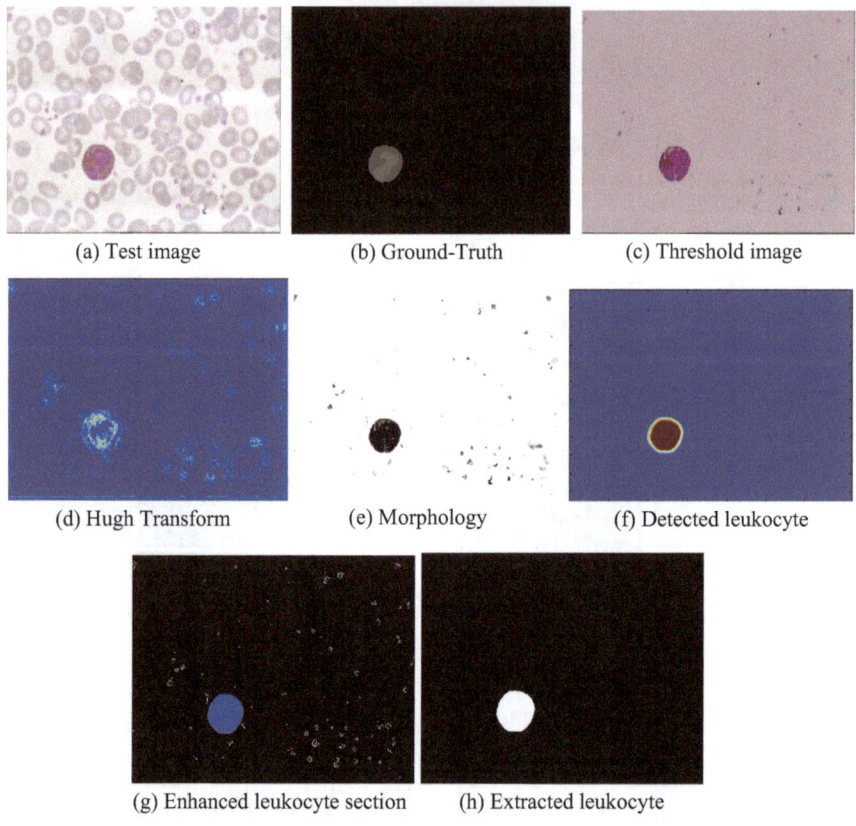

(a) Test image   (b) Ground-Truth   (c) Threshold image

(d) Hugh Transform   (e) Morphology   (f) Detected leukocyte

(g) Enhanced leukocyte section   (h) Extracted leukocyte

Fig. 3 Sample results obtained with the proposed method on eosinophil image

Actual Class

$TP=6514$ (pixels)	$FP=0$ (pixels)	$TP_{rate}=0.8408$	$TP_{rate}+FN_{rate}=1$
$FN=1233$ (pixels)	$TN=406973$ (pixels)	$FN_{rate}=0.1592$	
$P=TP+FN$ (pixels)	$N=FP+TN$ (pixels)	$TN_{rate}=1$	$TN_{rate}+FP_{rate}=1$
414720 (Total pixels=P+N)		$FP_{rate}=0$	

Identified Class

Fig. 4 Confusion matrix for the test picture

Table 1 Validation of AIET with other existing approaches

Method	Accuracy (%)	Precision (%)	Sensitivity (%)	Specificity (%)	F1 score (%)
Shannon's entropy-based Chan-Vese [17]	99.52	93.85	94.03	99.19	98.28
Proposed AIET (Basophil)	98.96	94.27	94.86	99.26	97.88
Shannon's entropy-based level-set [18]	98.14	94.62	90.28	97.63	96.24
Proposed AIET (LISC)	97.96	94.68	91.42	98.41	97.28

matrix provides all the essential IPPs computed in the proposed work. Figure 4 presents the IPPs attained for the chosen test image of Fig. 3.

Similar results are obtained with other images existing in LISC database, and its mean IPPs are then compared against the existing methods in the literature, and the values are presented in Table 1.

The results of Table 1 confirm that proposed AIET offers better precision, sensitivity, and specificity compared to Shannon's entropy-based Chan-Vese extraction technique for the basophil class images. This table also confirms that the overall performance of the proposed tool is better compared to the average result of Shannon's entropy-based level-set on the considered LISC database. These results verify that the proposed tool is an automated technique and offered better overall result compared to the semi-automated approaches available in the literature.

5 Future Research Direction

The work discussed in this chapter implemented a BFA-based AIET to examine the benchmark hematological images of the LISC database. The current work incorporates a pre- and post-processing image processing system to attain a better accuracy as discussed in earlier works [17, 18]. The chief task implemented in this work is the mining of the leukocyte section and comparing it with the GT.

This work considered only the BFA [56, 57] and the performance of AIET is confirmed based on the IPPs. In future, the performance of the BFA can be evaluated based on (i) computation time, (ii) implementation issue compared to other algorithms, (iii) optimization accuracy, and (iv) probability of receiving optimal results during multi-thresholding.

In this work, the preprocessing is executed using KE. In future, a study can be implemented to assess the performance of the KE with other techniques, like Otsu, Tsallis, and Shannon, and the best suitable thresholding technique for the LISC dataset can be confirmed.

After extracting the possible leukocyte section, the texture, as well as shape features, can be extracted using a chosen pattern recognition technique. A classifier system can be incorporated to categorize the LISC dataset into various classes.

6 Conclusion

In this chapter, an automated leukocyte examination procedure is proposed using the most successful nature-inspired metaheuristic technique. The developed AIET is experimentally validated using the benchmark LISC dataset. This work initially employs a preprocessing procedure based on the BFA and KE thresholding. This preprocessing enhances the stained leukocyte section by eliminating other sections existing in the considered hematology picture. Later, a HT is then implemented to identify and extract the leukocyte segment. The traditional HT detected all the existing pixels in the preprocessed image, which may result in a false outcome. Hence, the HT is then combined with other procedures, like Haralick texture feature extraction, morphological dilation, enhancement, and extraction of the leukocyte. The performance of the AIET is then established by calculating the essential IPPs with a relative study among the leukocyte and GT. Finally, the outcomes of the proposed technique are validated against other techniques obtainable in the literature. The experimental outcome verifies that proposed AIET offered better overall result compared to the semiautomated techniques existing in the literature. The results of AIET also confirm its clinical significance, and in future, the AIET can be considered to appraise the medical grade hematology images.

References

1. Ashour AS et al (2015) Computed tomography image enhancement using cuckoo search: a log transform based approach. J Signal Inf Process 6(3):244–257. https://doi.org/10.4236/jsip.2015.63023
2. Dey N et al (2013) Retention of electrocardiogram features insignificantly devalorized as an effect of watermarking for a multimodal biometric authentication system. Adv Biom Secur Hum Authentication Recognit 175
3. Kar R, Saha S, Bera SK, Kavallieratou E, Bhateja V, Sarkar R (2019) Novel approaches towards slope and slant correction for tri-script handwritten word images. Imaging Sci J 67(3):159–170
4. Koubarakis M et al (2018) AI in Greece: the case of research on linked geospatial data. AI Mag 39(2):91–96
5. Karampidis K, Kavallieratou E, Papadourakis G (2018) A review of image steganalysis techniques for digital forensics. J Inf Secur Appl 40:217–235
6. Kavallieratou E, Likforman-Sulem L, Vasilopoulos N (2018) Slant removal technique for historical document images. J Imaging 4(6):80
7. Satapathy SC, Rajinikanth V (2018) Jaya algorithm guided procedure to segment tumor from brain MRI. J Optim 2018:12. https://doi.org/10.1155/2018/3738049

8. Raja NSM, Rajinikanth V, Fernandes SL, Satapathy SC (2017) Segmentation of breast thermal images using Kapur's entropy and hidden Markov random field. J Med Imaging Health Inform 7(8):1825–1829

9. Fernandes SL, Rajinikanth V, Kadry S (2019) A hybrid framework to evaluate breast abnormality. IEEE Consum Electron Mag. https://doi.org/10.1109/MCE.2019.2905488

10. Wang Y et al (2019) Classification of mice hepatic granuloma microscopic images based on a deep convolutional neural network. Appl Soft Comput 74:40–50. https://doi.org/10.1016/j.asoc.2018.10.006

11. Wang Y et al (2019) Morphological segmentation analysis and texture-based support vector machines classification on mice liver fibrosis microscopic images. Curr Bioinform 14(4):282–294. https://doi.org/10.2174/1574893614666190304125221

12. Rajinikanth V, Dey N, Kumar R, Panneerselvam J, Raja NSM (2019) Fetal head periphery extraction from ultrasound image using jaya algorithm and Chan-Vese segmentation. Procedia Comput Sci 152:66–73. https://doi.org/10.1016/j.procs.2019.05.028

13. Rajinikanth V, Dey N, Satapathy SC, Ashour AS (2018) An approach to examine magnetic resonance angiography based on Tsallis entropy and deformable snake model. Futur Gener Comput Syst 85:160–172

14. Dey N et al (2014) Firefly algorithm for optimization of scaling factors during embedding of manifold medical information: an application in ophthalmology imaging. J Med Imaging Health Inform 4(3):384–394. https://doi.org/10.1166/jmihi.2014.1265

15. Dey N, Rajinikanth V, Ashour AS, Tavares JMRS (2018) Social group optimization supported segmentation and evaluation of skin melanoma images. Symmetry 10(2):51. https://doi.org/10.3390/sym10020051

16. Moraru L, Obreja CD, Dey, N, Ashour AS (2018) Dempster-shafer fusion for effective retinal vessels' diameter measurement. Soft Comput Based Med Image Anal 149–160

17. Dey N, Shi F, Rajinikanth V (2019) Leukocyte nuclei segmentation using entropy function and Chan-Vese approach. Inf Technol Intell Transp Syst 314:255–264. https://doi.org/10.3233/978-1-61499-939-3-255

18. Raja NSM, Arunmozhi S, Lin H, Dey N, Rajinikanth V (2019) A study on segmentation of leukocyte image with Shannon's entropy. Histopathol Image Anal Med Decis Mak, 1–27. https://doi.org/10.4018/978-1-5225-6316-7.ch001

19. Sghaier S, Farhat W, Souani C (2018) Novel technique for 3D face recognition using anthropometric methodology. Int J Ambient Comput Intell 9(1):60–77. https://doi.org/10.4018/ijaci.2018010104

20. Hemalatha S, Anouncia SM (2016) A computational model for texture analysis in images with fractional differential filter for texture detection. Int J Ambient Comput Intell 7(2):93–113. https://doi.org/10.4018/IJACI.2016070105

21. Hu J, Fan XP, Liu S, Huang L (2019) Robust target tracking algorithm based on superpixel visual attention mechanism: robust target tracking algorithm. Int J Ambient Comput Intell 10(2):1–17. https://doi.org/10.4018/IJACI.2019040101

22. Yang XS (2010) Engineering optimization: an Introduction with metaheuristic applications. Wiley & Sons, New Jersey

23. Yang XS (2010) Firefly algorithm, Lévy flights and global optimization. In: Bramer M, Ellis R, Petridis M (eds) Research and development in intelligent systems XXVI. Springer, London. https://doi.org/10.1007/978-1-84882-983-1_15

24. Gandomi AH, Yang X-S, Talatahari S, Alavi AH (2013) Firefly algorithm with chaos. Commun Nonlinear Sci Numer Simul 18(1):89–98

25. Tilahun SL, Ngnotchouye JMT (2017) Firefly algorithm for discrete optimization problems: A survey. KSCE J Civ Eng 21(2):535–545

26. Fister I, Yang X-S, Brest J (2013) A comprehensive review of firefly algorithms. Swarm Evol Comput 13:34–46. https://doi.org/10.1016/j.swevo.2013.06.001

27. Fister I, Yang X-S, Fister D, Fister I (2014) Firefly algorithm: a brief review of the expanding literature. In: Cuckoo search and firefly algorithm. Springer. pp 347–360. https://doi.org/10.1007/978-3-319-02141-6-17

28. Yang XS, He X. Why the Firefly Algorithm Works? https://doi.org/10.1007/978-3-319-67669-2_11arXiv:1806.01632%5bcs.NE

29. Dey N (ed) (2017) Advancements in applied metaheuristic computing. IGI Global

30. Raja NSM, Manic KS, Rajinikanth V (2013) Firefly algorithm with various randomization parameters: an analysis. Lect Notes Comput Sci 8297:110–121. https://doi.org/10.1007/978-3-319-03753-0_11

31. Raja NSM, Rajinikanth V, Latha K (2014) Otsu based optimal multilevel image thresholding using firefly algorithm. Model Simul Eng, 2014:17. Article ID 794574

32. Xu L, Oja E (1993) Randomized hough transform (RHT): basic mechanisms, algorithms, and computational complexities. CVGIP: Image Underst 57(2):131–154. https://doi.org/10.1006/ciun.1993.1009

33. Xu L, Oja E, Kultanen K (1990) A new curve detection method: randomized hough transform (RHT). Pattern Recogn Lett 11(5):331–338. https://doi.org/10.1016/0167-8655(90)90042-Z

34. Illingworth J, Kittler J (1988) A survey of the Hough transform. Comput Vis, Graph, Image Process 44(1):87–116. https://doi.org/10.1016/S0734-189X(88)80033-1

35. Mukhopadhyay P, Chaudhuri BB (2015) A survey of Hough transform. Pattern Recogn 48(3):993–1010

36. Venkatalakshmi B, Thilagavathi K (2013) Automatic red blood cell counting using Hough transform. In. IEEE conference on information and communication technologies, pp 267–271. https://doi.org/10.1109/cict.2013.6558103

37. Bagui OK, Zoueu JT (2014) Red blood cells counting by circular Hough transform using multispectral images. J Appl Sci 14:3591–3594. https://doi.org/10.3923/jas.2014.3591.3594

38. Cuevas E, Díaz M, Manzanares M, Zaldivar D, Perez-Cisneros M (2013) An improved computer vision method for white blood cells detection. Comput Math Methods Med 2013:14. Article ID 137392. http://dx.doi.org/10.1155/2013/137392

39. Prinyakupt J, Pluempitiwiriyawej C (2015) Segmentation of white blood cells and comparison of cell morphology by linear and naïve Bayes classifiers. BioMed Eng OnLine 14:63. https://doi.org/10.1186/s12938-015-0037-1

40. Rezatofighi SH, Soltanian-Zadeh H (2011) Automatic recognition of five types of white blood cells in peripheral blood. Comput Med Imaging Graph 35(4):333–343

41. LISC. http://users.cecs.anu.edu.au/~hrezatofighi/Data/Leukocyte%20Data.htm (Accessed on: 10 Apr 2019)

42. Yang XS (2008) Nature-inspired metaheuristic algorithms, Luniver Press

43. Alomari YM, Abdullah SNHA, Azma RZ, Omar K (2014) Automatic detection and quantification of WBCs and RBCs using iterative structured circle detection algorithm. Comput Math Methods Med 2014:979302. https://doi.org/10.1155/2014/979302

44. Metzler R, Klafter J (2000) The random walk's guide to anomalous diffusion: a fractional dynamics approach. Phys Rep 339(1):1–77

45. Nurzaman SG, Matsumoto Y, Nakamura Y, Shirai K, Koizumi S, Ishiguro H (2011) From L´evy to Brownian: a computational model based on biological fluctuation. PLoS ONE 6(2). Article ID e16168

46. Raja NSM, Rajinikanth V (2014) Brownian distribution guided bacterial foraging algorithm for controller design problem. Adv Intell Syst Comput 248:141–148. https://doi.org/10.1007/978-3-319-03107-1_17

47. Rajinikanth V, Satapathy SC, Dey N, Fernandes SL, Manic KS (2019) Skin melanoma assessment using Kapur's entropy and level set—A study with bat algorithm. Smart Innov, Syst Technol 104:193–202. https://doi.org/10.1007/978-981-13-1921-1_19

48. Shriranjani D, Tebby SG, Satapathy SC, Dey N, Rajinikanth V (2018) Kapur's entropy and active contour-based segmentation and analysis of retinal optic disc. Lect Notes Electr Eng 490:287–295. https://doi.org/10.1007/978-981-10-8354-9_26

49. Rajinikanth V, Satapathy SC, Fernandes SL, Nachiappan S (2017) Entropy based segmentation of tumor from brain MR images–A study with teaching learning based optimization. Pattern Recogn Lett 94:87–95. https://doi.org/10.1016/j.patrec.2017.05.028

50. Rajinikanth V, Satapathy SC, Dey N, Lin H (2018) Evaluation of ischemic stroke region from CT/MR images using hybrid image processing techniques. In: Intelligent multidimensional data and image processing. pp 194–219. https://doi.org/10.4018/978-1-5225-5246-8.ch007

51. Kapur JN, Sahoo PK, Wong AKC (1985) A new method for gray-level picture thresholding using the entropy of the histogram. Comput Vis Graph Image Process 29:273–285

52. Manic KS, Priya RK, Rajinikanth V (2016) Image multithresholding based on Kapur/Tsallis entropy and firefly algorithm. Indian J Sci Technol 9(12):89949

53. Cherabit N, Chelali FZ, Djeradi A (2012) Circular hough transform for iris localization. Sci Technol 2(5):114–121. https://doi.org/10.5923/j.scit.20120205.02

54. Likforman-Sulem L, Kavallieratou E (2018) Document image processing. J Imaging 4(7):84

55. Duda RO, Hart PE (1972) Use of the Hough transformation to detect lines and curves in pictures. Commun ACM 15:11–15

56. Manic KS, Rajinikanth V, Ananthasivam S, Suresh U (2015) Design of controller in double feedback control loop–an analysis with heuristic algorithms. Chem Prod Process Model 10(4):253–262. https://doi.org/10.1515/cppm-2015-0005

57. Jagatheesan K, Anand B, Samanta S, Dey N, Ashour AS, Balas VE (2019) Design of a proportional-integral-derivative controller for an automatic generation control of multi-area power thermal systems using firefly algorithm. IEEE/CAA J Autom Sin 6(2):503–515. https://doi.org/10.1109/jas.2017.7510436

Chapter 11
Effect of Population Size Over Parameter-less Firefly Algorithm

Krishna Gopal Dhal, Samarendu Sahoo, Arunita Das and Sanjoy Das

1 Introduction

In the last decade, researchers are incessantly developing nature-inspired optimization algorithms (NIOAs) under the inspiration of different natural and biological phenomena [1, 2]. NIOAs are exceptionally efficient because of their capability of providing promising solutions to the complex real-world problems in logical time [1, 2]. The aim of developing this kind of clever algorithms is to produce alternative and efficient solutions to complex real-life optimization problems. Recently, NIOAs are carrying out a vital role in many fields such as computational intelligence (CI), artificial intelligence (AI), machine learning (ML), and so on. The mathematical formulations of the major number of NIOA have been performed depending on the intelligent swarming behavior of animals in nature, and as a result, this subclass of NIOA is called swarm intelligence (SI). Some well-known examples of SI algorithms are Particle Swarm Optimizer (PSO) [1, 3], where the swarming behaviors of fish

K. G. Dhal (✉)
Department of Computer Science and Application, Midnapore College (Autonomous),
Paschim Medinipur, West Bengal, India
e-mail: krishnagopal.dhal@midnaporecollege.ac.in

S. Sahoo
Department of Computer Science, Vidyasagar University, Paschim Medinipur, West Bengal, India
e-mail: samarendusahoo@gmail.com

A. Das
Department of Information Technology, Kalyani Government Engineering College, Kalyani,
West Bengal, India
e-mail: arunita17@gmail.com

S. Das
Department of Engineering & Technological Studies, University of Kalyani, Kalyani,
West Bengal, India
e-mail: dassanjoy0810@hotmail.com

© Springer Nature Singapore Pte Ltd. 2020   237
N. Dey (ed.), *Applications of Firefly Algorithm and its Variants*,
Springer Tracts in Nature-Inspired Computing,
https://doi.org/10.1007/978-981-15-0306-1_11

and birds have been formulated; Firefly Algorithm (FA) [1, 4] has been built under the inspiration of flashing based attraction behavior of fireflies, Cuckoo Search (CS) [1, 5] algorithm is devised by mathematically formulating the brooding parasitism behavior of some cuckoo species, whereas Bat Algorithm (BA) [1, 6] is developed by formulating the echolocation of foraging behavior of bats. SI algorithms show their effective performance over different complex problem domain due to their some fascinating characteristics such as information sharing among solutions which facilitate learning while searching, population of possible solutions which is very much beneficial for parallelization. As an outcome, SI algorithms guarantee the promising set of solutions in large-scale optimization field.

The common and most important characteristics which control the efficiency of each NIOA are (a) exploration (global search) and exploitation (local search), (b) adaption and diversity, and (c) algorithm's operators [7, 8]. Proper balancing between exploration and exploitation plays a very crucial role in NIOA, and it purely depends on the implementation structure, incorporated search mechanism, and algorithm's parameters [9]. After that, appropriate adaption and diversity maintaining techniques of the concerned NIOA make it a successful one. Different approaches to achieve a balanced exploration and exploitation property in any NIOA are the key concepts of adaption [9, 10]. For example, the formulation of automatic strategies for finding the appropriate population size and parameter values for a specific NIOA are the different forms of adaption. Whereas, generating a well-varied initial population using randomization is the paradigm of diversity which helps to trounce premature convergence. Therefore, there is a correlation between adaption and diversity. The third characteristic, i.e., algorithm's operators are the main keys for controlling the diversity, balances between exploration and exploitation, etc. Evolutionary algorithms like Genetic Algorithm (GA) and Differential Evolution (DE) have three powerful operators, namely crossover (recombination), mutation, and selection [9]. Crossover and mutation generate new solution from the present solutions and select operator technically picks the solutions on which crossover and mutation will work. It is reported that mutation is responsible for global search (exploration), and crossover is a combination of global and local search. In addition to that when select operator chooses only the best quality solution then it helps in exploitation and it facilitates exploration when picks only poor quality solutions or a mixture of poor and best quality of solutions. As a consequence, it can be said that a well-balanced crossover and mutation is needed to develop a good variant GA. But the quantification of this balancing factor significantly depends on the problem type, values of the algorithm's parameters, and other factors.

Hence, in a broad sense, the above three discussed characteristics provide the adjustment capability to an NIOA over any problem type and objective space [9]. Literature also reports some NIOA's efficacy enhancement strategies like parameter adaption [11], use of different random numbers (e.g., uniform distribution [1], Gaussian distribution [1], Lévy distribution [1], chaotic sequence [12]), hybridization among NIOAs [12, 13, 15], inertia weight [16, 17], global–local search strategies [18, 19], and so on. Besides all the improvement strategies, the proper setting of control parameters including population size is very challenging as it varies based on the problem under consideration. To overcome that problem, this study focuses on the development of parameter-less variants of one important NIOA called Firefly Algorithm (FA). Implementation of a simple, adaptive, parameter-less optimization algorithm is needed to search optimal solutions in less time and without any parameter-setting specialist. In addition to that researchers do not have to bother about the complete knowledge or experience over the problem space as well as the influence of different parameters over the concerned NIOA. Proposed parameter-less FA variants have the following features: There is no parameter-setting problem, good optimization capability, and stability. The structure of the remaining chapter is as follows: Sect. 2 discusses the related work. Section 3 demonstrates the original FA algorithm and their extension to parameter-less variants. Experimental results are demonstrated and discussed in Sect. 4. Conclusions have been made in Sect. 5.

2   Related Work

In the literature, some improved and adaptive FA variants [20–22] had been developed and applied in different domains such as image processing [23, 24], medical imaging field [25, 26], electrical engineering [27], and so on. Baykasoglu and Ozsoydan [20] developed one adaptive FA (AFA) by incorporating adaptivity over parameters which provided it good exploitation ability. A chaotic version of the AFA had also been developed by the authors. Experimental results showed that proposed AFA outperformed traditional FA in terms of optimization ability and stability. Same authors also proposed an improved FA (IFA) [21] and multipopulation based FA (MPFA) [22] for solving dynamic optimization problems. Experimental results showed that IFA and MPFA provided superior outcomes to FA, GA, and DE. In conjunction with that parameter-less variants of some popular NIOA, such as the Bat Algorithm (BA) [28], Genetic Algorithm [29, 30] and Differential Evolution (DE) [31], Harmony Search (HS) [32, 33], Cuckoo Search (CS) [34], and Artificial Bee Colony (ABC) [35] had been developed and they demonstrated their competitive performance over diverse optimization fields. BA had been extended to its parameter-less variants by tuning frequency parameter depending on the range of the search space, population size, and random number. Other parameters like loudness and pulse rate are assigned constant values from the experience [28]. Population size took the value within [10, 1280] such that population size got doubled in each run starting with 10. Therefore, eight variants of the parameter-less BA (PLBA) indicated as PL-1 to PL-8 were obtained. PLBA variants were only compared among each other, and experimental results showed that higher population size is better for BA. Dhal et al. [33] proposed another eight variants of HS by assigning the constant values to the control parameters, varying the population size within [10, 1280] same as [28], and two stopping condition had been utilized which were a number of iteration and maximum fitness evaluation (MAX_FE). Experimental study demonstrated that in iteration based stopping criterion, lower population size was well-suited for HS with good stability and large execution time while solving Shannon's entropy based multilevel thresholding problem. The larger population reduced the execution time, but could perform premature convergence. On the MAX_FE based stopping criterion, population size in between 40 and 160 provided competitive and consistent outcomes. Here also, large population affected the stability issue. Silva et al. [32] proposed another parameter-setting-free HS algorithm by assigning the constant values to the control parameters. Experimental study proved that proposed HS gave promising results when it was applied to set the control parameters of Restricted Boltzmann Machines. Teo and Hamid [31] developed one parameter-less DE by incorporating a self-adaptive strategy to its parameters including population size. It was found that proposed self-adapting DE provided superior results to traditional DE in terms of objective function and stability. In the same way, parameter-less evolutionary algorithms like G A had been built and showed their effective performance over several engineering optimization fields [29, 30]. Self-adaptive strategy based parameter-less CS [34] and ABC [35] had been developed in image segmentation and enhancement domains, respectively. Experimental results proved that the self-adaptive strategy of control parameters significantly enhanced the efficiency of CS and ABC algorithms.

The experiments are performed inspired by methodologies same as in [28, 33] where the effect of different population size is also investigated because varying the population size is also one kind of significant adaption mechanism. Self-adaptive control parameters have been formulated by considering the population size, dimension, and range of the search space. To measure the performance of parameter-less FA (PLFA), the CEC-2014 benchmark function suite [36] has been utilized. Experimental results prove that self-adaptive control parameters are effective for the enhancement of FA, whereas population size within 40–80 provides the best results. The study also shows that population size 40 is best when the dimension increases.

3   Firefly Algorithm (FA)

Firefly Algorithm via Lévy flight had been developed by X. S. Yang, inspired by the flashing and attraction behaviors of fireflies [1, 4, 37]. Firefly flashes for performing two purposes: (a) to attract partners or (b) to evoke potential prey. Yang followed three principles to formulate mathematically the FA via Lévy flight algorithm:

(i)   Firefly is a unisex beetle, and hence, one firefly attracted to others despite their sex;
(ii) Attractiveness is correlated to brightness, and so, less bright firefly attracts to the dazzling fireflies.
(iii) Brightness positively correlated to the objective/fitness function in the case of the maximization problem.

In FA, two major concerns are (a) difference of glowing intensity and (b) creation of the attractiveness.

For a specified medium with a fixed light absorption coefficient γ, the light intensity I varies with the distance r following the equation:

$$I = I_0 e^{-\gamma r} \tag{1}$$

where I indicates actual light intensity.

As it is known that attractiveness is positively correlated to the light intensity, the attractiveness β may be formulated as:

$$\beta = \beta_0 e^{-\gamma r^2} \tag{2}$$

where β_0 represents attractiveness at $r = 0$. When $\beta_0 = 0$, then firefly walks randomly and information sharing among solutions do not occur. Based on the above discussion, the pseudocode of the traditional FA with Lévy flight is given as Fig. 1.

In FA, the initial population is created randomly by the following equation:

$$X_i = \text{low} + (\text{up} - \text{low}) \times \partial \tag{3}$$

X_i is the ith individual. up and low indicate the upper and lower bound of the search space. ∂ is a random variable within the range [0, 1] and usually created using uniform distribution.

The Cartesian distance between X_i and X_j is as follows:

Take objective function $f(X_i)$,

where, $X_i = (x_1, x_2, \dots \dots \dots, x_d)^T$.

Initial population of size n has been generated $X = \{X_i \mid i = 1, 2, \dots, n\}$ as per Eq.(3) by uniform distribution, where, X_i is the ith firefly.

Compute fitness of each solution i.e. $f(X_i)$.

Then define light absorption coefficient γ.

While *termination condition* does not meet Do

        for $i = 1$: n (all n fireflies)

            for $j = 1$: n (all n fireflies)

                if $(f(X_i) < f(X_j))$, then

        i^{th} firefly moves towards j^{th} firefly using Eq.(5)

            End if

        Modification of attraction w.r.t. distance r via

        $e^{-\gamma r^2}$ using Eq.(2)

        Calculate the fitness of new solutions and modify the light intensity

            End for j

        End for i

Sort the fireflies according to the fitness and find the current global best (*gbest*)

End While

Create Post-process request

Fig. 1   Pseudocode of traditional Firefly Algorithm

$$r_{ij} = \|X_i - X_j\| = \sqrt{\sum_{k=1}^{d}(x_{i,k} - x_{j,k})^2} \tag{4}$$

where $x_{i,k}$ is the kth component of the spatial coordinate X_i of ith firefly.

In FA, firefly X_i move toward X_j via Lévy flight using the expression given below:

$$X_i = X_i + \beta_0 e^{-\gamma r_{ij}^2}(X_i - X_j) + \alpha.\text{Lévy}(.) \tag{5}$$

where α is the scaling factor and Lévy() is the random walk generated from Lévy distribution which is capable of exploring a large amount of search space. Lévy distribution which has infinite variance with infinite mean is formulated below [18]

$$\text{Lévy} \sim u = t^{-\tau}, \; 1 < \tau \leq 3 \tag{6}$$

Depending on the above mathematical formulation, the FA via Lévy flight algorithm is given as Fig. 1.

3.1 Development of Parameter-less Variants of FA

Performance of a specific NIOA such as FA crucially relies on its parameters [29, 30]. The adaptive mechanism can also be incorporated into these parameters so that they modify automatically. In the literature, it can be noticed that self-adaptive mechanisms are very efficient to enhance the efficacy of any NIOA. There is a great influence of these parameters over the balance between exploration and exploitation, adaption, diversity, and algorithm's operators. But, the self-adaptive mechanism formulation is very critical due to its dependency over problem type and algorithm's structure. Furthermore, appropriate population size (n) has a significant influence over execution-time competence and effectiveness. A lower-sized population may affect the diversity, whereas the large-sized population can suffer from superior convergence speed. Therefore, it can be said that the small-sized populations have a propensity toward faster convergence but also enhance the possibility of local trapping. While, the larger-sized populations may converge slower but has the ability to explore the search space significantly. Therefore, the adaptive modification rules for three controlling parameters, i.e., β_0, γ, and α are formulated based on the population size, the dimension of the problem, and range of the search space. The rules are given as follows:

$$\alpha = \frac{n}{D \times (U - L)} \tag{7}$$

$$\beta_0 = \begin{cases} \frac{(U-L)\times\text{rand}(0,1)}{n} & \text{if } n \le 100 \\ \frac{(U-L)\times\text{rand}(0,1)}{2\times\sqrt{n}} & \text{if } n > 100 \end{cases} \tag{8}$$

$$\gamma = \begin{cases} \frac{D\times\text{rand}(0,1)}{n} & \text{if } D \le 30 \text{ and } n \le 80 \\ \frac{D\times\text{rand}(0,1)}{\sqrt{n}} & \text{if } D > 30 \text{ and } n > 80 \end{cases} \tag{9}$$

where n = population size, D = dimension of the search space, rand$(0, 1)$ is the random number within $[0, 1]$ generated using a uniform distribution, and U and L are the upper and lower bound of the concerned decision variable, respectively. In this study, $n \in [10, 1280]$, $D \in [10, 50]$ and $U = 100$, $L = -100$. Considering the values of n, D, U and L, it can be easily seen that

$$\alpha \in [0.001, 0.64],$$

$$\beta_0 \in \begin{cases} [0, 20] & \text{if } n \le 100 \\ [0, 9.95] & \text{if } n > 100 \end{cases},$$

$$\gamma \in \begin{cases} [0, 5] & \text{if } D \le 30 \text{ and } n \le 80 \\ [0, 5.56] & \text{if } D > 30 \text{ and } n > 80 \end{cases}$$

Basically, α controls the affect of the Lévy flight based random walk over fireflies' current position. When the population size is less, then α takes a smaller value which helps in exploitation, and when a large population size has been taken, then α becomes large and performs exploration. Therefore, population size influences the exploration and exploitation efficiency of the FA significantly. In FA, there is a critical relationship between β and γ. When $\gamma \to 0$, then $\beta = \beta_0$ and for $\gamma \to \infty$, β is almost zero. Therefore, there is a trade-off between β and γ. In this study, this trade-off is gently taken into consideration and depending on n, D, U, and L, and the values of β and γ have been deduced. From the equation of β_0 and γ, it is verified that β is large for small population size which significantly assists the exploration capability of FA. Whereas, in the case of larger population size, β helps in exploitation by taking small value. So, the above parameter-setting strategy clearly demonstrates that α, β and γ take a well combination of values for enhancing the ability of FA by balancing the exploration and exploitation trade-off. In Lévy distribution, τ has been taken as 1.5 which is same as in [18].

A plenty number of studies have been published to measure the impact of the population size by varying it during the entire search process. Although adaptive population variation techniques are fruitful, they generally defined by some complex metrics that entail supplementary meta-control parameters, which are also tough to set [38–42]. Therefore, the population size (n) of considered one algorithms is varied within the size [10, 1280] same as in [28]. Here, population size gets doubled in each run starting with $n = 10$. Therefore, eight instances of FA have been executed, and the best one is considered by the user over different classes of functions. The termination condition for all parameter-less variants of FA has been taken as number of fitness evaluations (FEs), and the maximum number of FEs (i.e., MAX_FE) has been set as $10,000 \times D$, where D indicates the dimension of the search space.

4   Experimental Results

The experiments are done using CEC'14 benchmark suite [36] (Table 1) which contains 30 benchmark functions which are classified into four classes:

(1) **Unimodal functions (Func1–Func3)**

Each unimodal function has a single global optimum and no local optima. They are also non-separable and rotated.

(2) **Simple multimodal functions (Func4–Func16)**

Multimodal functions are either separable or non-separable. In addition, they are also rotated or/and shifted.

(3) **Hybrid functions (Func17–Func22)**

Hybrid functions are developed by dividing the variables randomly into a number of subparts which are constructed by utilizing different basic functions.

(4) **Composition functions (Func23–Func30)**

Composition functions have been developed by summing up of two or more basic functions. Here, hybrid functions are utilized as the basic functions to create composition functions.

The efficiency of the proposed PLFA variants has been evaluated by computing the subsequent measures:

(a) Best solution (best) found over 30 runs,
(b) Worst solution (worst) found over 30 runs,
(c) Mean objective function (mean) value over 3 0 runs,
(d) Standard deviation (std.) over 30 runs, and
(e) Median of 30 runs.

Experimental results are given over 16 functions out of total 30 functions of CEC'14 benchmark suite. Three unimodal functions (*Func1–Func3*), five multimodal functions (*Func4–Func8*), four hybrid functions (*Func17–Func20*), and four composition functions (*Func23–Func26*) are taken into consideration in this study.

4.1   *Analysis of the Results*

Values of best, worst, mean, standard deviation, and median of PLFA variants over unimodal functions for dimensions 10, 30, and 50 are given in Tables 2, 3, and 4,

Table 1 Summary of the CEC'14 test functions

Func. type	Class	Sl. No.	Func. name	$F'_j = F_j(x')$ Global optima
Unimodal functions	1	Func1	Rotated High Conditioned Elliptic (RHCE)	1.00×10^2
		Func2	Rotated Bent Cigar (RBC)	2.00×10^2
		Func3	Rotated Discus (RD)	3.00×10^2
Simple multimodal functions	2	Func4	Shifted and Rotated Rosenbrock (SRR)	4.00×10^2
		Func5	Shifted and Rotated Ackley (SRA)	5.00×10^2
		Func6	Shifted and Rotated Weierstrass (SRW)	6.00×10^2
		Func7	Shifted and Rotated Griewank (SRG)	7.00×10^2
		Func8	Shifted Rastrigin (SR)	8.00×10^2
		Func9	Shifted and Rotated Rastrigin (SRR)	9.00×10^2
		Func10	Shifted Schwefel (SS)	10.00×10^2
		Func11	Shifted and Rotated Schwefel (SRS)	11.00×10^2
		Func12	Shifted and Rotated Katsuura (SRK)	12.00×10^2
		Func13	Shifted and Rotated Happy Cat (SRHC)	13.00×10^2
		Func14	Shifted and Rotated HGBat (SRHGB)	14.00×10^2
		Func15	Shifted and Rotated Expanded Griewank's plus Rosenbrock (SREGR)	15.00×10^2
		Func16	Shifted and Rotated Expanded Scaffer's F6 (SRESF6)	16.00×10^2
Hybrid functions	3	Func17	Hybrid Function 1 (HF1) ($N = 3$)	17.00×10^2
		Func18	Hybrid Function 2 (HF2) ($N = 3$)	18.00×10^2
		Func19	Hybrid Function 3 (HF3) ($N = 4$)	19.00×10^2
		Func20	Hybrid Function 4 (HF4) ($N = 4$)	20.00×10^2

(continued)

Table 1 (continued)

Func. type	Class	Sl. No.	Func. name	$F_j' = F_j(x')$ Global optima
		Func21	Hybrid Function 5 (HF5) ($N = 5$)	21.00×10^2
		Func22	Hybrid Function 6 (HF6) ($N = 5$)	22.00×10^2
Composition functions	4	Func23	Composition Function 1 (CF1) ($N = 5$)	23.00×10^2
		Func24	Composition Function 2 (CF2) ($N = 3$)	24.00×10^2
		Func25	Composition Function 3 (CF3) ($N = 3$)	25.00×10^2
		Func26	Composition Function 4 (CF4) ($N = 5$)	26.00×10^2
		Func27	Composition Function 5 (CF5) ($N = 5$)	27.00×10^2
		Func28	Composition Function 6 (CF6) ($N = 5$)	28.00×10^2
		Func29	Composition Function 7 (CF7) ($N = 3$)	29.00×10^2
		Func30	Composition Function 8 (CF8) ($N = 3$)	30.00×10^2

Search range: $[-100, 100]^D$
Dimension (D) = 10, 30, and 50

respectively. Whereas Tables 5, 6, and 7 represent the results over simple multimodal functions for dimensions 10, 30, and 50, respectively, Tables 8, 9, and 10 demonstrate the results over hybrid functions for dimensions 10, 30, and 50 respectively, and Tables 11, 12, and 13 represent the results over composite functions for dimensions 10, 30, and 50, respectively.

According to the Friedman ranking [43] based on mean in Table 14, PLFA-3 is the best for unimodal functions, PLFA-4 is the best for simple multimodal functions, hybrid functions, and composition functions, and PLFA-4 is the best FA variant over all the classes of functions which is revealed by the total rank depending on mean for dimension 10. All the ranking values corresponding to dimension 10 of Table 14 are graphically represented in Figs. 2a and 3, where Fig. 2a represents class wise ranking and Fig. 3 represents the total ranking. Therefore, it can be said that population size 80 is the best for FA in terms of mean value. Whereas, when standard deviation of objective functions for dimension 10 is considered, then according to the ranking in Table 15, PLFA-3 is the best for unimodal functions, PLFA-5 is the best for simple multimodal functions, PLFA-4 is the best for hybrid functions, PLFA-8 is the best for composition functions, and PLFA-4 is the best FA variant over all the classes of functions which is revealed by the total rank. Ranking depending on standard

Table 2 Results over unimodal functions with dimension $D = 10$ for PLFA variants

Func.	Algo.	Best	Worst	Mean	Med.	Std.
F1	PLFA-1	$\mathbf{1.03 \times 10^3}$	5.55×10^6	$\mathbf{4.00 \times 10^4}$	1.18×10^5	1.07×10^6
	PLFA-2	1.75×10^3	$\mathbf{1.48 \times 10^5}$	6.68×10^4	$\mathbf{6.55 \times 10^4}$	$\mathbf{4.87 \times 10^4}$
	PLFA-3	1.63×10^4	2.37×10^5	1.00×10^5	7.48×10^4	6.80×10^4
	PLFA-4	7.87×10^4	4.84×10^5	2.24×10^5	2.24×10^5	1.01×10^5
	PLFA-5	9.99×10^5	5.83×10^6	2.76×10^6	2.68×10^6	1.06×10^6
	PLFA-6	4.80×10^6	4.34×10^7	2.00×10^7	2.02×10^7	8.84×10^6
	PLFA-7	5.83×10^6	2.26×10^7	1.25×10^7	1.26×10^7	4.17×10^6
	PLFA-8	4.40×10^6	1.41×10^7	9.17×10^6	8.80×10^6	2.30×10^6
F2	PLFA-1	$\mathbf{2.28 \times 10^2}$	$\mathbf{1.19 \times 10^4}$	5.99×10^3	6.49×10^3	$\mathbf{4.35 \times 10^3}$
	PLFA-2	3.07×10^2	1.22×10^4	$\mathbf{4.32 \times 10^3}$	$\mathbf{2.25 \times 10^3}$	4.45×10^3
	PLFA-3	1.15×10^5	2.91×10^5	2.01×10^5	2.09×10^5	4.98×10^4
	PLFA-4	4.93×10^5	1.97×10^6	1.19×10^6	1.20×10^6	3.59×10^5
	PLFA-5	4.49×10^7	1.53×10^8	9.96×10^7	1.08×10^8	2.79×10^7
	PLFA-6	1.31×10^9	3.92×10^9	2.71×10^9	2.86×10^9	7.06×10^8
	PLFA-7	7.25×10^8	2.71×10^9	1.93×10^9	2.07×10^9	5.36×10^8
	PLFA-8	8.69×10^8	2.02×10^9	1.43×10^9	1.47×10^9	3.18×10^8
F3	PLFA-1	$\mathbf{3.09 \times 10^2}$	1.36×10^4	4.97×10^3	4.24×10^3	3.80×10^3
	PLFA-2	3.14×10^2	1.37×10^4	2.87×10^3	1.39×10^3	3.40×10^3
	PLFA-3	3.60×10^2	$\mathbf{7.13 \times 10^3}$	$\mathbf{1.73 \times 10^3}$	$\mathbf{1.06 \times 10^3}$	1.75×10^3
	PLFA-4	7.52×10^2	7.28×10^3	3.02×10^3	2.80×10^3	$\mathbf{1.46 \times 10^3}$
	PLFA-5	8.12×10^3	1.50×10^4	1.13×10^4	1.14×10^4	1.86×10^3
	PLFA-6	7.60×10^3	2.47×10^4	1.71×10^4	1.83×10^4	4.56×10^3
	PLFA-7	6.76×10^3	2.11×10^4	1.43×10^4	1.43×10^4	3.74×10^3
	PLFA-8	5.37×10^3	1.40×10^4	9.62×10^3	9.66×10^3	2.39×10^3

Best results obtained are given in bold

deviation for dimension 10 is graphically shown as Figs. 4a and 5, where Figs. 4a and 5 represent class wise ranking and total ranking, respectively. Thus, population size 80 is the best for FA by considering optimization capability (i.e., mean) and robustness (i.e., standard deviation) over any classes of functions.

In the case of dimension 30, Friedman ranking of PLFA variants depending on mean over different classes of functions is done in Table 14 which clearly demonstrates that PLFA-2 is the best for unimodal functions, PLFA-3 is the best for simple multimodal and hybrid functions, and PLFA-3 and 4 are the best variants for composition functions. However, PLFA-3 outperformed all the other FA variants over all classes of functions according to the minimum total rank value. All the ranking values corresponding to dimension 30 of Table 14 are graphically represented in Figs. 2b and 3, where Fig. 2b represents class wise ranking and Fig. 3 represents the total ranking. But when robustness has been considered which is measured by calculating

Table 3 Results over unimodal functions with dimension $D = 30$ for PLFA variants

Func.	Algo.	Best	Worst	Mean	Med.	Std.
F1	PLFA-1	3.34×10^6	6.22×10^7	1.82×10^7	1.49×10^7	1.33×10^7
	PLFA-2	$\mathbf{1.05 \times 10^5}$	$\mathbf{2.54 \times 10^6}$	$\mathbf{1.06 \times 10^6}$	$\mathbf{8.35 \times 10^5}$	$\mathbf{6.68 \times 10^5}$
	PLFA-3	5.50×10^5	5.55×10^6	2.05×10^6	1.59×10^6	1.28×10^6
	PLFA-4	2.72×10^6	2.05×10^7	8.28×10^6	6.89×10^6	4.56×10^6
	PLFA-5	2.63×10^7	8.37×10^7	4.89×10^7	4.68×10^7	1.58×10^7
	PLFA-6	1.52×10^8	3.42×10^8	2.71×10^8	2.71×10^8	4.36×10^7
	PLFA-7	4.64×10^8	9.29×10^8	7.18×10^8	7.09×10^8	1.29×10^8
	PLFA-8	3.67×10^8	8.25×10^8	5.64×10^8	5.64×10^8	1.06×10^8
F2	PLFA-1	2.07×10^2	3.37×10^7	2.20×10^6	1.16×10^4	2.02×10^9
	PLFA-2	$\mathbf{2.02 \times 10^2}$	$\mathbf{2.64 \times 10^4}$	$\mathbf{5.36 \times 10^3}$	$\mathbf{1.25 \times 10^3}$	3.45×10^4
	PLFA-3	1.60×10^3	2.75×10^4	8.78×10^3	3.76×10^3	$\mathbf{3.44 \times 10^4}$
	PLFA-4	1.09×10^7	1.67×10^7	1.31×10^7	1.33×10^7	8.45×10^4
	PLFA-5	1.17×10^9	2.15×10^9	1.65×10^9	1.63×10^9	1.29×10^7
	PLFA-6	3.18×10^{10}	4.44×10^{10}	3.69×10^{10}	3.68×10^{10}	8.99×10^8
	PLFA-7	9.09×10^{10}	1.13×10^{11}	1.01×10^{11}	1.01×10^{11}	2.52×10^{10}
	PLFA-8	1.14×10^{11}	1.67×10^{11}	1.45×10^{11}	1.47×10^{11}	7.24×10^{10}
F3	PLFA-1	9.05×10^3	8.43×10^4	3.13×10^4	2.55×10^4	2.02×10^4
	PLFA-2	$\mathbf{3.57 \times 10^2}$	1.14×10^4	2.60×10^3	1.88×10^3	2.26×10^3
	PLFA-3	5.00×10^2	$\mathbf{3.16 \times 10^3}$	$\mathbf{1.49 \times 10^3}$	$\mathbf{1.42 \times 10^3}$	$\mathbf{7.50 \times 10^2}$
	PLFA-4	3.53×10^3	1.78×10^4	8.78×10^3	9.02×10^3	3.59×10^3
	PLFA-5	2.97×10^4	4.59×10^4	3.65×10^4	3.61×10^4	3.90×10^3
	PLFA-6	5.68×10^4	1.12×10^5	8.87×10^4	8.91×10^4	1.08×10^4
	PLFA-7	9.44×10^4	1.42×10^5	1.15×10^5	1.15×10^5	1.23×10^4
	PLFA-8	6.52×10^4	1.18×10^5	9.88×10^4	9.97×10^4	1.05×10^4

Best results obtained are given in bold

standard deviation, then PLFA-3 is the best for unimodal functions and PLFA-4 is the best variants for all other classes of functions. Therefore, PLFA-4 gets the best total rank in Table 15, and PLFA-3 gets the second best rank. Ranking depending on standard deviation for dimension 30 is graphically shown as Figs. 4b and 5, where Figs. 4b and 5 represent class wise ranking and total ranking, respectively. So, it can be said that population size resides within 40–80 will give promising results according to better optimization ability and robustness.

The ranking according to mean over dimension 50 has been performed in Table 14. Where PLFA-2 is the best for unimodal functions, PLFA-3 is the best for simple multimodal functions and hybrid functions, PLFA-4 is the best for composition functions, and PLFA-3 is the best FA variant over all the classes of functions by confirming the minimum total rank. All the ranking values corresponding to mean value over dimension 50 of Table 14 are graphically represented in Figs. 2c and 3, where Fig. 2c

Table 4 Results over unimodal functions with dimension $D = 50$ for PLFA variants

Func.	Algo.	Best	Worst	Mean	Med.	Std.
F1	PLFA-1	7.80×10^6	6.03×10^7	3.32×10^7	3.03×10^7	1.30×10^7
	PLFA-2	$\mathbf{1.08 \times 10^6}$	$\mathbf{3.87 \times 10^6}$	$\mathbf{2.34}$	$\mathbf{2.24 \times 10^6}$	$\mathbf{7.29 \times 10^5}$
	PLFA-3	1.88×10^6	5.68×10^6	3.30	3.09×10^6	1.04×10^6
	PLFA-4	6.98×10^6	1.98×10^7	1.16×10^7	1.06×10^7	3.30×10^6
	PLFA-5	4.18×10^7	1.29×10^8	7.52×10^7	6.62×10^7	2.45×10^7
	PLFA-6	3.51×10^8	6.02×10^8	4.98×10^8	4.92×10^8	6.42×10^7
	PLFA-7	7.94×10^8	1.57×10^9	1.17×10^9	1.18×10^9	1.91×10^8
	PLFA-8	1.46×10^9	2.53×10^9	2.08×10^9	2.08×10^9	2.64×10^8
F2	PLFA-1	2.07×10^2	3.37×10^7	2.20×10^6	1.16×10^4	7.99×10^6
	PLFA-2	$\mathbf{2.02 \times 10^2}$	$\mathbf{2.64 \times 10^4}$	$\mathbf{5.36 \times 10^3}$	$\mathbf{1.25 \times 10^3}$	$\mathbf{8.20 \times 10^3}$
	PLFA-3	1.60×10^3	2.75×10^4	8.78×10^3	3.76×10^3	8.43×10^3
	PLFA-4	1.09×10^7	1.67×10^7	1.31×10^7	1.33×10^7	1.34×10^6
	PLFA-5	1.17×10^9	2.15×10^9	1.65×10^9	1.63×10^9	2.54×10^8
	PLFA-6	3.18×10^{10}	4.44×10^{10}	3.69×10^{10}	3.68×10^{10}	3.02×10^9
	PLFA-7	9.09×10^{10}	1.13×10^{11}	1.01×10^{11}	1.01×10^{11}	6.21×10^9
	PLFA-8	1.14×10^{11}	1.67×10^{11}	1.45×10^{11}	1.47×10^{11}	1.33×10^{10}
F3	PLFA-1	3.33×10^4	1.37×10^5	8.57×10^4	8.07×10^4	2.76×10^4
	PLFA-2	$\mathbf{2.46 \times 10^3}$	3.16×10^4	1.25×10^4	1.18×10^4	8.41×10^3
	PLFA-3	2.56×10^3	$\mathbf{1.78 \times 10^4}$	$\mathbf{8.23 \times 10^3}$	$\mathbf{7.14 \times 10^3}$	$\mathbf{4.41 \times 10^3}$
	PLFA-4	1.43×10^4	5.26×10^4	3.30×10^4	3.26×10^4	9.05×10^3
	PLFA-5	6.75×10^4	1.05×10^5	8.56×10^4	8.38×10^4	9.95×10^3
	PLFA-6	1.42×10^5	1.85×10^5	1.62×10^5	1.64×10^5	1.19×10^4
	PLFA-7	1.10×10^5	2.18×10^5	1.87×10^5	1.91×10^5	2.20×10^4
	PLFA-8	1.62×10^5	2.30×10^5	2.02×10^5	2.00×10^5	1.63×10^4

Best results obtained are given in bold

represents class wise ranking and Fig. 3 represents the total ranking. Now by considering standard deviation ranking over dimension 50 has been done in Table 15 which clearly revealed that PLFA-2 is the best for unimodal functions, PLFA-4 is the best for simple multimodal functions, PLFA-3 is the best for hybrid functions and composition functions, and PLFA-3 is the best FA variant over all the classes of functions which is clear by its minimum total rank. Ranking depending on standard deviation for dimension 50 is graphically shown as Figs. 4c and 5, where Figs. 4c and 5 represent class wise ranking and total ranking, respectively. Therefore, PLFA-3 is the finest variant of FA over dimension 50 in terms of robustness, consistency, and optimization ability. It can also be seen that when dimension D increases from 30 to 50, PLFA-3 provides best outcomes by considering mean objective function and standard deviation. Therefore, it can be said that PLFA-3 is the best PLFA variant for complex optimization problem.

Table 5 Results over simple multimodal functions with dimension $D = 10$ for PLFA variants

Func.	Algo.	Best	Worst	Mean	Med.	Std.
F4	PLFA-1	**4.00×10^2**	4.50×10^2	4.25×10^2	4.35×10^2	1.69×10
	PLFA-2	**4.00×10^2**	**4.35×10^2**	4.19×10^2	4.35×10^2	1.67×10
	PLFA-3	**4.00×10^2**	**4.35×10^2**	4.21×10^2	4.35×10^2	1.66×10
	PLFA-4	4.01×10^2	**4.35×10^2**	**4.17×10^2**	**4.08×10^2**	1.49×10
	PLFA-5	4.19×10^2	4.55×10^2	4.38×10^2	4.40×10^2	**8.270**
	PLFA-6	4.84×10^2	8.15×10^2	6.41×10^2	6.48×10^2	7.86×10
	PLFA-7	4.90×10^2	6.37×10^2	5.56×10^2	5.56×10^2	3.89×10
	PLFA-8	4.67×10^2	5.68×10^2	5.27×10^2	5.33×10^2	2.77×10
F5	PLFA-1	5.20×10^2	5.21×10^2	5.20×10^2	**5.20×10^2**	1.87×10^{-1}
	PLFA-2	**5.00×10^2**	5.21×10^2	**5.19×10^2**	5.20×10^2	5.170
	PLFA-3	5.02×10^2	**5.20×10^2**	**5.19×10^2**	**5.20×10^2**	5.670
	PLFA-4	5.04×10^2	5.21×10^2	**5.19×10^2**	**5.20×10^2**	4.010
	PLFA-5	5.20×10^2	5.21×10^2	5.20×10^2	**5.20×10^2**	9.55×10^{-2}
	PLFA-6	5.20×10^2	5.21×10^2	5.20×10^2	**5.20×10^2**	8.93×10^{-2}
	PLFA-7	5.20×10^2	**5.20×10^2**	5.20×10^2	**5.20×10^2**	**8.76×10^{-2}**
	PLFA-8	5.20×10^2	**5.20×10^2**	5.20×10^2	**5.20×10^2**	1.67×10^{-1}
F6	PLFA-1	6.01×10^2	6.06×10^2	6.03×10^2	6.03×10^2	1.610
	PLFA-2	**6.00×10^2**	6.03×10^2	**6.01×10^2**	**6.01×10^2**	9.43×10^{-1}
	PLFA-3	**6.00×10^2**	6.04×10^2	**6.01×10^2**	**6.01×10^2**	1.00
	PLFA-4	6.01×10^2	**6.02×10^2**	**6.01×10^2**	**6.01×10^2**	**3.89×10^{-1}**
	PLFA-5	6.04×10^2	6.06×10^2	6.05×10^2	6.05×10^2	5.60×10^{-1}
	PLFA-6	6.07×10^2	6.10×10^2	6.09×10^2	6.09×10^2	6.57×10^{-1}
	PLFA-7	6.07×10^2	6.09×10^2	6.08×10^2	6.08×10^2	4.83×10^{-1}
	PLFA-8	6.06×10^2	6.08×10^2	6.07×10^2	6.07×10^2	6.34×10^{-1}
F7	PLFA-1	**7.00×10^2**	**7.00×10^2**	**7.00×10^2**	**7.00×10^2**	3.39×10^{-2}
	PLFA-2	**7.00×10^2**	**7.00×10^2**	**7.00×10^2**	**7.00×10^2**	**2.83×10^{-2}**
	PLFA-3	**7.00×10^2**	7.01×10^2	7.01×10^2	7.01×10^2	7.04×10^{-2}
	PLFA-4	**7.00×10^2**	7.01×10^2	7.01×10^2	7.01×10^2	8.47×10^{-2}
	PLFA-5	7.02×10^2	7.05×10^2	7.03×10^2	7.03×10^2	6.10×10^{-1}
	PLFA-6	7.21×10^2	7.71×10^2	7.48×10^2	7.48×10^2	1.05×10
	PLFA-7	7.24×10^2	7.59×10^2	7.36×10^2	7.36×10^2	8.870
	PLFA-8	7.16×10^2	7.39×10^2	7.28×10^2	7.27×10^2	6.290
F8	PLFA-1	8.12×10^2	8.51×10^2	8.23×10^2	8.20×10^2	9.440
	PLFA-2	8.08×10^2	8.30×10^2	8.16×10^2	8.15×10^2	5.940
	PLFA-3	**8.04×10^2**	**8.15×10^2**	8.09×10^2	8.09×10^2	3.520
	PLFA-4	**8.04×10^2**	**8.15×10^2**	**8.08×10^2**	**8.08×10^2**	**2.890**
	PLFA-5	8.16×10^2	8.35×10^2	8.28×10^2	8.28×10^2	4.440
	PLFA-6	8.50×10^2	8.77×10^2	8.63×10^2	8.64×10^2	5.660
	PLFA-7	8.34×10^2	8.64×10^2	8.54×10^2	8.54×10^2	6.690
	PLFA-8	8.35×10^2	8.55×10^2	8.46×10^2	8.47×10^2	5.390

Best results obtained are given in bold

Table 6 Results over simple multimodal functions with dimension $D = 30$ for PLFA variants

Func.	Algo.	Best	Worst	Mean	Med.	Std.
F4	PLFA-1	4.71×10^2	6.43×10^2	5.30×10^2	5.34×10^2	**9.280**
	PLFA-2	4.05×10^2	5.42×10^2	**4.81×10^2**	**4.75×10^2**	2.74×10
	PLFA-3	**4.00×10^2**	5.73×10^2	5.04×10^2	5.06×10^2	4.10×10
	PLFA-4	4.75×10^2	**5.58×10^2**	5.19×10^2	5.22×10^2	2.03×10
	PLFA-5	5.67×10^2	6.77×10^2	6.24×10^2	6.21×10^2	2.42×10
	PLFA-6	1.55×10^3	3.20×10^3	2.45×10^3	2.51×10^3	3.42×10^2
	PLFA-7	6.29×10^3	1.26×10^4	9.19×10^3	9.13×10^3	1.50×10^3
	PLFA-8	4.29×10^3	1.03×10^4	7.57×10^3	7.69×10^3	1.46×10^3
F5	PLFA-1	**5.20×10^2**	5.21×10^2	**5.21×10^2**	**5.21×10^2**	2.54×10^{-1}
	PLFA-2	**5.20×10^2**	5.21×10^2	**5.21×10^2**	**5.21×10^2**	1.70×10^{-1}
	PLFA-3	5.21×10^2	**5.21×10^2**	**5.21×10^2**	**5.21×10^2**	6.69×10^{-2}
	PLFA-4	5.21×10^2	**5.21×10^2**	**5.21×10^2**	**5.21×10^2**	5.44×10^{-2}
	PLFA-5	5.21×10^2	**5.21×10^2**	**5.21×10^2**	**5.21×10^2**	4.84×10^{-2}
	PLFA-6	5.21×10^2	**5.21×10^2**	**5.21×10^2**	**5.21×10^2**	4.95×10^{-2}
	PLFA-7	5.21×10^2	**5.21×10^2**	**5.21×10^2**	**5.21×10^2**	**4.51×10^{-2}**
	PLFA-8	5.21×10^2	**5.21×10^2**	**5.21×10^2**	**5.21×10^2**	4.69×10^{-2}
F6	PLFA-1	6.09×10^2	6.28×10^2	6.19×10^2	6.18×10^2	4.130
	PLFA-2	6.04×10^2	6.18×10^2	6.12×10^2	6.12×10^2	3.070
	PLFA-3	**6.03×10^2**	6.15×10^2	6.09×10^2	**6.08×10^2**	3.170
	PLFA-4	6.05×10^2	**6.12×10^2**	**6.08×10^2**	**6.08×10^2**	1.930
	PLFA-5	6.15×10^2	6.24×10^2	6.20×10^2	6.21×10^2	2.240
	PLFA-6	6.33×10^2	6.37×10^2	6.34×10^2	6.34×10^2	9.05×10^{-1}
	PLFA-7	6.37×10^2	6.41×10^2	6.39×10^2	6.39×10^2	8.92×10^{-1}
	PLFA-8	6.36×10^2	6.40×10^2	6.38×10^2	6.38×10^2	**8.45×10^{-1}**
F7	PLFA-1	**7.00×10^2**	**7.00×10^2**	**7.00×10^2**	**7.00×10^2**	1.42×10^{-2}
	PLFA-2	**7.00×10^2**	**7.00×10^2**	**7.00×10^2**	**7.00×10^2**	**5.05×10^{-3}**
	PLFA-3	**7.00×10^2**	**7.00×10^2**	**7.00×10^2**	**7.00×10^2**	1.28×10^{-2}
	PLFA-4	7.01×10^2	7.01×10^2	7.01×10^2	7.01×10^2	1.16×10^{-2}
	PLFA-5	7.06×10^2	7.10×10^2	7.08×10^2	7.08×10^2	1.010
	PLFA-6	8.54×10^2	9.42×10^2	8.96×10^2	8.93×10^2	2.39×10
	PLFA-7	1.15×10^3	1.34×10^3	1.25×10^3	1.26×10^3	5.62×10
	PLFA-8	1.07×10^3	1.33×10^3	1.19×10^3	1.19×10^3	5.62×10
F8	PLFA-1	8.59×10^2	9.84×10^2	9.11×10^2	9.10×10^2	2.83×10
	PLFA-2	8.42×10^2	9.73×10^2	8.80×10^2	8.73×10^2	2.80×10
	PLFA-3	8.22×10^2	8.91×10^2	8.57×10^2	8.61×10^2	1.45×10
	PLFA-4	8.24×10^2	8.80×10^2	8.47×10^2	8.46×10^2	1.26×10
	PLFA-5	9.10×10^2	9.69×10^2	9.44×10^2	9.44×10^2	1.57×10
	PLFA-6	9.84×10^2	1.06×10^3	1.04×10^3	1.05×10^3	1.55×10
	PLFA-7	1.11×10^3	1.18×10^3	1.15×10^3	1.15×10^3	1.72×10
	PLFA-8	**6.36×10^2**	**6.40×10^2**	**6.38×10^2**	**6.38×10^2**	**8.45×10^{-1}**

Best results obtained are given in bold

Table 7 Results over simple multimodal functions with dimension $D = 50$ for PLFA variants

Func.	Algo.	Best	Worst	Mean	Med.	Std.
F4	PLFA-1	5.06×10^2	8.88×10^2	6.66×10^2	6.46×10^2	8.98×10
	PLFA-2	$\mathbf{4.45 \times 10^2}$	5.97×10^2	$\mathbf{5.05 \times 10^2}$	$\mathbf{4.95 \times 10^2}$	3.39×10
	PLFA-3	4.74×10^2	6.11×10^2	5.06×10^2	4.98×10^2	$\mathbf{2.54 \times 10}$
	PLFA-4	4.94×10^2	$\mathbf{6.03 \times 10^2}$	5.18×10^2	5.06×10^2	2.75×10
	PLFA-5	6.40×10^2	8.35×10^2	6.96×10^2	6.90×10^2	4.14×10
	PLFA-6	4.00×10^3	5.65×10^3	4.83×10^3	4.90×10^3	4.73×10^2
	PLFA-7	1.53×10^4	2.49×10^4	1.94×10^4	1.93×10^4	2.48×10^3
	PLFA-8	2.70×10^4	4.39×10^4	3.76×10^4	3.84×10^4	4.55×10^3
F5	PLFA-1	$\mathbf{5.20 \times 10^2}$	$\mathbf{5.21 \times 10^2}$	$\mathbf{5.21 \times 10^2}$	$\mathbf{5.21 \times 10^2}$	2.28×10^{-1}
	PLFA-2	5.21×10^2	$\mathbf{5.21 \times 10^2}$	$\mathbf{5.21 \times 10^2}$	$\mathbf{5.21 \times 10^2}$	4.23×10^{-2}
	PLFA-3	5.21×10^2	$\mathbf{5.21 \times 10^2}$	$\mathbf{5.21 \times 10^2}$	$\mathbf{5.21 \times 10^2}$	$\mathbf{2.95 \times 10^{-2}}$
	PLFA-4	5.21×10^2	$\mathbf{5.21 \times 10^2}$	$\mathbf{5.21 \times 10^2}$	$\mathbf{5.21 \times 10^2}$	3.78×10^{-2}
	PLFA-5	5.21×10^2	$\mathbf{5.21 \times 10^2}$	$\mathbf{5.21 \times 10^2}$	$\mathbf{5.21 \times 10^2}$	3.46×10^{-2}
	PLFA-6	5.21×10^2	$\mathbf{5.21 \times 10^2}$	$\mathbf{5.21 \times 10^2}$	$\mathbf{5.21 \times 10^2}$	3.28×10^{-2}
	PLFA-7	5.21×10^2	$\mathbf{5.21 \times 10^2}$	$\mathbf{5.21 \times 10^2}$	$\mathbf{5.21 \times 10^2}$	3.96×10^{-2}
	PLFA-8	5.21×10^2	$\mathbf{5.21 \times 10^2}$	$\mathbf{5.21 \times 10^2}$	$\mathbf{5.21 \times 10^2}$	4.26×10^{-2}
F6	PLFA-1	6.31×10^2	6.51×10^2	6.41×10^2	6.42×10^2	4.720
	PLFA-2	6.19×10^2	6.45×10^2	6.29×10^2	6.29×10^2	5.210
	PLFA-3	$\mathbf{6.07 \times 10^2}$	6.32×10^2	6.21×10^2	6.21×10^2	5.920
	PLFA-4	6.12×10^2	$\mathbf{6.28 \times 10^2}$	$\mathbf{6.19 \times 10^2}$	$\mathbf{6.18 \times 10^2}$	3.960
	PLFA-5	6.29×10^2	6.38×10^2	6.34×10^2	6.34×10^2	2.190
	PLFA-6	6.60×10^2	6.65×10^2	6.62×10^2	6.62×10^2	1.570
	PLFA-7	6.64×10^2	6.69×10^2	6.68×10^2	6.68×10^2	$\mathbf{1.140}$
	PLFA-8	6.68×10^2	6.73×10^2	6.71×10^2	6.70×10^2	1.230
F7	PLFA-1	$\mathbf{7.00 \times 10^2}$	7.01×10^2	$\mathbf{7.00 \times 10^2}$	$\mathbf{7.00 \times 10^2}$	2.88×10^{-1}
	PLFA-2	$\mathbf{7.00 \times 10^2}$	$\mathbf{7.00 \times 10^2}$	$\mathbf{7.00 \times 10^2}$	$\mathbf{7.00 \times 10^2}$	3.99×10^{-3}
	PLFA-3	$\mathbf{7.00 \times 10^2}$	$\mathbf{7.00 \times 10^2}$	$\mathbf{7.00 \times 10^2}$	$\mathbf{7.00 \times 10^2}$	$\mathbf{3.06 \times 10^{-3}}$
	PLFA-4	7.01×10^2	7.01×10^2	7.01×10^2	7.01×10^2	1.74×10^{-2}
	PLFA-5	7.10×10^2	7.18×10^2	7.15×10^2	7.15×10^2	1.970
	PLFA-6	9.89×10^2	1.11×10^3	1.05×10^3	1.05×10^3	2.95×10
	PLFA-7	1.55×10^3	1.81×10^3	1.67×10^3	1.67×10^3	5.88×10
	PLFA-8	1.90×10^3	2.29×10^3	2.12×10^3	2.12×10^3	8.78×10
F8	PLFA-1	9.59×10^2	1.17×10^3	1.07×10^3	1.08×10^3	5.75×10
	PLFA-2	9.22×10^2	1.02×10^3	9.72×10^2	9.73×10^2	3.07×10
	PLFA-3	8.75×10^2	9.78×10^2	9.21×10^2	9.20×10^2	2.79×10
	PLFA-4	$\mathbf{8.59 \times 10^2}$	$\mathbf{9.46 \times 10^2}$	$\mathbf{9.07 \times 10^2}$	$\mathbf{9.05 \times 10^2}$	2.38×10
	PLFA-5	9.92×10^2	1.07×10^3	1.03×10^3	1.03×10^3	1.81×10
	PLFA-6	1.20×10^3	1.28×10^3	1.25×10^3	1.26×10^3	$\mathbf{1.60 \times 10}$
	PLFA-7	1.34×10^3	1.41×10^3	1.39×10^3	1.39×10^3	2.20×10
	PLFA-8	1.41×10^3	1.53×10^3	1.49×10^3	1.50×10^3	2.84×10

Best results obtained are given in bold

Table 8 Results over hybrid functions with dimension $D = 10$ for PLFA variants

Func.	Algo.	Best	Worst	Mean	Med.	Std.
F17	PLFA-1	2.06×10^3	1.86×10^4	4.22×10^4	1.79×10^4	6.34×10^4
	PLFA-2	$\mathbf{1.73 \times 10^3}$	1.86×10^4	5.44×10^3	4.08×10^3	3.94×10^3
	PLFA-3	2.26×10^3	$\mathbf{1.17 \times 10^4}$	5.89×10^3	$\mathbf{3.57 \times 10^3}$	4.19×10^3
	PLFA-4	2.65×10^3	3.03×10^4	$\mathbf{4.93 \times 10^3}$	4.38×10^3	$\mathbf{2.14 \times 10^3}$
	PLFA-5	4.20×10^3	3.41×10^5	1.42×10^4	1.21×10^4	6.57×10^3
	PLFA-6	3.03×10^4	7.09×10^4	1.02×10^5	6.59×10^4	8.16×10^4
	PLFA-7	8.33×10^3	3.91×10^4	3.74×10^4	3.74×10^4	1.70×10^4
	PLFA-8	5.86×10^3	1.86×10^4	1.38×10^4	1.24×10^4	7.07×10^3
F18	PLFA-1	1.89×10^3	3.64×10^4	1.40×10^4	1.01×10^4	1.09×10^4
	PLFA-2	$\mathbf{1.87 \times 10^3}$	3.70×10^4	1.48×10^4	1.13×10^4	1.22×10^4
	PLFA-3	1.92×10^3	3.53×10^4	9.14×10^3	5.68×10^3	8.35×10^3
	PLFA-4	2.25×10^3	$\mathbf{1.17 \times 10^4}$	$\mathbf{4.09 \times 10^3}$	$\mathbf{3.41 \times 10^3}$	$\mathbf{2.17 \times 10^3}$
	PLFA-5	2.59×10^3	1.34×10^4	6.50×10^3	6.36×10^3	2.92×10^3
	PLFA-6	6.18×10^3	6.51×10^5	1.75×10^5	1.16×10^5	1.65×10^5
	PLFA-7	3.03×10^3	9.37×10^4	3.94×10^4	3.42×10^4	2.59×10^4
	PLFA-8	3.01×10^3	3.45×10^4	1.86×10^4	1.99×10^4	3.49×10^4
F19	PLFA-1	$\mathbf{1.90 \times 10^3}$	$\mathbf{1.90 \times 10^3}$	$\mathbf{1.90 \times 10^3}$	$\mathbf{1.90 \times 10^3}$	0.16×10
	PLFA-2	$\mathbf{1.90 \times 10^3}$	1.91×10^3	$\mathbf{1.90 \times 10^3}$	$\mathbf{1.90 \times 10^3}$	0.990
	PLFA-3	$\mathbf{1.90 \times 10^3}$	$\mathbf{1.90 \times 10^3}$	$\mathbf{1.90 \times 10^3}$	$\mathbf{1.90 \times 10^3}$	0.550
	PLFA-4	$\mathbf{1.90 \times 10^3}$	$\mathbf{1.90 \times 10^3}$	$\mathbf{1.90 \times 10^3}$	$\mathbf{1.90 \times 10^3}$	$\mathbf{0.210}$
	PLFA-5	$\mathbf{1.90 \times 10^3}$	$\mathbf{1.90 \times 10^3}$	$\mathbf{1.90 \times 10^3}$	$\mathbf{1.90 \times 10^3}$	0.400
	PLFA-6	1.91×10^3	1.91×10^3	1.91×10^3	1.91×10^3	0.15×10
	PLFA-7	$\mathbf{1.90 \times 10^3}$	1.91×10^3	1.91×10^3	1.91×10^3	0.990
	PLFA-8	$\mathbf{1.90 \times 10^3}$	1.91×10^3	1.91×10^3	1.91×10^3	0.820
F20	PLFA-1	2.15×10^3	3.17×10^4	1.06×10^4	5.23×10^3	1.06×10^3
	PLFA-2	$\mathbf{2.02 \times 10^3}$	2.67×10^4	5.04×10^3	3.00×10^3	5.26×10^3
	PLFA-3	2.06×10^3	1.02×10^4	3.22×10^3	$\mathbf{2.37 \times 10^3}$	1.91×10^3
	PLFA-4	2.07×10^3	1.14×10^4	2.78×10^3	$\mathbf{2.37 \times 10^3}$	6.70×10^2
	PLFA-5	2.13×10^3	5.09×10^3	3.00×10^3	2.92×10^3	7.59×10^2
	PLFA-6	2.55×10^3	1.61×10^4	$\mathbf{5.74 \times 10^2}$	5.94×10^3	9.05×10^2
	PLFA-7	2.25×10^3	7.18×10^3	4.13×10^3	4.09×10^3	3.71×10^2
	PLFA-8	2.24×10^3	$\mathbf{3.47 \times 10^3}$	2.68×10^3	2.62×10^3	$\mathbf{3.70 \times 10^2}$

Best results obtained are given in bold

Table 9 Results over hybrid functions with dimension $D = 30$ for PLFA variants

Func.	Algo.	Best	Worst	Mean	Med.	Std.
F17	PLFA-1	3.12×10^4	5.59×10^6	$\mathbf{1.10 \times 10^5}$	9.42×10^5	1.09×10^6
	PLFA-2	$\mathbf{2.89 \times 10^4}$	$\mathbf{9.37 \times 10^5}$	3.45×10^5	2.90×10^5	2.59×10^5
	PLFA-3	3.53×10^4	1.17×10^6	3.26×10^5	2.45×10^5	$\mathbf{2.46 \times 10^5}$
	PLFA-4	4.91×10^4	1.54×10^6	5.98×10^5	$\mathbf{5.23 \times 10^5}$	4.27×10^5
	PLFA-5	3.32×10^5	1.63×10^6	8.97×10^5	8.47×10^5	3.96×10^5
	PLFA-6	2.38×10^6	1.02×10^7	5.69×10^6	5.53×10^6	1.81×10^6
	PLFA-7	6.74×10^6	3.37×10^7	1.97×10^7	1.96×10^7	6.07×10^6
	PLFA-8	4.88×10^6	2.52×10^7	1.48×10^7	1.50×10^7	4.26×10^6
F18	PLFA-1	1.89×10^3	3.64×10^4	1.40×10^4	1.01×10^4	1.09×10^4
	PLFA-2	$\mathbf{1.87 \times 10^3}$	3.70×10^4	1.48×10^4	1.13×10^4	1.22×10^4
	PLFA-3	1.92×10^3	3.53×10^4	9.14×10^3	5.68×10^3	8.35×10^3
	PLFA-4	2.25×10^3	$\mathbf{1.17 \times 10^4}$	$\mathbf{4.09 \times 10^3}$	$\mathbf{3.41 \times 10^3}$	$\mathbf{2.17 \times 10^3}$
	PLFA-5	2.59×10^3	1.34×10^4	6.50×10^3	6.36×10^3	2.92×10^3
	PLFA-6	6.18×10^3	6.51×10^5	1.75×10^5	1.16×10^5	1.65×10^5
	PLFA-7	3.03×10^3	9.37×10^4	3.94×10^4	3.42×10^4	2.59×10^4
	PLFA-8	3.01×10^3	3.45×10^4	1.86×10^4	1.99×10^4	9.27×10^3
F19	PLFA-1	1.91×10^3	2.04×10^3	1.92×10^3	$\mathbf{1.91 \times 10^3}$	2.96×10
	PLFA-2	$\mathbf{1.90 \times 10^3}$	1.97×10^3	$\mathbf{1.91 \times 10^3}$	$\mathbf{1.91 \times 10^3}$	1.10×10
	PLFA-3	$\mathbf{1.90 \times 10^3}$	$\mathbf{1.91 \times 10^3}$	$\mathbf{1.91 \times 10^3}$	$\mathbf{1.91 \times 10^3}$	0.13×10
	PLFA-4	1.91×10^3	$\mathbf{1.91 \times 10^3}$	$\mathbf{1.91 \times 10^3}$	$\mathbf{1.91 \times 10^3}$	$\mathbf{0.930}$
	PLFA-5	1.91×10^3	1.92×10^3	$\mathbf{1.91 \times 10^3}$	$\mathbf{1.91 \times 10^3}$	$\mathbf{0.930}$
	PLFA-6	1.95×10^3	2.03×10^3	1.99×10^3	1.98×10^3	1.84×10
	PLFA-7	2.08×10^3	2.23×10^3	2.15×10^3	2.15×10^3	3.74×10
	PLFA-8	2.02×10^3	2.17×10^3	2.11×10^3	2.10×10^3	3.08×10
F20	PLFA-1	8.21×10^3	8.77×10^4	4.02×10^4	3.86×10^4	1.72×10^4
	PLFA-2	2.97×10^3	2.82×10^4	1.11×10^4	1.01×10^4	1.11×10^4
	PLFA-3	$\mathbf{2.19 \times 10^3}$	2.19×10^4	$\mathbf{4.66 \times 10^3}$	$\mathbf{4.11 \times 10^3}$	3.51×10^3
	PLFA-4	2.93×10^3	$\mathbf{1.46 \times 10^4}$	5.70×10^3	4.81×10^3	$\mathbf{2.69 \times 10^3}$
	PLFA-5	5.01×10^3	1.67×10^4	1.05×10^4	1.05×10^4	3.40×10^3
	PLFA-6	1.12×10^4	4.65×10^4	2.85×10^4	2.69×10^4	8.47×10^3
	PLFA-7	2.48×10^4	1.52×10^5	7.05×10^4	6.77×10^4	2.53×10^4
	PLFA-8	1.50×10^4	9.23×10^4	4.55×10^4	4.36×10^4	1.72×10^4

Best results obtained are given in bold

Table 10 Results over hybrid functions with dimension $D = 50$ for PLFA variants

Func.	Algo.	Best	Worst	Mean	Med.	Std.
F17	PLFA-1	4.76×10^5	7.17×10^6	3.77×10^6	3.95×10^6	1.73×10^6
	PLFA-2	1.32×10^5	$\mathbf{9.97 \times 10^5}$	$\mathbf{4.27 \times 10^5}$	$\mathbf{3.99 \times 10^5}$	$\mathbf{2.15 \times 10^5}$
	PLFA-3	$\mathbf{8.47 \times 10^4}$	1.56×10^6	6.19×10^5	5.73×10^5	3.47×10^5
	PLFA-4	1.47×10^5	4.39×10^6	1.33×10^6	9.06×10^5	1.10×10^6
	PLFA-5	9.63×10^5	8.20×10^6	3.43×10^6	3.25×10^6	1.79×10^6
	PLFA-6	1.16×10^7	3.43×10^7	2.15×10^7	2.20×10^7	6.08×10^6
	PLFA-7	2.64×10^7	7.23×10^7	5.15×10^7	5.53×10^7	1.25×10^7
	PLFA-8	6.59×10^7	1.64×10^8	1.21×10^8	1.23×10^8	2.80×10^7
F18	PLFA-1	2.02×10^3	1.73×10^4	5.32×10^3	4.28×10^3	3.73×10^3
	PLFA-2	$\mathbf{1.90 \times 10^3}$	1.73×10^4	5.71×10^3	$\mathbf{3.82 \times 10^3}$	4.20×10^3
	PLFA-3	2.80×10^3	$\mathbf{1.31 \times 10^4}$	$\mathbf{5.19 \times 10^3}$	4.30×10^3	$\mathbf{2.74 \times 10^3}$
	PLFA-4	6.62×10^4	2.08×10^5	1.34×10^5	1.35×10^5	3.45×10^4
	PLFA-5	2.40×10^5	1.00×10^6	6.26×10^5	6.08×10^5	1.62×10^5
	PLFA-6	2.58×10^7	1.13×10^8	7.37×10^7	7.27×10^7	2.32×10^7
	PLFA-7	3.02×10^8	1.56×10^9	1.01×10^9	9.99×10^8	2.87×10^8
	PLFA-8	4.73×10^8	1.10×10^9	7.86×10^8	8.08×10^8	2.03×10^8
F19	PLFA-1	1.92×10^3	1.99×10^3	1.96×10^3	1.98×10^3	2.35×10
	PLFA-2	$\mathbf{1.91 \times 10^3}$	1.98×10^3	1.93×10^3	$\mathbf{1.92 \times 10^3}$	1.66×10
	PLFA-3	$\mathbf{1.91 \times 10^3}$	1.98×10^3	1.92×10^3	$\mathbf{1.92 \times 10^3}$	1.26×10
	PLFA-4	$\mathbf{1.91 \times 10^3}$	$\mathbf{1.95 \times 10^3}$	1.93×10^3	1.94×10^3	$\mathbf{0.96 \times 10}$
	PLFA-5	1.93×10^3	1.97×10^3	1.95×10^3	1.95×10^3	1.31×10
	PLFA-6	2.03×10^3	2.08×10^3	2.05×10^3	2.05×10^3	1.22×10
	PLFA-7	2.12×10^3	2.47×10^3	2.35×10^3	2.35×10^3	7.98×10
	PLFA-8	2.16×10^3	2.79×10^3	$\mathbf{1.74 \times 10^2}$	2.85×10^3	3.05×10^3
F20	PLFA-1	$\mathbf{6.35 \times 10^2}$	1.42×10^5	7.69×10^4	7.50×10^4	2.95×10^4
	PLFA-2	3.31×10^3	3.22×10^4	1.54×10^4	1.34×10^4	8.41×10^3
	PLFA-3	2.80×10^3	$\mathbf{1.37 \times 10^4}$	$\mathbf{5.73 \times 10^3}$	$\mathbf{4.82 \times 10^3}$	$\mathbf{2.50 \times 10^3}$
	PLFA-4	3.22×10^3	1.51×10^4	7.53×10^3	6.84×10^3	3.00×10^3
	PLFA-5	1.24×10^4	2.90×10^4	1.92×10^4	1.82×10^4	4.78×10^3
	PLFA-6	2.70×10^4	8.22×10^4	5.50×10^4	5.67×10^4	1.29×10^4
	PLFA-7	5.65×10^4	1.49×10^5	1.04×10^5	1.03×10^5	2.10×10^4
	PLFA-8	7.83×10^4	2.68×10^5	1.62×10^5	1.58×10^5	4.35×10^4

Best results obtained are given in bold

Table 11 Results over composition functions with dimension $D = 10$ for PLFA variants

Func.	Algo.	Best	Worst	Mean	Med.	Std.
F23	PLFA-1	**2.63 × 10³**	**2.63 × 10³**	**2.63 × 10³**	**2.63 × 10³**	7.36×10^{-2}
	PLFA-2	**2.63 × 10³**	**2.63 × 10³**	**2.63 × 10³**	**2.63 × 10³**	**1.21 × 10⁻⁶**
	PLFA-3	**2.63 × 10³**	**2.63 × 10³**	**2.63 × 10³**	**2.63 × 10³**	1.50×10^{-3}
	PLFA-4	**2.63 × 10³**	**2.63 × 10³**	**2.63 × 10³**	**2.63 × 10³**	8.19×10^{-3}
	PLFA-5	**2.63 × 10³**	**2.63 × 10³**	**2.63 × 10³**	**2.63 × 10³**	8.54×10^{-1}
	PLFA-6	2.65×10^3	2.70×10^3	2.67×10^3	2.67×10^3	1.13×10
	PLFA-7	2.65×10^3	2.67×10^3	2.66×10^3	2.66×10^3	0.74×10
	PLFA-8	2.65×10^3	2.66×10^3	2.65×10^3	2.65×10^3	3.60×10
F24	PLFA-1	2.52×10^3	**2.62 × 10³**	2.54×10^3	2.54×10^3	3.06×10^3
	PLFA-2	**2.51 × 10³**	**2.62 × 10³**	2.53×10^3	2.52×10^3	1.77×10
	PLFA-3	**2.51 × 10³**	**2.62 × 10³**	**2.51 × 10³**	2.52×10^3	0.73×10
	PLFA-4	**2.51 × 10³**	**2.62 × 10³**	**2.51 × 10³**	**2.51 × 10³**	0.37×10
	PLFA-5	2.53×10^3	**2.62 × 10³**	2.54×10^3	2.54×10^3	0.51×10
	PLFA-6	2.56×10^3	**2.62 × 10³**	2.57×10^3	2.58×10^3	0.88×10
	PLFA-7	2.56×10^3	**2.62 × 10³**	2.57×10^3	2.57×10^3	0.56×10
	PLFA-8	2.56×10^3	**2.62 × 10³**	2.56×10^3	2.56×10^3	**0.36 × 10**
F25	PLFA-1	2.64×10^3	2.70×10^3	2.69×10^3	2.70×10^3	1.84×10
	PLFA-2	2.64×10^3	2.70×10^3	2.69×10^3	2.70×10^3	1.97×10
	PLFA-3	2.62×10^3	2.70×10^3	2.68×10^3	2.70×10^3	3.19×10
	PLFA-4	**2.61 × 10³**	2.70×10^3	**2.61 × 10³**	**2.64 × 10³**	3.06×10
	PLFA-5	2.65×10^3	**2.67 × 10³**	2.67×10^3	2.67×10^3	1.17×10
	PLFA-6	2.67×10^3	2.70×10^3	2.69×10^3	2.69×10^3	0.72×10
	PLFA-7	2.66×10^3	2.69×10^3	2.68×10^3	2.68×10^3	0.77×10
	PLFA-8	2.66×10^3	2.69×10^3	2.67×10^3	2.68×10^3	**0.5 × 10**
F26	PLFA-1	**2.70 × 10³**	2.80×10^3	**7.04 × 10²**	2.70×10^3	1.82×10
	PLFA-2	**2.70 × 10³**	**2.70 × 10³**	2.70×10^3	2.70×10^3	0.040
	PLFA-3	**2.70 × 10³**	**2.70 × 10³**	2.70×10^3	2.70×10^3	0.020
	PLFA-4	**2.70 × 10³**	**2.70 × 10³**	2.70×10^3	2.70×10^3	0.040
	PLFA-5	**2.70 × 10³**	**2.70 × 10³**	2.70×10^3	2.70×10^3	0.060
	PLFA-6	**2.70 × 10³**	**2.70 × 10³**	2.70×10^3	2.70×10^3	0.300
	PLFA-7	**2.70 × 10³**	**2.70 × 10³**	2.70×10^3	2.70×10^3	0.210
	PLFA-8	**2.70 × 10³**	**2.70 × 10³**	2.70×10^3	**0.010**	**0.010**

Best results obtained are given in bold

Table 12 Results over composition functions with dimension $D = 30$ for PLFA variants

Func.	Algo.	Best	Worst	Mean	Med.	Std.
F23	PLFA-1	**2.62 × 10³**	**2.62 × 10³**	**2.62 × 10³**	**2.62 × 10³**	0.12 × 10
	PLFA-2	**2.62 × 10³**	**2.62 × 10³**	**2.62 × 10³**	**2.62 × 10³**	**2.65 × 10⁻⁹**
	PLFA-3	**2.62 × 10³**	**2.62 × 10³**	**2.62 × 10³**	**2.62 × 10³**	**2.65 × 10⁻⁹**
	PLFA-4	**2.62 × 10³**	**2.62 × 10³**	**2.62 × 10³**	**2.62 × 10³**	0.030
	PLFA-5	**2.62 × 10³**	**2.62 × 10³**	**2.62 × 10³**	**2.62 × 10³**	0.940
	PLFA-6	2.67 × 10³	2.74 × 10³	2.71 × 10³	2.71 × 10³	1.29 × 10
	PLFA-7	2.86 × 10³	3.07 × 10³	2.98 × 10³	2.99 × 10³	5.95 × 10⁴
	PLFA-8	2.76 × 10³	2.98 × 10³	2.90 × 10³	2.90 × 10³	4.81 × 10
F24	PLFA-1	2.64 × 10³	**2.62 × 10³**	2.65 × 10³	2.65 × 10³	**2.65 × 10⁻⁹**
	PLFA-2	**2.62 × 10³**	**2.62 × 10³**	2.64 × 10³	2.64 × 10³	0.030
	PLFA-3	**2.62 × 10³**	**2.62 × 10³**	**2.63 × 10³**	**2.63 × 10³**	0.940
	PLFA-4	**2.62 × 10³**	**2.62 × 10³**	**2.63 × 10³**	**2.63 × 10³**	1.29 × 10
	PLFA-5	2.65 × 10³	**2.62 × 10³**	2.65 × 10³	2.65 × 10³	5.95 × 10⁴
	PLFA-6	2.70 × 10³	**2.62 × 10³**	2.71 × 10³	2.71 × 10³	0.56 × 10
	PLFA-7	2.76 × 10³	**2.62 × 10³**	2.79 × 10³	2.80 × 10³	1.42 × 10
	PLFA-8	2.75 × 10³	2.79 × 10³	2.77 × 10³	2.78 × 10³	2.77 × 10³
F25	PLFA-1	2.71 × 10³	2.75 × 10³	2.71 × 10³	2.72 × 10³	0.880
	PLFA-2	**2.70 × 10³**	2.72 × 10³	**2.70 × 10³**	**2.71 × 10³**	0.30 × 10
	PLFA-3	**2.70 × 10³**	**2.71 × 10³**	**2.70 × 10³**	**2.71 × 10³**	0.180
	PLFA-4	**2.70 × 10³**	**2.71 × 10³**	**2.70 × 10³**	**2.71 × 10³**	**0.150**
	PLFA-5	2.71 × 10³	2.72 × 10³	2.71 × 10³	**2.71 × 10³**	0.23 × 10
	PLFA-6	2.72 × 10³	2.74 × 10³	2.73 × 10³	2.73 × 10³	0.38 × 10
	PLFA-7	2.74 × 10³	2.78 × 10³	2.75 × 10³	2.76 × 10³	0.93 × 10
	PLFA-8	2.74 × 10³	2.76 × 10³	2.74 × 10³	2.75 × 10³	0.61 × 10
F26	PLFA-1	**2.70 × 10³**	2.80 × 10³	2.73 × 10³	**2.70 × 10³**	4.79 × 10
	PLFA-2	**2.70 × 10³**	2.80 × 10³	**2.70 × 10³**	**2.70 × 10³**	4.06 × 10
	PLFA-3	**2.70 × 10³**	**2.70 × 10³**	**2.70 × 10³**	**2.70 × 10³**	0.060
	PLFA-4	**2.70 × 10³**	**2.70 × 10³**	**2.70 × 10³**	**2.70 × 10³**	**0.040**
	PLFA-5	**2.70 × 10³**	**2.70 × 10³**	**2.70 × 10³**	**2.70 × 10³**	0.060
	PLFA-6	**2.70 × 10³**	**2.70 × 10³**	**2.70 × 10³**	**2.70 × 10³**	0.240
	PLFA-7	2.71 × 10³	2.71 × 10³	2.71 × 10³	2.71 × 10³	0.750
	PLFA-8	2.71 × 10³	2.71 × 10³	**2.70 × 10³**	2.71 × 10³	0.390

Best results obtained are given in bold

Table 13 Results over composition functions with dimension $D = 50$ for PLFA variants

Func.	Algo.	Best	Worst	Mean	Med.	Std.
F23	PLFA-1	$\mathbf{2.64 \times 10^3}$	2.67×10^3	2.65×10^3	2.65×10^3	0.58×10
	PLFA-2	$\mathbf{2.64 \times 10^3}$	$\mathbf{2.64 \times 10^3}$	$\mathbf{2.64 \times 10^3}$	$\mathbf{2.64 \times 10^3}$	$\mathbf{1.03 \times 10^{-9}}$
	PLFA-3	$\mathbf{2.64 \times 10^3}$	$\mathbf{2.64 \times 10^3}$	$\mathbf{2.64 \times 10^3}$	$\mathbf{2.64 \times 10^3}$	2.14×10^{-5}
	PLFA-4	$\mathbf{2.64 \times 10^3}$	$\mathbf{2.64 \times 10^3}$	$\mathbf{2.64 \times 10^3}$	$\mathbf{2.64 \times 10^3}$	0.030
	PLFA-5	2.65×10^3	2.65×10^3	2.65×10^3	2.65×10^3	0.760
	PLFA-6	2.76×10^3	2.85×10^3	2.80×10^3	2.80×10^3	2.050
	PLFA-7	3.11×10^3	3.36×10^3	3.25×10^3	3.26×10^3	6.240
	PLFA-8	3.51×10^3	3.96×10^3	3.71×10^3	3.72×10^3	1.07×10^2
F24	PLFA-1	2.69×10^3	2.74×10^3	2.71×10^3	2.71×10^3	1.04×10^2
	PLFA-2	2.68×10^3	2.70×10^3	2.69×10^3	2.68×10^3	0.56×10
	PLFA-3	$\mathbf{2.67 \times 10^3}$	2.69×10^3	2.68×10^3	2.68×10^3	0.40×10
	PLFA-4	$\mathbf{2.67 \times 10^3}$	$\mathbf{2.68 \times 10^3}$	$\mathbf{2.67 \times 10^3}$	$\mathbf{2.67 \times 10^3}$	0.27×10
	PLFA-5	2.70×10^3	2.71×10^3	2.70×10^3	2.70×10^3	$\mathbf{0.23 \times 10}$
	PLFA-6	2.79×10^3	2.81×10^3	2.80×10^3	2.80×10^3	0.64×10
	PLFA-7	2.89×10^3	2.96×10^3	2.93×10^3	2.93×10^3	1.55×10
	PLFA-8	3.00×10^3	3.07×10^3	3.02×10^3	3.02×10^3	1.74×10
F25	PLFA-1	2.72×10^3	2.77×10^3	$\mathbf{7.590}$	2.74×10^3	1.18×10
	PLFA-2	$\mathbf{2.71 \times 10^3}$	$\mathbf{2.71 \times 10^3}$	2.06×10^3	2.72×10^3	0.58×10
	PLFA-3	$\mathbf{2.71 \times 10^3}$	2.72×10^3	1.96×10^3	$\mathbf{2.71 \times 10^3}$	$\mathbf{0.280}$
	PLFA-4	$\mathbf{2.71 \times 10^3}$	2.72×10^3	1.82×10^3	$\mathbf{2.71 \times 10^3}$	0.29×10
	PLFA-5	2.72×10^3	2.73×10^3	1.86×10^3	2.72×10^3	0.40×10
	PLFA-6	2.76×10^3	2.80×10^3	1.89×10^3	2.78×10^3	0.94×10
	PLFA-7	2.81×10^3	2.84×10^3	3.13×10^3	2.83×10^3	0.97×10
	PLFA-8	2.84×10^3	2.90×10^3	6.62×10^3	2.88×10^3	1.70×10
F26	PLFA-1	$\mathbf{2.70 \times 10^3}$	2.80×10^3	$\mathbf{7.04 \times 10^2}$	$\mathbf{2.70 \times 10^3}$	1.82×10
	PLFA-2	$\mathbf{2.70 \times 10^3}$	$\mathbf{2.70 \times 10^3}$	2.70×10^3	$\mathbf{2.70 \times 10^3}$	0.040
	PLFA-3	$\mathbf{2.70 \times 10^3}$	$\mathbf{2.70 \times 10^3}$	2.70×10^3	$\mathbf{2.70 \times 10^3}$	$\mathbf{0.020}$
	PLFA-4	$\mathbf{2.70 \times 10^3}$	$\mathbf{2.70 \times 10^3}$	2.70×10^3	$\mathbf{2.70 \times 10^3}$	0.040
	PLFA-5	$\mathbf{2.70 \times 10^3}$	$\mathbf{2.70 \times 10^3}$	2.70×10^3	$\mathbf{2.70 \times 10^3}$	0.060
	PLFA-6	$\mathbf{2.70 \times 10^3}$	$\mathbf{2.70 \times 10^3}$	2.70×10^3	$\mathbf{2.70 \times 10^3}$	0.300
	PLFA-7	$\mathbf{2.70 \times 10^3}$	$\mathbf{2.70 \times 10^3}$	2.70×10^3	$\mathbf{2.70 \times 10^3}$	0.210
	PLFA-8	2.71×10^3	2.74×10^3	2.71×10^3	2.71×10^3	6.830

Best results obtained are given in bold

Table 14 Final ranking based on mean objective function over $D = 10, 30,$ and 50

Algorithm	Rank of the algorithm														
	Unimodal function			Simple multimodal function			Hybrid function			Composition function			Total rank		
	10	30	50	10	30	50	10	30	50	10	30	50	10	30	50
PLFA-1	10	11	14	17	16	16	21	20	13	12	17	11	60	64	54
PLFA-2	8	*4*	*4*	8	11	9	18	12	8	12	6	12	46	33	33
PLFA-3	7	5	5	10	*10*	*8*	12	*6*	*7*	8	*4*	10	37	*25*	*30*
PLFA-4	10	10	10	7	12	10	*6*	7	13	*5*	*4*	*6*	*28*	33	39
PLFA-5	16	16	16	24	23	20	13	10	19	9	10	13	62	59	68
PLFA-6	24	20	20	36	27	25	19	25	25	23	19	18	102	91	88
PLFA-7	21	18	18	32	27	29	26	29	30	20	31	23	99	105	100
PLFA-8	17	21	21	28	23	33	19	27	24	16	22	32	80	93	110

Best rank obtained are given in bold and italic

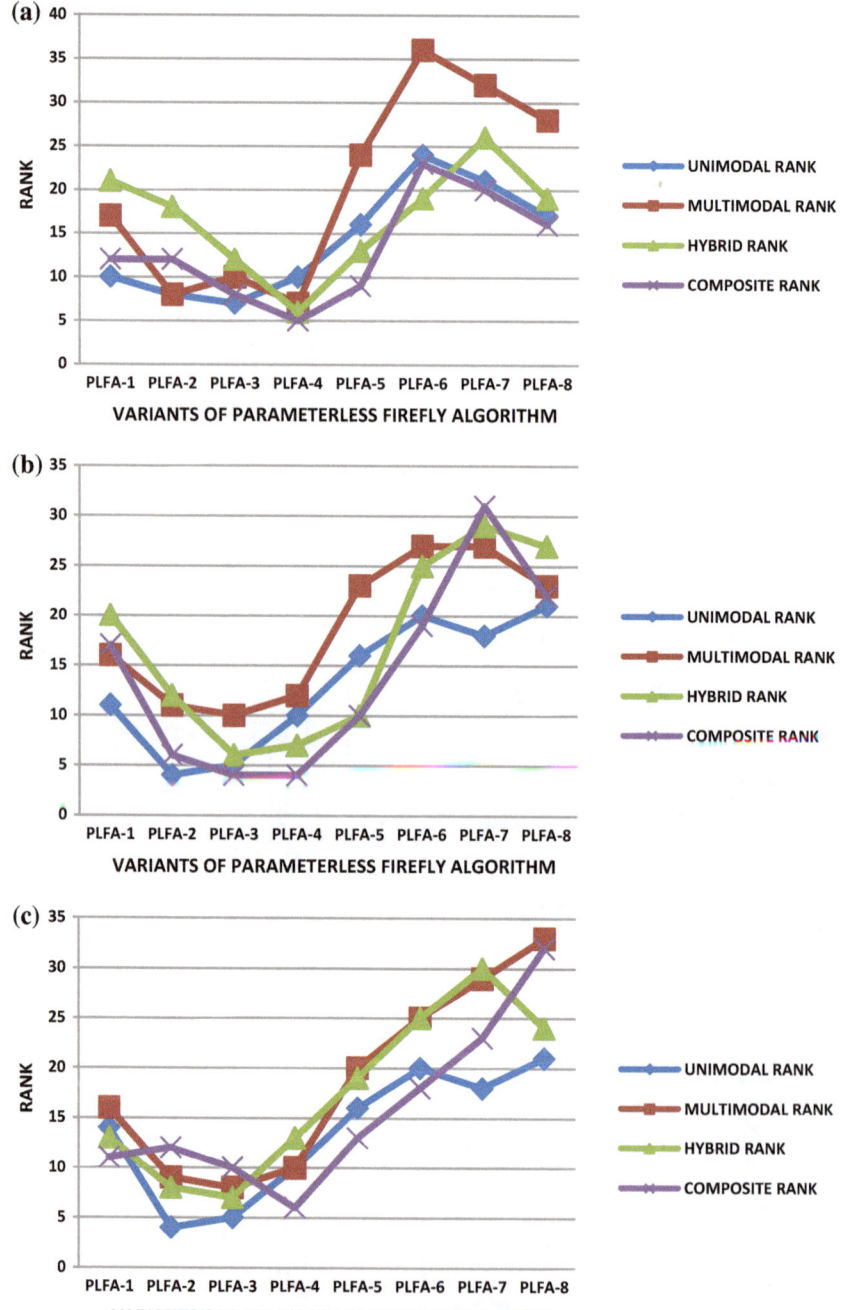

Fig. 2 Ranking of PLFA variants based on mean objective function: **a** class wise ranking for $D = 10$; **b** class wise ranking for $D = 30$; and **c** class wise ranking for $D = 50$

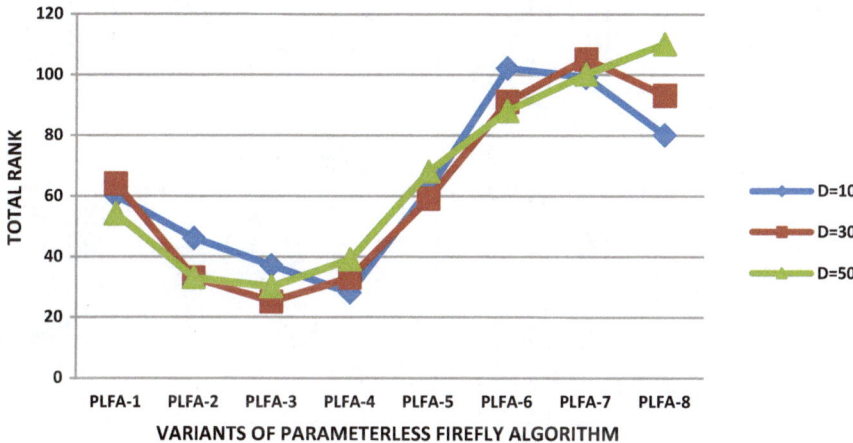

Fig. 3 Total ranking of PLFA variants based on mean objective function for $D = 10$, $D = 30$, and $D = 50$

So, from the above discussion, it can be concluded that population size 40–80 is the most suitable for FA by taking robustness, optimization ability into account over any classes of functions and over any dimension. Additionally, the entries which are highlighted in boldface indicate the best performance results.

5   Conclusion

The efficiency of the FA considerably relies on the proper setting of the algorithm's control parameters. But the proper setting of these parameters is very time-consuming and problem-specific. Therefore, this study focuses on the development of parameter-less variants of FA (PLFA). The three significant parameters of FA have been set by considering the population size, the range of the search space and the dimensionality of the concerned problem. Eight instances of PLFA variants have been developed based on population size. A complete study over experimental results proves that according to mean objective function values, PLFA-4 is the best variants for parameter-less FA for dimension 10. For dimension 30 and 50, PLFA-3 is the best variants for parameter-less FA. On the other hand, PLFA-4 is the best variant in terms of standard deviation over dimension 10 and 30. PLFA-3 is the best variant over dimension 50 when the standard deviation is considered only. Therefore, it can be easily concluded that population size 40–80 is best for FA over any dimension and over any class of objective function. The adaptive formulation of the parameters depending on population size also influences the capability of FA significantly. Hence, it is true that the development of robust adaptive nature-inspired optimization algorithms is very crucial where all parameters including population for a set of problems are automatically adapted.

Table 15 Final ranking based on standard deviation over $D = 10$, 30, and 50

Algorithm	Rank of the algorithm														
	Unimodal function			Simple multimodal function			Hybrid function			Composition function			Total rank		
	10	30	50	10	30	50	10	30	50	10	30	50	10	30	50
PLFA-1	13	18	16	29	26	31	25	22	18	25	22	29	92	88	94
PLFA-2	08	05	**04**	23	26	20	20	17	13	16	18	11	67	66	48
PLFA-3	**07**	**04**	05	22	18	23	16	10	**08**	18	08	**07**	63	40	**43**
PLFA-4	08	09	09	**16**	**13**	**18**	**06**	**07**	10	**14**	**07**	09	**44**	**36**	46
PLFA-5	12	13	14	22	22	20	12	08	17	17	11	13	53	54	64
PLFA-6	24	18	17	22	20	22	28	23	19	23	20	20	97	81	78
PLFA-7	20	23	21	29	29	21	19	31	28	19	29	23	87	112	93
PLFA-8	16	19	22	27	25	25	17	24	31	11	27	31	71	95	109

Best rank obtained are given in bold and italic

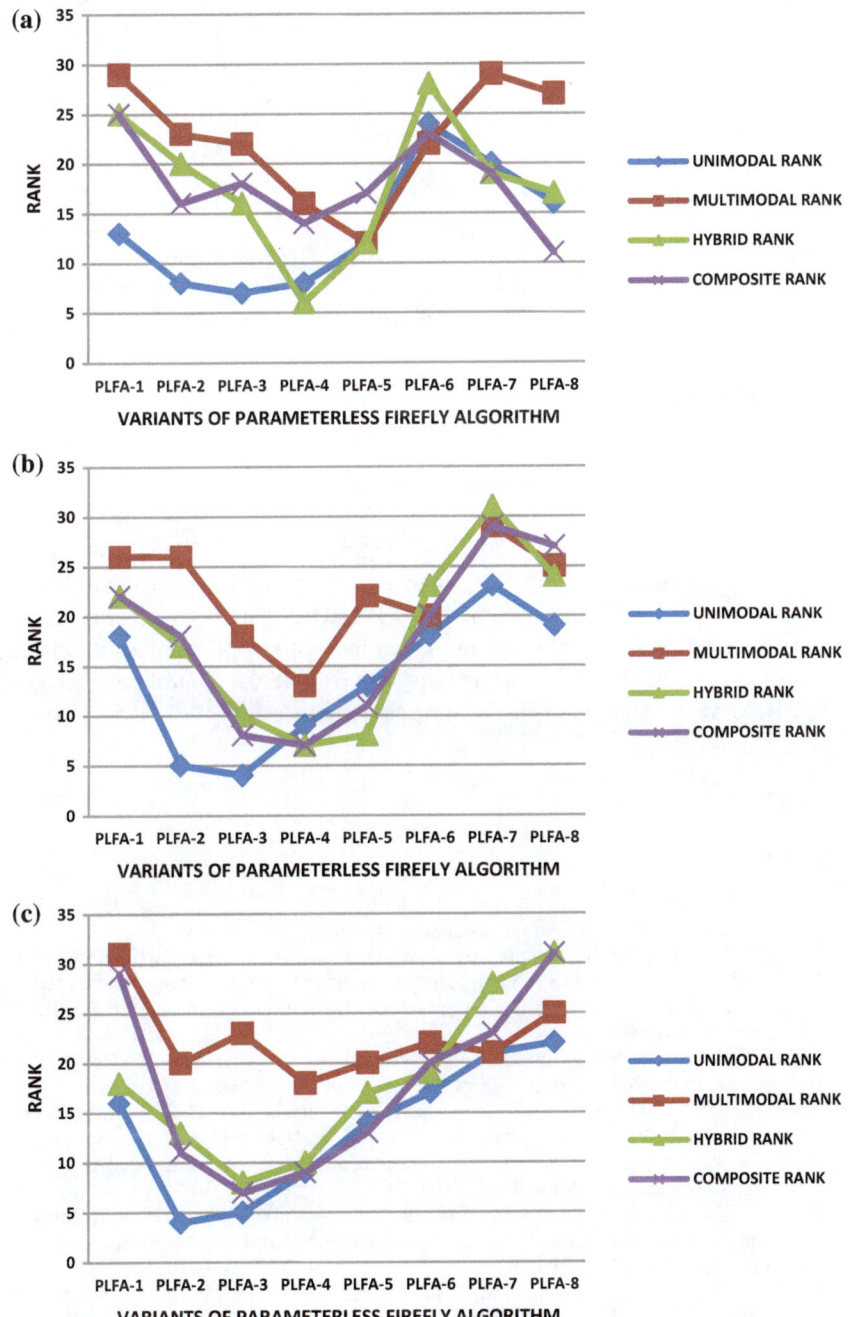

Fig. 4 Ranking of PLFA variants based on standard deviation: **a** class wise ranking for $D = 10$; **b** class wise ranking for $D = 30$; and **c** class wise ranking for $D = 50$

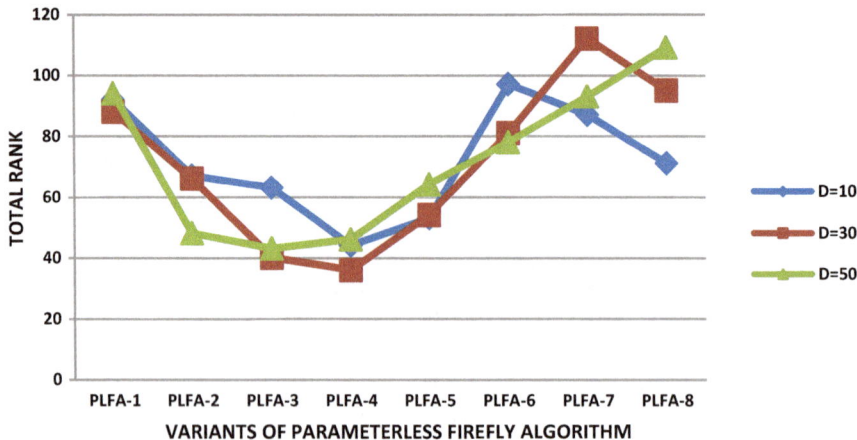

Fig. 5 Total Ranking of PLFA variants based on standard deviation for $D = 10$, $D = 30$, and $D = 50$

The main future direction of this paper is to perform an extensive study and systematic analysis of the influence of different NIOA's parameters over its performance over the diverse type of problems. Other motivating opportunities may be the utilization of the parameter-less as an approach to tune the control parameters of other NIOA. Measuring the impact of the population size over NIOA can also be a challenging task.

References

1. Yang XS (2010) Nature-inspired metaheuristic algorithms. Luniver press
2. Yang XS (2010) Engineering optimization: an introduction to metaheuristic applications. Wiley
3. Eberhart R, Kennedy J (1995 Oct) A new optimizer using particle swarm theory. In: Proceedings of the sixth international symposium on micro machine and human science, 1995. MHS'95. IEEE, New York, pp 39–43
4. Yang XS (2010) Firefly algorithm, Lévy flights, and global optimization. In: Research and development in intelligent systems XXVI. Springer, London, pp 209–218
5. Yang XS, Deb S (2009 Dec) Cuckoo search via Lévy flights. In: NaBIC 2009. world congress on nature & biologically inspired computing, 2009. IEEE, New York, pp 210–214
6. Yang XS (2010) A new metaheuristic bat-inspired algorithm. Nature inspired cooperative strategies for optimization (NICSO 2010), 65–74
7. Dhal KG, Das A, Ray S, Gálvez J, Das S (2019) Nature-inspired optimization algorithms and their application in multi-thresholding image segmentation. Arch Comput Methods Eng, 1–34. https://doi.org/10.1007/s11831-019-09334-y
8. Dhal KG, Ray S, Das A, Das S (2018) A survey on nature-inspired optimization algorithms and their application in image enhancement domain. Arch Comput Methods Eng, 1–32. https://doi.org/10.1007/s11831-018-9289-9
9. Yang XS, He X (2016) Nature-inspired optimization algorithms in engineering: overview and applications. In: Nature-inspired computation in engineering. Springer International Publishing, pp 1–20

10. Booker L (ed) (2005) Perspectives on adaptation in natural and artificial systems (vol 8). Oxford University Press on Demand
11. Valdez F, Melin P, Castillo O (2014) A survey on nature-inspired optimization algorithms with fuzzy logic for dynamic parameter adaptation. Expert Syst Appl 41(14):6459–6466
12. Sheikholeslami R, Kaveh A (2013) A survey of chaos embedded meta-heuristic algorithms. Int J Optim Civil Eng 3(4):617–633
13. Črepinšek M, Liu SH, Mernik M (2013) Exploration and exploitation in evolutionary algorithms: a survey. ACM Comput Surv (CSUR) 45(3):35
14. Črepinšek M, Mernik M, Liu SH (2011) Analysis of exploration and exploitation in evolutionary algorithms by ancestry trees. Int J Innov Comput Appl 3(1):11–19
15. Eiben AE, Schippers CA (1998) On evolutionary exploration and exploitation. Fundam Inf 35(1–4):35–50
16. Bansal JC, Singh PK, Saraswat M, Verma A, Jadon SS, Abraham A (2011 Oct) Inertia weight strategies in particle swarm optimization. In: 2011 third world congress on nature and biologically inspired computing (NaBIC). IEEE, pp 633–640
17. Yang X, Yuan J, Yuan J, Mao H (2007) A modified particle swarm optimizer with dynamic adaptation. Appl Math Comput 189(2):1205–1213
18. Wang H, Wu Z, Rahnamayan S (2011) Particle swarm optimisation with simple and efficient neighbourhood search strategies. Int J Innov Comput Appl 3:97–104
19. Wang H, Cui Z, Sun H, Rahnamayan S, Yang XS, Randomly attracted firefly algorithm with neighborhood search and dynamic parameter adjustment mechanism. Soft Comput, pp 1–15. https://doi.org/10.1007/s00500-016-2116-z
20. Baykasoğlu A, Ozsoydan FB (2015) Adaptive firefly algorithm with chaos for mechanical design optimization problems. Appl Soft Comput 36:152–164
21. Baykasoğlu A, Ozsoydan FB (2014) An improved firefly algorithm for solving dynamic multidimensional knapsack problems. Expert Syst Appl 41(8):3712–3725
22. Ozsoydan FB, Baykasoglu A (2015 Dec) A multi-population firefly algorithm for dynamic optimization problems. In: 2015 IEEE international conference on evolving and adaptive intelligent systems (EAIS). IEEE, pp 1–7
23. Samanta S, Mukherjee A, Ashour AS, Dey N, Tavares JMR, Abdessalem Karâa WB, … Hassanien AE (2018) Log transform based optimal image enhancement using firefly algorithm for autonomous mini unmanned aerial vehicle: an application of aerial photography. Int J Image Gr 18(04):1850019
24. Dhal KG, Quraishi MI, Das S (2016) Development of firefly algorithm via chaotic sequence and population diversity to enhance the image contrast. Nat Comput 15(2):307–318
25. Dey N, Samanta S, Chakraborty S, Das A, Chaudhuri SS, Suri JS (2014) Firefly algorithm for optimization of scaling factors during embedding of manifold medical information: an application in ophthalmology imaging. J Med Imaging Health Inf 4(3):384–394
26. Dhal KG, Das S (2018) Colour retinal images enhancement using modified histogram equalisation methods and firefly algorithm. Int J Biomed Eng Technol 28(2):160–184
27. Jagatheesan K, Anand B, Samanta S, Dey N, Ashour AS, Balas VE (2017) Design of a proportional-integral-derivative controller for an automatic generation control of multi-area power thermal systems using firefly algorithm. IEEE/CAA J Automat Sin
28. Fister I Jr, Mlakar U, Yang X-S, Fister I (2016) Parameterless bat algorithm and its performance study. Nat Inspir Comput Eng Stud Comput Intell 637:267–276
29. Lobo FG, Goldberg DE (2003) An overview of the parameterless genetic algorithm. In: Proceedings of the 7th joint conference on information sciences (Invited paper), pp 20–23
30. Papa G (2013) Parameter-less algorithm for evolutionary-based optimization For continuous and combinatorial problems. Comput Optim Appl 56:209–229
31. Teo J, Hamid MY (2005) A parameterless differential evolution optimizer. In: Proceedings of the 5th WSEAS/IASME international conference on systems theory and scientific computation, pp 330–335
32. De-Silva LA, da-Costa KAP, Ribeiro PB, Rosa G, Papa JP (2015) Parameter-setting free harmony search optimization of restricted Boltzmann machines and its applications to spam detection. In: 12th international conference applied computing, pp 143–150

33. Dhal KG, Fister I Jr, Das S (2017) Parameterless harmony search for image multi-thresholding. In: 4th student computer science research conference (StuCosRec-2017), pp 5–12
34. Dhal KG, Sen M, Das S (2018) Multi-thresholding of histopathological images using Fuzzy entropy and parameterless Cuckoo Search. In: Critical developments and application of swarm intelligence (IGI-GLOBAL), pp 339–356
35. Dhal KG, Sen M, Ray S, Das S (2018) Multi-thresholded histogram equalization based on parameterless artificial bee colony. In: Incorporating of nature-inspired paradigms in computational applications, (IGI-GLOBAL), pp 108–126
36. Liang J, Qu B, Suganthan P, Problem definitions and evaluation criteria for the CEC 2014 special session and competition on single objective real parameter numerical optimization. Computational Intelligence Laboratory, Zhengzhou University, Zhengzhou China and Technical Report, Nanyang Technological University, Singapore
37. Durbhaka GK, Selvaraj B, Nayyar A (2019) Firefly swarm: metaheuristic swarm intelligence technique for mathematical optimization. In: Data management, analytics and innovation. Springer, Singapore, pp 457–466
38. Röhler AB, Chen S (2011 Dec) An analysis of sub-swarms in multi-swarm systems. In: Australasian joint conference on artificial intelligence. Springer, Berlin, pp 271–280
39. Lanzarini L, Leza V, De Giusti A (2008 June) Particle swarm optimization with variable population size. In: International conference on artificial intelligence and soft computing. Springer, Berlin, pp 438–449
40. Zhu W, Tang Y, Fang JA, Zhang W (2013) Adaptive population tuning scheme for differential evolution. Inf Sci 223:164–191
41. Chen D, Zhao C (2009) Particle swarm optimization with adaptive population size and its application. Appl Soft Comput 9(1):39–48
42. Piotrowski AP (2017) Review of differential evolution population size. Swarm Evolut Comput 32:1–24
43. Derrac J, Garcia S, Molina D, Herrera F (2011) A practical tutorial on the use of nonparametric statistical tests as a methodology for comparing evolutionary and swarm intelligence algorithms. Swarm Evolut Comput 1:3–18